D0026015

Exploring Engineering: An Introduction for Freshman to Engineering and to the Design Process

"...it is engineering that changes the world."

Issac Asimov, *Isaac Asimov's Book of Science and Nature Quotations* (Simon and Schuster, 1970)

"...engineering is the art of doing that well with one dollar which any bungler can do with two."

Arthur Wellington, *Economic Theory of the Location of Railways*, 2nd ed. (Wiley, New York, 1887)

Exploring Engineering: An Introduction for Freshman to Engineering and to the Design Process

P. G. Kosky, G. Wise, R. T. Balmer,
and
W. D. Keat

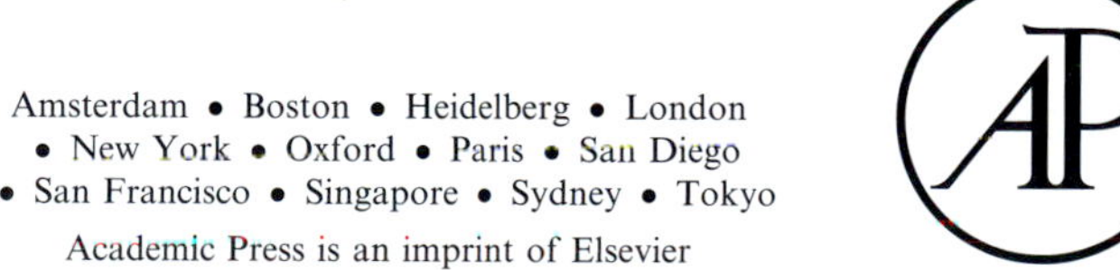
Amsterdam • Boston • Heidelberg • London
• New York • Oxford • Paris • San Diego
• San Francisco • Singapore • Sydney • Tokyo
Academic Press is an imprint of Elsevier

Academic Press is an imprint of Elsevier
30 Corporate Drive, Suite 400, Burlington, MA 01803, USA
525 B Street, Suite 1900, San Diego, California 92101-4495, USA
84 Theobald's Road, London WC1X 8RR, UK

This book is printed on acid-free paper. ♾

Library of Congress Cataloging-in-Publication Data

British Library Cataloguing-in-Publication Data

A catalogue record for this book is available from the British Library.

ISBN 13: 978-0-12-369405-8
ISBN 10: 0-12-369405-1

For information on all Academic Press publications
visit our Web site at www.books.elsevier.com

Printed in the United States of America
07 08 09 10 9 8 7 6 5 4 3 2

Contents

Foreword

This text was originally written for a first-year engineering course at Union College, Schenectady, New York. It is divided into two parts: Part I is an Introduction to Engineering, and Part II covers the Design Process. Broadly speaking, Part I deals with the "minds-on" tools and Part II with the "hands-on" tools one needs as an engineer. Chapters in Part 1 are organized around just one or two principles and have several worked examples and about 20 exercises at a generally increasing level of complexity at the end of the chapter. Answers are given to several selected problems to encourage the students to work toward self-proficiency.

Part II of this text is a "Design Studio" and is associated with the design of engineering *systems*. It is just as essential and challenging as the minds-on aspects covered in Part I. It is also, for most students, a lot more fun. Few things are more satisfying than seeing a machine, an electronic device, or a computer program you have designed and built doing exactly what you intended it to do. Such initial successes may sound simple, but they provide the basis of a rigorous system that will enable an engineering graduate, as part of a team of engineers, to achieve the even greater satisfaction of designing a system that can provide new means of transportation, information access, medical care, energy supply, and so on, and can change for the better the lives of people around the world.

We cover *introductory* material explicitly from the following engineering subdisciplines: bioengineering, chemical engineering, civil engineering, computer and electronic engineering, control systems engineering, electrical engineering, materials engineering, and mechanical engineering. The text is organized around the theme of a "smart car" because it affords a relatively familiar basis that provides an entry into each of these engineering subdisciplines.

Figure 1 shows the sinews that connect the theme to core technical topics. These topics are kept to a level compatible with the background of first-year students. Some topics obviously are closer to the core material in one subdiscipline of engineering than to another, but some are generic to all. In order to cover such broad, and sometimes relatively advanced, subject matter, we have taken some liberties in simplifying those topics. Instructors should expect to find some shortcuts that will pain the purists. We have tried, nevertheless, to be accurate as to basic principles.

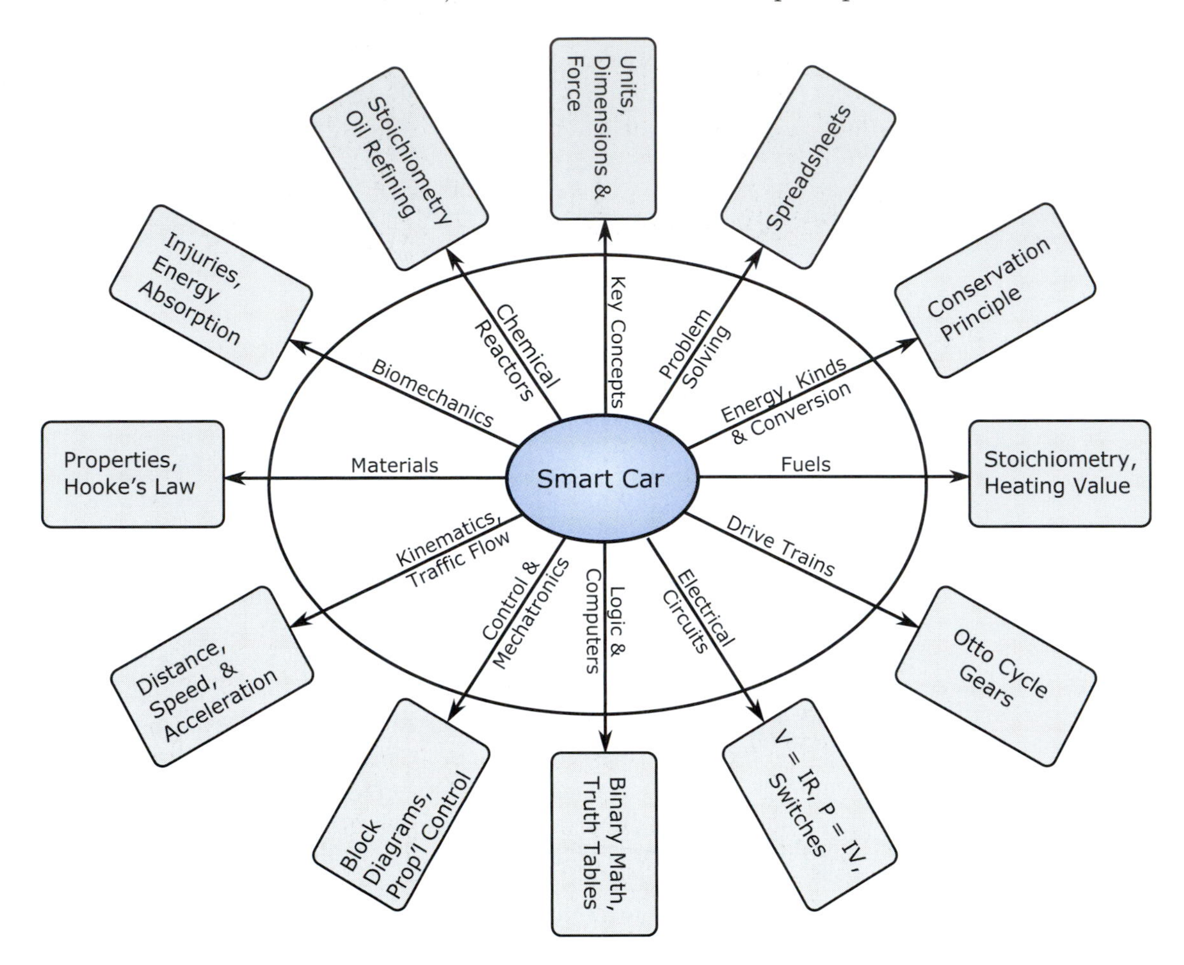

We physically separated the two parts of this text to emphasize the different character of their content. Each chapter of the minds-on section has about the equivalent amount of new ideas and principles. Our experience is that any chapter can be sufficiently covered in about two hours of class time so that the students can complete the rest of the chapter unaided. On the other hand, the Design Studio needs up to three contiguous laboratory hours per week to do it justice. It culminates in a team-orientated competition. Typically the student teams build a small model "vehicle" that is on wheels, or walks, or floats, that may or may not be autonomous, and so forth. Students then compete head-to-head against other teams from the course with the same design goals plus an offensive and defensive strategy to overcome all the other teams in the competition. Our experience is that it is highly motivating for the students.

There is probably too much material, as well as too broad coverage, in this text provided for just one introductory course. Given the necessary breaks for testing and for a final examination, typically a class will cover several chapters of Part 1. Exactly which chapters depends on which disciplines are offered in each engineering school. Certainly we think the more fundamental chapters need to be included. Suggested Part I coverage includes the basics in Chapters 1–4, the summary Chapter 14, plus several other chapters that can be selected for suitability for particular students' subdisciplines. Part 2 of this text can be thought of as independent of Part 1, but at Union College we taught Part 2 as an integral part of our first-year engineering course and that is our recommendation for a full introduction to engineering at any school where there is room to accommodate it.

The approach taken in this first-year text is unique, in part because of the atypical character of authorship. Two of the authors have industrial backgrounds, mostly at the GE Research Center in Niskayuna, New York, one in engineering research and applied science and the other in industrial communications. The other two authors have followed more usual academic career paths and have the appropriate academic credentials to draw upon. We believe the synergy of the combined authorship provides a fresh perspective for first-year engineering education. Specifically, while elementary in coverage, this textbook parallels the combined authors' wide experience that *engineering is not a spectator sport.* We therefore do not avoid the introduction of relatively advanced topics in this otherwise elementary text. Here are some of the nonstandard approaches to familiar engineering topics:

1. We introduce spreadsheets early in the text, and almost every chapter of Part 1 has one or more spreadsheet exercises.
2. We rigorously enforce the use of appropriate significant figures throughout the text. For example, we always try to differentiate between 60. and 60 (notice the decimal point or its absence). We obviously recognize often it appears to be clumsy to write numbers such as 6.00×10^1, but we do so to discourage bad habits such as electronic calculator answers to undeserved significant figures.
3. We develop all of our Exercise solutions in a rigorous format using a simple mnemonic "Need to Know How to Solve" to discourage the student who thinks he or she knows the answer and writes down the wrong one (or even the correct one!). This, too, can appear clumsy in its usage, but it is invaluable in training a young engineer to leave an audit trail of his or her methods, a good basic work habit of practicing engineers.
4. We recognize that the Engineering English unit system of lbf, lbm, and g_c system will be used throughout the careers of many, if not most, of today's young engineers. A clear exposition is used to develop it and to use it so we can avoid the terrible results of a factor of 32.2 that should or shouldn't be there!

5. Conservation principles, particularly energy and mass, are introduced early in the text, as well as emphasis on the use of control boundaries that focus on the essential problem at hand.
6. The use of tables is a powerful tool, both in the hands of students and of qualified engineers. We have developed a number of tabular methods for stoichiometric and thermodynamic problems that should eliminate the problem of the wrong stoichiometric coefficients and of sign errors respectively. Methods based on tables are also fundamental to design principles as taught in the Design Studio section of the book.
7. We have emphasized the power of electrical switches as vital elements of computer design and their mathematical logic analogues.
8. Since standard mathematical control theory is far too advanced for our intended audience, we have used spreadsheet methods that graphically show the effects of feedback gains, paralleling the results of the standard mathematical methods.
9. We have developed the solution method to standard kinematics problems using a visual/geometric technique of speed-time graphs rather than the standard equations. We believe this is a usefully visual way to deal with multielement kinematics problems. We have also quoted, but not developed, the standard kinematics equations because they are derived in every introductory college textbook, and their use does not increase basic understanding of kinematics *per se*.
10. The design methodology in the Design Studio is presented in a stepwise manner to help lead student and instructor through a "hands-on" design project.
11. Pacing of the hands-on projects is accomplished through gradable "design milestones." These are general time-tested project assignments that we believe are the most powerful tool in getting a freshman design course to work well.
12. The many design examples were selected from past student projects, ranging from the freshman to the senior year, to appeal to and be readily grasped by the beginning engineering student. In one chapter we present a typical first-year design project and follow the evolution of one team's design from clarification of the task to detailed design.
13. The culmination of the hands-on Design Studio is a head-to-head team competition, and it is recommended that all first-year engineering courses based on this text should strive to include it.
14. The Accreditation Board for Engineering and Technology (ABET) sets curriculum criteria that require students to have "an understanding of professional and ethical responsibility." We agree totally with the need for this requirement, but we do not see much structure in the way that ethics problems are handled in most engineering textbooks. In the first chapter, we discuss canons of ethics *and how to apply them*. This is a structured approach that we trust will bring discipline to this subject

for first-year engineers. Each chapter, except Chapters 1 and 14, has ethics exercises pertinent to its particular content, and some provide possible answers. We believe it is more useful to infuse ethics continually during the term rather than as a single arbitrarily inserted lecture.

15. The last chapter of Part 1 looks to the near-term future technological development of a "smart car." Pedagogically, it is best included as an "extra" in the last week of the term, since it offers neither new concepts nor testable material. We have found that it can be fitted in while the major activity for the last week is a review of the course as a whole and, for those who use the Design Studio portion of the text, the finals of its competition and submission of final reports.

Union College, Schenectady, New York,

PGK, GW, RTB, & WDK July 2006

A companion web site for this textbook with material for instructors and for students is available at:

http://books.elsevier.com/companions/0123694051

It has links for:

(a) **Instructors** – containing a solution manual, design contest material, student feedback forms, and a textbook feedback link.
(b) **Students** – containing time management and study skills information, links to unit conversion programs, and practice exercises with solutions.

Acknowledgments

We wish to acknowledge the help, suggestions, and advice of several Union colleagues and especially from coteachers for the Union freshman engineering course: Dean Cherrice Traver and Professors Brad Bruno, James Hedrick, Thomas Jewell, John Rogers, John Spinelli, Frank Wicks, and Andrew Wolfe.

In addition we have received advice, assistance, and, most important, individual chapter reviews from Professors Nicholas Krouglicof and Thomas Jewell.

The competition-based hands-on approach to teaching design was inspired by Professor Michael C. Larson, Mechanical Engineering Department, Tulane University, New Orleans, Louisiana, and by Mr. Daniel Retajczyk, then a graduate student at Clarkson University, New York.

Union undergraduates Andrew Krauss (Mechanical Engineering/Philosophy) checked the solution set to Part 1, and Craig Ferguson (Computer Science/Mechanical Engineering) developed the student design for the "A Bridge Too Far" example in Part 2.

The graphic illustrations were produced by Ted Balmer at March Twenty Productions (http://www.marchtwenty.com/)

Part 1

Minds-on

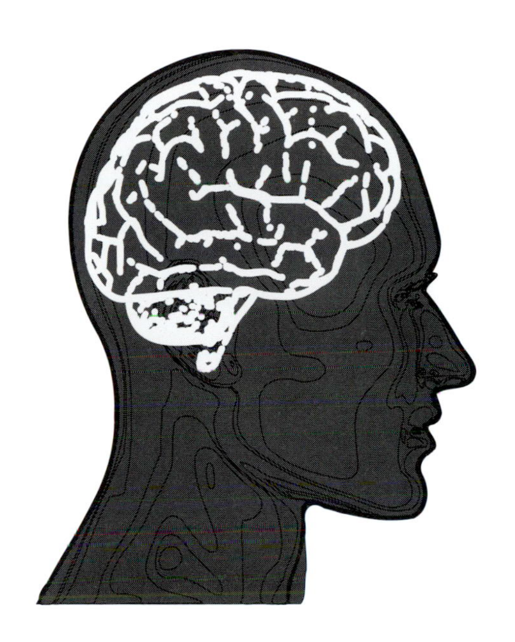

Chapter 1

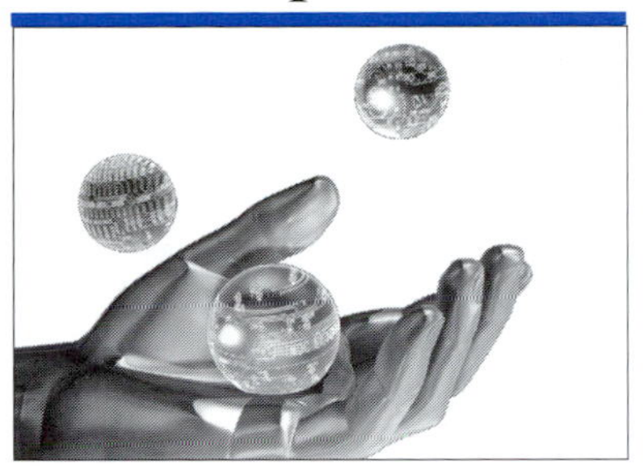

© iStockphoto.com/Antonis Papantoniou

What Engineers Do

Introduction

What is an engineer, and what does he or she do? You can get a good answer to this question by just looking at the word itself. The word *engine* comes from the Latin *ingenerare*, meaning "to create." About 2000 years ago, the Latin word *ingenium* ("the product of genius") was used to describe the design of a new machine. Soon after, the word *ingen* was used to describe all machines. In English, "ingen" was spelled "engine," and people who designed creative things were known as "engine-ers." In French, German, and Spanish today the word for *engineer* is *ingenieur*, and in Italian it is *ingegnere*.

So—again—what is an engineer?
Answer: *An engineer is a creative, ingenious person.*

What does an engineer do?
Answer: *Engineers create (i.e., design) ingenious solutions to societal problems.*

What makes a "good" engineer?
This is actually a difficult question to answer because the knowledge and skills required to be an engineer (i.e., to create ingenious solutions) is a moving target. The factors that will lead to your career success are not the same as they were 20 years ago. In this book we illustrate the key characteristics of a successful 21st-century engineer by exploring the multidisciplinary creative engineering process required to produce a "good" competitive modern automobile.

This book is *not* about automotive engineering; it *is* linked by the theme of a "smart" car. It is about what almost all kinds of engineers do and how they do it. The theme itself is not a discipline; it's just a thread to pull in many disciplines in science and in engineering.

So just what *is* a smart car? It's the *convergence* of many technologies and engineering systems. It's the car of today and of tomorrow, in which computers, sensors, accident controls, protective systems, and modern alloys and plastics are as important as is a continuing expertise in the traditional engineering disciplines. This book is intended to appeal to a number of aspects of modern engineering subdisciplines. The smart car theme embraces a surprising number of them.

Obviously, in a beginning engineering text, we can discuss only a small segment of those subdisciplines:

- **Bioengineers** deal with such elements as a seat belt, a method of preventing forces on your body from causing you injury during accidents. More generally, bioengineers deal with the engineering analysis of living systems.
- **Chemical engineers** deal with the gas tank and what goes into it, and with the processes that refine that gasoline from crude oil. They also handle the combustion process that occurs within the engine when the gasoline is ignited. More generally, chemical engineers deal with the way atoms and molecules link up and how those connections shape the properties of materials, and the conversion of energy from one form to another.
- **Civil engineers** design the roads that the car rides on. More generally, civil engineers design and analyze large-scale structures such as buildings, bridges, water treatment systems, and so forth.
- **Computer and electronic engineers** design the dozens of embedded computers and electronic systems that are essential for the operation of a modern automobile. More generally, computer and electronic engineers design and analyze electronic systems for computation and control.
- **Control system engineers** develop cruise controls for smooth riding. More generally, control systems engineers design and analyze systems that sense the environment and provide responses that enable people to accomplish goals safely.
- **Electrical engineers** design the systems that provide headlights, taillights, ignition systems, and so forth. More generally, electrical engineers design and analyze systems that apply electrical energy.
- **Materials engineers** develop the dozens of materials found in today's automobiles. These include the steel in the body, the aluminum in the engine block, the plastic in the bumpers, and the glass in the windows, as well as specialized fabrics, insulating materials, and engine exhaust catalysts. More generally, materials engineers select and apply materials to enhance the performance of engineered systems.

- **Mechanical engineers** develop better drive systems for converting the chemical energy of fuels into the kinetic energy of motion. More generally, mechanical engineers design and analyze systems for the manipulation of mechanical energy. It is one of the most pervasive engineering disciplines, with multiple applications from transport systems such as aircraft, spacecraft, trains, cars, and so on.

What This Book Covers

In your mind, what makes a "good" automobile? If you were in the market to purchase one, you might want one that has high performance and good gas mileage and is roomy, safe, and stylish. Or you might describe it in categories: new or used; sedan, sports car, or SUV; two doors or four doors. Or maybe you would only be interested in the price tag.

As a consumer making a decision about purchasing a car, it is enough to use these words, categories, and questions to reach a decision. But engineers think differently. They design and analyze, and consequently they must have a different set of words, categories, and questions. In order to design and analyze, engineers ask precise questions that can be answered with **variables, numbers,** and **units.** They do it to accomplish a safe and reliable product. From this point of view, an automobile is an engineer's answer to the question "What's a good way to move people safely and reliably?"

The purpose of this book is to introduce you to the engineering profession. It does so by introducing you to the way engineers think, ask, and answer questions like: What makes an automobile—or a computer, or an airplane, or a washing machine, or a bridge, or a prosthetic limb, or an oil refinery, or a space satellite—*good*?

Again, we must emphasize that the choice of automobiles as the theme for this text is strictly for convenience. It no more and no less expresses the essence of engineering than would a theme based on a computer, an airplane, a washing machine, a bridge, a prosthetic limb, an oil refinery, or a space satellite. In each case, the essence of the text would focus on the creative use of energy, materials, motion, and information to serve human needs, so a more detail-oriented engineer might answer our original question like this:

> A good 21st-century automobile employs stored energy (on the order of 100 million joules), complex materials (on the order of 1000 kilograms of steel, aluminum, glass, and plastics), and information (on the order of millions of bits processed every second) to produce a human transport vehicle capable of high speed (on the order of 30 meters/second), low cost (a few tens of cents per mile), low pollution (a few grams of pollutants per mile), and high safety.

That's a long and multidimensional answer, but an engineer would be unapologetic about that. Engineering is *inherently* multidimensional and multidisciplinary. It needs to be multidimensional to create compromises among conflicting criteria, and it needs to be multidisciplinary to understand the technical impact of the compromises. Making a car heavier, for example, might make it safer, but it would also be less fuel efficient. Engineers often deal with such competing factors. They break down general issues into concrete questions. They then answer those questions with design variables, units, and numbers.

Engineering is not a spectator sport. It is a *hands-on* and *minds-on* activity. In this book, you will be asked to participate in a "Design Studio." This is the part of the course that is "hands on"—and, it's *fun!* But you will still learn the principles of good design practice (irrespective of your intended major in engineering), and you will have to integrate skills learned in construction, electrical circuits, logic, and computers in building a vehicle ("vehicle" in the widest sense of the word) that will have to compete against vehicles built by other young engineers whose motivation is to stop your vehicle from succeeding in achieving the same goals! You will learn how to organize data and the vital importance of good communication skills. You will present your ideas and your designs orally and in written format. In the design studio you will design and build increasingly complex engineering systems, starting with the tallest tower made from a single sheet of paper and ending with a controlled vehicle combining many parts into a system aimed at achieving complex goals.

As a start to the "minds-on" portion of the book, can you mentally take apart and put back together an imaginary automobile called the conceptual car? Instead of using wrenches and screwdrivers, your tools will be mental and computerized tools for engineering thought.

EXAMPLE 1

Figure 1 shows a generic car with numbered parts. Fill in the correct number corresponding to the parts in each of the blanks.[1]

As visually appealing as this figure is, an engineer would consider it inadequate because it fails to express the functional connections among the various parts. Expressing in visual form the elements and relationships involved in a problem is a crucial tool of engineering, called a **conceptual sketch.** A first step in an engineer's approach to a problem is to draw a conceptual sketch of the problem. Artistic talent is not an issue, nor is graphic accuracy. The engineer's conceptual sketch will not look exactly like the thing it portrays. Rather, it is intended to (1) help the engineer identify the elements in a problem, (2) see how groups of elements are connected together to form subsystems, and (3) understand how all those subsystems work together to create a system.

[1] 1-distributor, 2-transmission, 3-spare tire, 4-muffler, 5-gas tank, 6-starter motor, 7-exhaust manifold, 8-oil filter, 9-radiator, 10-alternator, 11-battery.

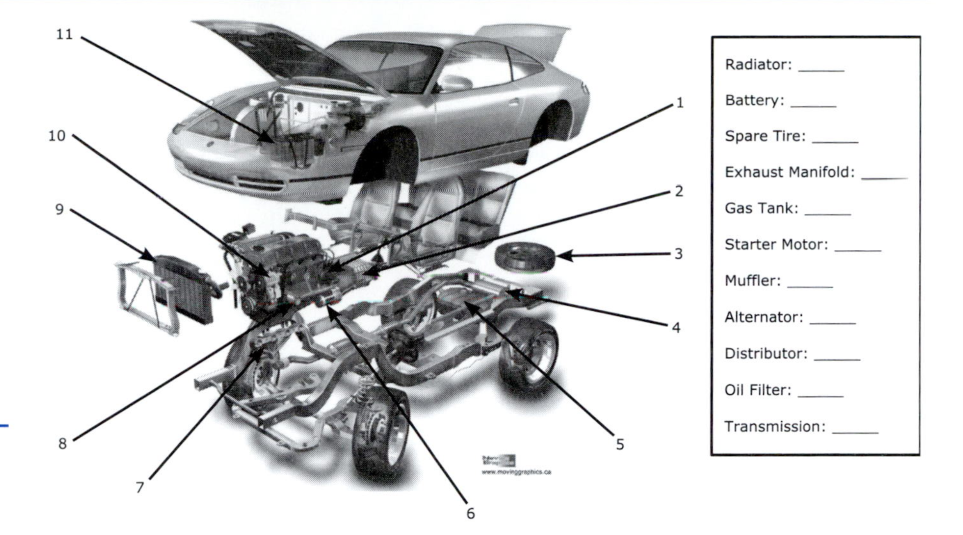

Figure 1

An exploded view of a modern automobile. © Moving Graphics

EXAMPLE 2

On a piece of paper draw a conceptual sketch of what happens when the driver of a car steps on the accelerator. Before you begin here are some questions you should think about:

1. What are the key components that connect the accelerator pedal to the engine?
2. Which ones are connected to each other?
3. How does doing something to one of the components affect the others?
4. What do those connections and changes have to do with accomplishing the task of accelerating the car?

Solution For *one* possible answer to this problem, see Figure 8 in Chapter 6. It shows that the accelerator controls the throttle, which in turn controls the flow of fuel and air into the engine.

For any engineering concept, many different conceptual sketches are possible. You are encouraged to draw conceptual sketches of each of the key points in the learning sections in this book.

Learning Sections

Table 1 contains a list of topics and some of the corresponding intellectual tools needed to answer questions about them (exclusive of the Design Studio, Part 2). Additionally, possible automotive questions and relevant parts of a car using these tools are provided. Most chapters of the book focus on one just topic, one question, and one tool.

This book uses the elements of a modern (smart) automobile as a medium to introduce you to the engineering profession. Engineering is

Table 1: Text Organization

Chapter	Topic	Typical Questions	Car Concept	Tools
1	**Introduction**	What makes a good automobile?	Conceptual car	**Conceptual sketch**
2	**Key concepts: Variables, units and numbers; force, weight, and mass**	How much force must be applied where the tire meets the road to cause a car to achieve "0 to 60 in seven seconds"?	Body, tire, road, etc.	**Dimensions and units method; significant figures; Newton's law of motion**
3	**Problem-solving methods**	Rental car analysis	System analysis	**Need to know how to solve; spreadsheets**
4	**Energy**	How much power must gasoline supply in order for a car to maintain a constant speed of 30 miles per hour?	Engine, radiator, wheels	**Energy conservation principle; control boundaries**
5	**Chemical energy and combustion**	How much carbon dioxide does a car emit for every mile it travels?	Fuel line, cylinder, air intake exhaust	**Molecules and moles; stoichiometry; chemical energy, A/F**
6	**Drive train**	How do the displacement, RPM, A/F, and gear ratio determine the power and torque a car delivers?	Cylinder, piston, valves, sparkplug, ignition, transmission	**Otto cycle analysis; gear train analysis**
7	**Electric circuits**	Why can a car battery supply enough power to start a car and run its lights, but not enough to run the car?	Battery, lights	**Circuit diagram; Ohm's Law; $P = I \times V$; switches**
8	**Logic and computers**	Why do cars have computers?	Seat belts, air flow sensor, fuel injector.	**Binary mathematics; truth tables**
9	**Control system design and mechatronics**	How does a cruise control work?	Accelerator, throttle, speed sensor, computer	**Block diagram; spreadsheets; feedback**
10	**Civil engineering: Kinematics and traffic flow**	How long should the on ramp to a superhighway be?	How many cars/mile of highway?	**Speed vs. time graph; highway capacity diagram**
11	**Introduction to materials engineering**	Of what material should a car bumper be made?	Bumper	**Stress-strain diagram**
12	**Introduction to bioengineering**	How does a seat belt save lives?	Seat belt	**Deceleration and bone stress; Gadd impact parameter**
13	**Introduction to chemical engineering**	How do you get gasoline from crude oil?	Fuel system, refinery principles	**Stoichiometry**
14	**Cars of tomorrow**	What will the cars of the future look like?		

creative design and analysis that uses energy, materials, motion, and information to serve human needs in innovative ways. Engineers express knowledge in the form of variables, numbers, and units. Engineers typically begin their attack on a problem with a conceptual sketch. There are many kinds of engineers, but all share the ideas and methods introduced in this book.

Ethics

What are personal ethics ... and what do they have to do with engineering?

Personal ethics are the standards of human behavior that individuals of different cultures have constructed to make moral judgments about personal or group situations. Ethical principles have developed as people have reflected on the intentions and consequences of their acts. Naturally, they vary over time and from culture to culture, resulting in conflict when what is acceptable in one culture is not in another. For example, the notion of privacy in U.S. culture is very strong, and a desk is considered an extension of that privacy, whereas in another culture, such as Japan, office space is open and one's desk would be considered public domain.

Suppose you are a passenger in a car driven by a close friend. The friend is exceeding the speed limit and has an accident. There are no witnesses, and his lawyer tells you that if you testify that your friend was not exceeding the speed limit, it will save him from a jail sentence. What do you do?

Lying is more accepted in cultures that stress human relationships, but it is less accepted in cultures that stress laws. People in cultures that emphasize human relationships would most likely lie to protect the relationship, whereas people in cultures that put greater value on laws would lie less in order to obey the law.

How do you reconcile a belief in certain moral absolutes such as "I will not kill anyone" with the reality that in some circumstances (e.g., war) it might be necessary to endanger or kill innocent people for the greater good? This issue gets particularly difficult if one denies tolerance to other faiths, yet the prevailing morality that most of us would describe as "good" is to extend tolerance to others.

The Five Cornerstones of Ethical Behavior

Here are some examples of codes of personal ethics. At this point you might want to compare your own personal code of ethics with the ones listed here.[2]

- Do what you say you will do.

[2]Manske, F.A., Jr., *Secrets of Effective Leadership*, Leadership Education and Development, Inc., 1987.

- Never divulge information given to you in confidence.
- Accept responsibility for your mistakes.
- Never become involved in a lie.
- Never accept gifts that compromise your ability to perform in the best interests of your organization.

Top Ten Questions You Should Ask Yourself When Making an Ethical Decision[3]

10. Could the decision become habit forming? ***If so, don't do it.***
9. Is it legal? ***If it isn't, don't do it.***
8. Is it safe? ***If it isn't, don't do it.***
7. Is it the right thing to do? ***If it isn't, don't do it.***
6. Will this stand the test of public scrutiny? ***If it won't, don't do it.***
5. If something terrible happened, could I defend my actions? ***If you can't, don't do it.***
4. Is it just, balanced, and fair? ***If it isn't, don't do it.***
3. How will it make me feel about myself? ***If it's lousy, don't do it.***
2. Does this choice lead to the greatest good for the greatest number? ***If it doesn't, don't do it.***

And the #1 question you should ask yourself when making an ethical decision:

1. Would I do this in front of my mother? ***If you wouldn't, don't do it.***

What Are *Professional* Ethics?

A professional code of ethics has the goal of ensuring that a profession serves the legitimate goals of *all* its constituencies: self, employer, profession, and public. The code protects the members of the profession from some undesired consequences of competition (for example, the pressure to cut corners to save money) while leaving the members of the profession free to benefit from the desired consequences of competition (for example, invention and innovation). Having a code of ethics enables an engineer to resist the pressure to produce substandard work by saying, "As a professional, I cannot ethically put business concerns ahead of professional ethics." It also enables the engineer to similarly resist pressures to allow concerns such as personal desires, greed, ideology, religion, or politics to override professional ethics.

[3]From: http://www.cs.bgsu.edu/maner/heuristics/1990Taylor.htm.

National Society of Professional Engineers (NSPE) Code of Ethics for Engineers

Engineering is an important and learned profession. As members of this profession, engineers are expected to exhibit the highest standards of honesty and integrity. Engineering has a direct and vital impact on the quality of life for all people. Accordingly, the services provided by engineers require honesty, impartiality, fairness, and equity, and must be dedicated to the protection of the public health, safety, and welfare. Engineers must perform under a standard of professional behavior that requires adherence to the highest principles of ethical conduct.[4]

Fundamental Canons[5]

Engineers, in the fulfillment of their professional duties, shall:

- Hold paramount the safety, health, and welfare of the public.
- Perform services only in areas of their competence.
- Issue public statements only in an objective and truthful manner.
- Act for each employer or client as faithful agents or trustees.
- Avoid deceptive acts.
- Conduct themselves honorably, responsibly, ethically, and lawfully so as to enhance the honor, reputation, and usefulness of the profession.

EXAMPLE 3

An Ethical Situation The following scenario is a common situation faced by engineering students. Read it and discuss how you would respond. What are your ethical responsibilities?

You and your roommate are both enrolled in the same engineering class. Your roommate spent the weekend partying and did not do the homework that is due on Monday. You did the homework, and your roommate asks to see it. You are afraid he/she will just copy it and turn it in as his/her own work. What are you ethically obligated to do?

a. Show your roommate the homework.
b. Show the homework but ask your roommate not to copy it.
c. Show the homework and tell the roommate that if the homework is copied, you will tell the professor.
d. Refuse to show the homework.
e. Refuse to show the homework but offer to spend time tutoring the roommate.

---

[4]See http://www.nspe.org/ethics/eh1-code.asp.
[5]Canons were originally church laws; the word has come to mean rules of acceptable behavior for specific groups.

Solution For the purposes of this course, the answer to an ethics question will consist of appropriately *applying a code of ethics.* In this example, The Five Cornerstones of Ethical Behavior will be used.

In subsequent chapters, the Code of Engineering Ethics will be used, but this does not constitute an endorsement of the code or any other particular code for personal ethics. Use of the Code of Engineering Ethics in subsequent answers, by contrast, *does* constitute a reminder that you must accept that code in your professional dealings if you want to be a professional engineer.

Let us see which of the Five Cornerstones apply here.

1. **Do what you say you will do.** If the teacher has made it clear that this is an individual assignment, then by participating in the assignment you have implicitly agreed to keep your individual effort private. Allowing one's homework to be copied means going back on this implicit promise. This implies that answer d) or e), "Refuse to show the homework," is at least part of the right answer.
2. **Never divulge information given to you in confidence.** Again, homework is implicitly a confidential communication between individual student and teacher. By solving the problem, you have created a confidential communication with the teacher. This is more support for choice d) or e).
3. **Accept responsibility for your mistakes.** Sharing your homework will enable your roommate to evade this standard. Being an accomplice in the violation of standards by others is itself an ethical violation. This is further support for choice d) or e).
4. **Never become involved in a lie.** Allowing your homework to be copied is participating in a lie: that the work the roommate turns in is his or her own work. This further supports choice d) or e).
5. **Never accept gifts that compromise your ability to perform in the best interests of your organization.** Since the roommate has not offered anything in exchange for the help, this standard appears not to apply in this case.

Four of the five cornerstones endorse choice d) or e), refuse to show the homework, while the fifth cornerstone is silent. These results indicate that your ethical obligation under this particular code of personal ethics is to refuse to show the homework.

Many people will find the Five Cornerstones to be incomplete because they lack a canon common to most of the world's ethical codes: the Golden Rule.[6] Including the Golden Rule would create the additional obligation to show some empathy for your roommate's plight, just as you would hope

[6]There are many versions of the Golden Rule in the world's major religions. Here's one attributed to Confucius: "Do not do to others what you would not like yourself."

to receive such empathy if you were in a similar situation. This suggests the appropriateness of choice e), offering to tutor the roommate in doing the homework. In much the same way, in subsequent exercises you may feel the need to supplement the Code of Engineering Ethics with elements from your own personal code of ethics. However, this must not take the form of *replacing* an element in the Code of Engineering Ethics with a personal preference.

What You Should Expect from This Book

The old joke goes something like this: "One term ago, I couldn't spell *injuneer*, but now I are one!" Well, you will *not* be an engineer at the end of this course, and, if anything you will learn at least that much. On the other hand, if you pay attention, you will learn the following.

1. Engineering is based on well-founded fundamental principles grounded in physics, chemistry, mathematics, and in logic, to name just a few skills.
2. Its most general principles include (1) definition of a Newtonian force unit, (2) conservation of energy, (3) conservation of mass, and (4) the use of control boundaries.
3. Engineering problems are multidisciplinary in approach, and the lines between each subdiscipline blur.
4. Engineering success is often based on successful teamwork.
5. The ability to carry out an introductory analysis in several engineering disciplines should be based on fundamental principles. It often depends on:

   a. Identifying the basic steps in the design process.
   b. Applying those basic steps to simple designs.
   c. Completing a successful team design project.

6. You will require sound thinking skills as well as practical hands-on skills.
7. The Design Studio will teach you that you will also need writing and oral presentation skills.
8. No project is complete without reporting what you have accomplished. Therefore, you will need to demonstrate effective communication skills.
9. Computer skills are essential to answer many kinds of practical engineering problems.
10. Engineering skills can be intellectually rewarding as well as demanding.
11. You should come away with some idea of what is meant by each subdiscipline of engineering and, for those who will continue to seek an engineering career, some idea of which of these subdisciplines most appeals to you.
12. We offer a practical way to ask if your behavior is ethical according to well-established engineering ethical canons. If you always act in

concordance, no matter the short-term temptations not to, you *will* come out ahead.

SUMMARY

Engineering is about changing the world by creating new solutions to society's problems. This text covers introductions to bioengineering, chemical engineering, civil engineering, computer and electronic engineering, control systems engineering, electrical engineering, materials engineering, and mechanical engineering.

What is common to the branches of engineering is their use of fundamental ideas involving variables, numbers, and units, and the creative use of energy, materials, motion, and information. Engineering is hands-on and minds-on. The hands-on activity for this book is the Design Studio, in which good design practices are used to construct a "vehicle" to compete against similar vehicles built by other students. You will learn how to keep a data book and how to protect your designs. You will use conceptual sketches to advance your designs.

Finally, you will take your first steps to learn the need for professional ethics in your career; this is an ongoing activity you will need irrespective of your specialization or job level in whatever direction your career takes you.

Chapter 2

© iStockphoto.com/Mari Mansikka

Key Elements of Engineering Analysis

The Elements of Engineering Analysis

Many physical problems of interest to engineers are modeled by mathematical analysis. In the following chapters you will learn about a few such models. All of those models and the analysis methods used to construct them will share five key elements. One of them, **numerical value,** is familiar to you. Answering a numerical question requires coming up with the right number. But in engineering that is only one part of answering such a question. This chapter introduces other core elements of engineering analysis: **variables, dimensions, units,** and **significant figures,** as well as a fail-safe method of dealing with units and dimensions.

The essential idea to take away from this chapter is that arriving at the right numerical value in performing an analysis or solving a problem is only one step in the engineer's task. The result of an engineering calculation must involve the appropriate variables; it must be expressed in the appropriate units; it must express the numerical value (with the appropriate number of digits; or significant figures); and it must be accompanied by an explicit method so that others can understand and evaluate the merits and defects of your analysis or solution.

There is one variable introduced in this chapter that also has a strong claim to appear in another chapter that deals with energy and related subjects. That variable is **force,** and it is the scaffold on which much of modern engineering, as well as "classical" physics relies. The definition of *force* and its associated units is crucial to what follows in much of this text.

It is the strongest example of the notion of units and dimensions that appears in this text and has thus been placed in this chapter.

In addition to the preceding concepts, modern engineers have computerized tools at their fingertips that were unavailable just a generation ago. Because these tools pervasively enhance an engineer's productivity, it is necessary for the beginning engineer to learn them as soon as possible in his or her career. Today, all written reports and presentations are prepared on a computer. But there is another comprehensive computer tool that all engineers use: **spreadsheets.** This tool is another computer language that the engineer must master. We put its study in Chapter 3 so you can soon get some practice in the use of this tool.

Variables

Engineers typically seek answers to such questions as "How hot will this get?" "How heavy will it be?" "What's the voltage?" Each of these questions involves a variable, a precisely defined quantity describing an aspect of nature. What an engineering calculation does is different from what a pure mathematical calculation might do; the latter usually focuses on the final numerical answer as the end product of an analysis. For example, $\pi = 3.1415926\ldots$ is a legitimate answer to the question "What is the value of π?" The question "How hot?" is answered using the variable "temperature." The question "How heavy?" uses the variable "weight." "What voltage?" uses the variable "electric potential."

For our purposes, variables will almost always be defined in terms of measurements made with familiar instruments such as thermometers, rulers, and clocks. Speed, for example, is defined as a ruler-measurement, distance, divided by a clock-measurement, time. This makes possible what a great engineer and scientist William Thomson, Lord Kelvin (1824–1907), described as the essence of scientific and engineering knowledge.

> I often say that when you can measure what you are speaking about, and express it in numbers, you know something about it; but when you cannot measure it, when you cannot express it in numbers, your knowledge is of a meager and unsatisfactory kind: it may be the beginning of knowledge, but you have scarcely, in your thoughts, advanced to the stage of science, whatever the matter may be.

But expressing something in numbers is only the beginning of engineering knowledge. In addition to variables based on measurements and expressed as numbers, achieving Lord Kelvin's aspiration requires a second key element of engineering analysis: units.

Units

What if you are stopped by the highway patrol on a Canadian highway and get a ticket saying you were driving at "100"? You would probably guess that the variable involved is speed. But it would also be of interest to know if the claim was that you were traveling at 100 *miles* per hour or 100 *kilometers* per hour, knowing that 100 kph is only 62 mph. Units can and do make a difference!

While the fundamental laws of nature are independent of the **system of units** we use with them, in engineering and the sciences a calculated quantity always has two parts: the numerical value *and* its associated units, if any.[1] Therefore, the result of any engineering calculation must always be correct in two separate categories: ***It must have the correct numerical value, and it must have the correct units.***

Units are a way of quantifying the underlying concept of *dimensions*. Dimensions are the fundamental quantities we perceive such as mass, length, and time. Units provide us with a numerical scale whereby we can carry out a measurement of a quantity in some dimension. On the other hand, units are established quite arbitrarily and are codified by civil law or cultural custom. How the dimension of length ends up being measured in units of feet or meters has nothing to do with any physical law. It is solely dependent on the creativity and ingenuity of people. Therefore, the basic tenets of units systems are often grounded in the complex roots of past civilizations and cultures.

The SI Unit System

The international standard in units is the SI system or, officially, "the International System of Units" (or "Le Système International d'Unités" in French), which has been abbreviated to "SI" in many languages. It is the standard of modern science and technology and is based on MKS units (meter, kilogram, second). The fundamental units in the SI system are:

The meter (m), the fundamental unit of length
The second (s), the fundamental unit of time
The kilogram (kg), the fundamental unit of mass
The degree kelvin (K), the fundamental unit of temperature
The mole (mol), the fundamental unit of quantity of particles
The ampere (A), the fundamental unit of electric current

Table 1 illustrates a variety of SI units that were all derived from proper names of scientists who made discoveries in each of the fields in which these units are used. All unit names are written without capitalization (unless of course they appear at the beginning of a sentence), regardless of whether they were derived from proper names. However, when the unit is to be

[1]Some engineering quantities legitimately have no associated units—for instance, a ratio of like quantities.

abbreviated, the abbreviation is capitalized if the unit was derived from a proper name. Note that abbreviations use two letters *only* when necessary to prevent them from being confused with other established unit abbreviations[2] (e.g., Wb for the magnetic field unit "weber" pronounced as in German, "veyber" to distinguish it from the more common W, the watt unit of power), or to express prefixes (e.g., kW for kilowatt). Also, a unit abbreviation is never pluralized, whereas the unit's name may be pluralized. For example, kilograms are abbreviated as kg, and *not* kgs, newtons as N and *not* Ns, and so forth. And finally, unit name abbreviations are *never written with a terminal period* unless they appear at the end of a sentence.

Table 1

Some SI Units and Their Abbreviations

ampere (A)	henry (H)	pascal (Pa)
becquerel (Bq)	hertz (Hz)	siemens (S)
celsius (°C)	joule (J)	tesla (T)
coulomb (C)	kelvin (K)	volt (V)
farad (F)	newton (N)	watt (W)
gray (Gy)	ohm (Ω)	weber (Wb)

All other units whose names were not derived from the names of historically important people are both written and abbreviated with lowercase letters—for example, meter (m), kilogram (kg), second (s), and so forth. The correct abbreviation of seconds is s, not sec. or secs., and so forth.

We have assumed you are familiar with **mass** and you may indeed think you are, but it is not a trivial concept. As you will soon learn, mass is *not* weight. In a gravitational field, mass certainly produces weight, but the mass is present even where there is no gravity such as in outer space. Mass is best considered as the quantity of matter; it is a property of the substance.

In examples that follow in this text, we introduce a failsafe method that will always allow you to develop the correct units. We will follow the numerical part of the question with square brackets [...] enclosing the unit conversions that are needed. For example, knowing there are 12 inches in one foot would produce the units conversion factor [12 inch/ft], so if we wanted to covert $12.7\,\text{ft}^2$ to in^2, we would write $12.7\ [\text{ft}^2][12\ \text{in/ft}]^2 = 12.7 \times 144\ [\text{ft}^2][\text{in/ft}]^2 = 1830\,\text{in}^2$.

While this may seem ponderous in this example, in examples that are more complicated **it is essential** to follow this methodology. We will attempt to be consistent in presenting solutions and follow this technique throughout this text.

[2]Non-SI unit systems do not generally follow this simple rule. For example, the English length unit, foot, could be abbreviated "f" rather than "ft." However, the latter abbreviation is well established within society, and changing it at this time would only cause confusion.

Table 2

Some Derived SI Units

Quantity	Name	Symbol	Formula	Fundamental Units
Frequency	hertz	Hz	1/s	s^{-1}
Force	newton	N	$kg \cdot m/s^2$	$m \cdot kg \cdot s^{-2}$
Energy	joule	J	$N \cdot m$	$m^2 \cdot kg \cdot s^{-2}$
Power	watt	W	J/s	$m^2 \cdot kg \cdot s^{-3}$
Electric charge	coulomb	C	$A \cdot s$	$A \cdot s$
Electric potential	volt	V	W/A	$m^2 \cdot kg \cdot s^{-3} \cdot A^{-1}$
Electric resistance	ohm	Ω	V/A	$m^2 \cdot kg \cdot s^{-3} \cdot A^{-2}$
Electric capacitance	farad	F	C/V	$m^{-2} \cdot kg^{-1} \cdot s^4 \cdot A^2$

Table 3

SI Unit Prefixes

Multiples	Prefixes	Symbols	Submultiples	Prefixes	Symbols
10^{18}	exa	E	10^{-1}	deci	d
10^{15}	peta	P	10^{-2}	centi	c
10^{12}	tera	T	10^{-3}	milli	m
10^{9}	giga	G	10^{-6}	micro	μ
10^{6}	mega	M	10^{-9}	nano	n
10^{3}	kilo	k	10^{-12}	pico	p
10^{2}	hecto	h	10^{-15}	femto	f
10^{1}	deka	da	10^{-18}	atto	a

There are many SI units pertaining to different quantities being measured and their multiples thereof (Tables 2 and 3, respectively). Some of the rationale for their fundamental units will become clearer as we proceed.

Table 2 has value beyond merely listing these units: It relates the unit's name to the fundamental MKS units—that is, the fact that a frequency is expressed in "hertz" may not be as useful as the fact that a hertz is nothing but the name of an inverse second, s^{-1}. In Table 3, multiples of these quantities are arranged in factors of 1,000 for convenience for very large and very small multiples thereof.[3]

Force, Weight, and Mass

Central to any scientific set of units is the definition of *force*. You have probably had some prior introduction to this concept in your high school physics classes. For example, for a constant mass system, **Newton's Second Law of Motion** is correctly stated as follows:

Force on a mass is proportional to the acceleration it produces.

[3]In recent years, a new subcategory of materials and technology known as "Nanotechnology" has arisen; it is so called because it deals with materials whose size is in the nanometer or 10^{-9} m range.

Newton's Second Law can be written in an equation form:

$$F \propto ma \tag{1}$$

where a is the **acceleration** of mass m. To convert Newton's force law into an "equality," we need to introduce a "constant of proportionality."

Suppose there exists a set of units for which the force F_1 accelerates the mass m_1 by a_1. Then Newton's Second Law can be written as:

$$F_1 \propto m_1 a_1 \tag{2}$$

If we now eliminate the proportionality by dividing the general force-defining equation (1) by equation (2)

$$\frac{F}{F_1} = \frac{m}{m_1}\frac{a}{a_1} \quad or \quad F = \left(\frac{F_1}{m_1 a_1}\right) ma \tag{3}$$

Clearly, the proportionality constant, now explicitly the ratio of a specific force to a specific mass and to a specific acceleration, is very important to the calculations made with this equation. We must choose both the magnitude and the dimensions of this ratio. Any consistent set of units will satisfy this equation. With this degree of flexibility, it is easy to see how a large number of different force systems have evolved.

EXAMPLE 1

Suppose you define a *mass* unit called the "slug"—a rather ugly word that is supposed to suggest the phrase "*gravitational* foot-pound-second system." The slug is the *mass* that accelerates by exactly 1 ft/s^2 when a force of exactly 1 lb force is applied to it.

What is the *weight* of a mass of 5.00 slug? (*Weight* is the term for a force in Earth's gravitational field, which produces an acceleration of 32.2 ft/s^2.)

Equation 3 is the applicable principle for this problem:

$$F = \left(\frac{F_1}{m_1 a_1}\right) ma \text{ in which } \left(\frac{F_1}{m_1 a_1}\right) = \frac{1}{1 \times 1} = 1\left[\frac{\text{lb force} \times \text{s}^2}{\text{slug} \times \text{ft}}\right]\text{; hence}$$

$$F = \left(\frac{F_1}{m_1 a_1}\right) mg = 1\left[\frac{\text{lb force} \times \text{s}^2}{\text{slug} \times \text{ft}}\right] \times 5.00[\text{slug}] \times 32.2\left[\frac{\text{ft}}{\text{s}^2}\right] = 161 \text{ lb force.}$$

In fact, the slug system is still preferred in the United States by some engineering specialties. Its obvious advantage is that the constant of proportionality for Newton's law is exactly unity so Newton's law can be written in the familiar form of $F = ma$, with the stipulation that mass m is in units of slugs. Its other disadvantage is that you probably have little "feel" for the size of a unit of slugs. ■

The SI system is somewhat similar to the slug system of units but with the choice for the constant of proportionality being both easy and logical: Pick the ratio $F_1/m_1 a_1$ so that its numerical value is exactly "1," and pick it so that it *effectively* has no (i.e., [0]) dimensions.[4] The unit of force thus identified is called the *newton* when the unit of mass is the kilogram and the acceleration is exactly 1 m/s^2. It other words, the force F_1 of 1 N is defined to be that which causes the acceleration a_1 of exactly 1 m/s^2 when acting on a mass m_1 of exactly 1 kg. Thus, in the SI system, $F_1/m_1a_1 \equiv 1$ and Newton's Second Law is written:

$$F = ma \tag{4}$$

This simplification leads to the familiar form of Newton's Law of Motion that most of you have seen before. Since the conversion factor F_1/m_1a_1,—that is, [N/(kg $\times$ m/s^2)]—is exactly unity, we have dropped it altogether.

EXAMPLE 2

What is the force in newtons on a body of mass 102 g (0.102 kg) that is accelerated at 9.81 m/s^2?

Equation 4 is the principle we use: $F = ma = 0.102 \times 9.81[\text{kg}][\text{m/s}^2] = 1.00[\text{kg-m/s}^2] = 1.00$ N. ■

The last two examples use an acceleration of special interest, which is that caused by Earth's gravity: $g = 32.2$ ft/s^2 = 9.81 m/s^2. Looking at equation 4, the SI force acting on 1 kg mass due to gravity is not 1 N but 9.81 N—a fact that causes distress to newcomers to the subject. Weight is thus just a special force—that due to gravity. In this sense, equation 4 can be modified for the acceleration of gravity to yield that special force we call "weight," W, by writing it as:

$$W = mg \tag{5}$$

EXAMPLE 3

What is the *weight* in newtons of a *mass* of 0.102 kg?
Equation 5 is the underlying principle; the arithmetic is identical to that of Example 2: the force on 0.102 kg of mass (which is about 4 oz in common English units) is just about 1.00 N; in other words, one newton is just about the weight of a small apple here on Earth! Perhaps this will help you mentally imagine the magnitude of a force stated in newtons. ■

In addition to SI units, a North American engineer must master at least one of the other systems that relates mass and force, one whose persistence in the United States is due more to custom than logic: it is the **Engineering English** system of units. In this system, the conversion factor between force and mass $\times$ acceleration is *not* unity. Because of this we must carry an explicit proportionality constant every time we use this unit system. And because the proportionality factor is not unity, it requires the use of an explicit set of units as well.

[4]Read [0] to mean "dimensionless."

This system has also evolved a rather unfortunate convention regarding both the "pound" unit and the definition of force. It was decided that the name "pound" would be used both for mass and weight (force). Since mass and force are distinctly different quantities, a modifier had to be added to the "pound" unit to distinguish which (mass or weight) was being used. This was solved by simply using the phrase "pound mass" or "pound force" with the associated abbreviations "lbm" and "lbf," respectively, to distinguish between them.

In the English Engineering system it was decided that a "pound mass" should weigh a "pound force" at standard gravity. (Standard gravity accelerates a mass by 32.174 ft/s^2.) This has the helpful convenience of allowing one the intuitive ability to understand immediately what is meant by, say, a force of 15 lbf. It would be the force you would experience if you picked up a rock of mass 15 lbm on the Earth's surface.

This convenience was accomplished by setting the ratio $(F_1/m_1a_1) = 1/32.174$ [lbf s^2/lbm ft]. In other words, 1 lbf is defined as the force that will accelerate exactly 1 lbm by exactly 32.174 ft/s^2. The designers of the Engineering English system cleverly decided to define the inverse of the proportionality constant as g_c given by:

$$g_c \equiv 32.174 \frac{lbm \cdot ft}{lbf \cdot s^2}$$

The g_c symbolism was originally chosen because the *numerical value* (but not the dimensions) of g_c is the same as that of the acceleration in standard gravity in the English Engineering units system. However, this is awkward because it tends to make you think that g_c is the same as (i.e., equal to) the acceleration due to local gravity, g, which it definitely is not. The constant g_c is nothing more than a proportionality constant with dimensions of mass $\times$ length/(force $\times$ time^2). Because the use of g_c is so widespread today in the United States and because it is important that you are able to recognize the meaning of g_c when you see it elsewhere, it will be used in the relevant equations in this course except when we are using the much more convenient (and universal) SI units. For example, in the English Engineering unit system, we will henceforth write Newton's Second Law as:

$$F = \frac{ma}{g_c} \tag{6}$$

The consequence of this choice for F_1, m_1, and a_1 as expressed by g_c is that you can easily calculate the force in lbf corresponding to an acceleration in ft/s^2 and a mass in lbm.

EXAMPLE 4

What is the force necessary to accelerate a mass of 65 lbm at a rate of 15.0 ft/s^2?

Since the problem is stated in English units, assume the answer is required in these units. Equation 6 is the principle used here:

$$F = \frac{ma}{g_c} = \frac{65 \times 15.0}{32.174} = [\text{lbm}]\left[\frac{\text{ft}}{\text{s}^2}\right]\left[\frac{\text{lbf s}^2}{\text{lbm ft}}\right] = 30.3 \text{ lbf}$$

Notice how the **units** as well as the value of g_c enter the problem; without g_c our "force" would be in the nonsense units of lbm/ft/s^2 and our calculated numerical value in those nonsense units would be 975.

There are another series of consequences, too. In a subsequent chapter of this book you will be introduced to the quantity "kinetic energy." An engineer using the SI system would define kinetic energy as $= ½\, mv^2$ (here the v stands for speed). However, an engineer using the Engineering English system would define kinetic energy as ½ $(m/g_c)v^2$. The convenient mnemonic for all applications of the Engineering English system is "**When you see a mass, divide it by g$_c$.**" Of course, it is logically safer to argue the units using the [..] convention previously introduced. For example, if the grouping ½ mv^2 were expressed in English units, it would be [lbm] [ft/s]^2 if g_c were ignored. This is a meaningless collection of units, but, using the g_c conversion factor, the definition of kinetic energy now would read dimensionally as [lbm] [lbf s^2/lbm ft][ft/s]^2 = [ft lbf], a legitimate unit of energy in the English system (see Chapter 3).

Understand that the force conversion constant g_c has the same value everywhere in the universe—that is, the value of 32.174 lbm ft/lbf s^2, even if its domain of acceptance is confined to the United States! In this regard, as previously stated, it should not be confused with the physical quantity g, the acceleration due to gravity, which has different numerical values at different locations (as well as different dimensions from g_c).

The concept of "weight" always has the notion of the *local* gravity associated with it. Weight thus varies with location—indeed only slightly over the face of the Earth but significantly on nonterrestrial bodies. To restate what has been learned: the **weight** of a body of mass m on the surface of the earth is the force experienced on it due to the acceleration of "g" due to gravity—that is, $W = (mg/g_c)$ (g = 32.174 ft/s^2) in English Engineering units, or $W = mg$ in SI units (g = 9.81 m/s^2). The weight of a body of mass m when the local gravity is g' is mg'/g_c in English units and mg' in SI. A person on the International Space Station experiences *micro*gravity, a much smaller g than we do on Earth, and a person on the Moon experiences about $\frac{1}{6}$ of g compared to a person on the surface of the Earth. However, if that person is an engineer using the Engineering English system of units, he/she must use the same numerical value for g_c wherever in the universe he/she may be.

In summary: When you see Newton's Second Law written as $F = ma$ (in physics books, for example), you must use a unit system in which the proportionality constant between force and mass × acceleration is unity and is also effectively dimensionless, such as in the SI (MKS) system.[5]

EXAMPLE 5

a) What is the weight on Earth in Engineering English units[6] of a 10.0 lbm?
b) What is the weight on Earth in SI units of a 10 kg mass?
c) What is the mass of a 10.0 lbm object on the Moon (local $g = \frac{1}{6}$ that of Earth's)?
d) What is the weight of that 10.0 lbm object on the Moon?

a) $W = mg/g_c$ in English units. Thus, $W = 10.0 \times 32.2/32.2$ [lbm][ft/s^2]/[lbm ft/lbf s^2] = 10.0 lbf.

b) $W = mg$ in SI units. Thus, $W = 10.0 \times 9.81$ [kg][m/s^2] = 98.1 [kgm/s^2] = 98.1 N. (See Table 2 for the definition of newtons in terms of MKS fundamental units.)

c) Mass is a property of the material. Thus, the object still has a mass of 10.0 lbm on Earth, on the Moon, or anywhere in the cosmos.

d) $W = mg/g_c$ in English units. On the Moon, $g = 32.2/6.0$ ft/s^2. Thus, on the Moon, $W = 10 \times (32.2/6.0)/32.2$ [lbm][ft/s^2]/[lbm ft/lbf s^2] = 10.0/6.0 = 1.67 lbf.

Until the mid-twentieth century, most English-speaking countries used one or more forms of the Engineering English units system. But because of world trade pressures and the worldwide acceptance of the SI system, many engineering textbooks today present examples and homework problems in both the Engineering English and the SI unit systems. The United States is *slowly* converting to common use of the SI system. However, it appears likely that this conversion will take at least a significant fraction of your lifetime. So to succeed as an engineer in the United States, you must learn the Engineering English system. Doing so will help you avoid future repetition of such disasters as NASA's embarrassing loss in 1999 of an expensive and scientifically important Mars Lander due to an improper conversion between Engineering English and SI units!

Significant Figures

Having defined your variable and specified its units, you now calculate its value. Your calculator will obediently spew out that value to as many digits as its display will hold. But how many of those digits really matter? How many of those digits actually contribute toward achieving the purpose of engineering, which is to design useful objects and systems and to understand, predict, and control their function in useful ways? This question introduces into engineering analysis a concept you have possibly seen in your high school science or mathematics courses: the concept of significant figures.

[5]Or in the "cgs" (centimeter-gram-second) system, the now outdated predecessor to SI.
[6]Generally, you should give your answers in the "natural" set of units that is suggested by the problem.

Even the greatest scientists have made the same kind of howling errors by quoting more significant figures than were justified. Newton wrote that "the mass of matter in the Moon will be to the mass of matter in the Earth as 1 to 39.788" (Principia, Book 3, proposition 37, problem 18). "As the present-day figure of the ratio between the mass of the Earth and the mass of the Moon is $M_e/M_m = 81.300588$, it is clear that Newton had gone wrong somewhere ... his quoted value of M_e/M_m to five significant figures being completely unjustified."[7]

The use of the proper number of significant figures in experimental work is an important part of the experimentation process. Reporting a measurement of, say, 10 meters, or as 10. meters (*notice the period after the zero*), or as 10.0 meters, or as 10.00 meters, implies something about how accurately the measurement was made. The implication of 10 meters as written is that the accuracy of our measuring rule is of the order of ± 10 meters. However, 10. meters implies the measurement was good to ± 1 meter. Likewise, 10.0 meters implies accuracy to 0.1 meter and 10.00 meters to 0.01 meters and so on, a convention we will try to maintain henceforth in this book. Unless they are integers, numbers such as 1, 30, 100 have one significant figure.

The concept of significant figures arises since arithmetic alone will not increase the accuracy of a measured quantity.[8] If arithmetic is applied indiscriminately it might actually decrease the accuracy of the result. The use of significant figures is a method to avoid such blunders as $10/6 = 1.666666667$ (as easily obtained on many electronic calculators), whereas the strict answer is 2! (six and 10 are apparently known only to 1 *significant* figure, and if they represent real physical measurements, they apparently have not been measured to the implied accuracy of the arithmetical operation that produced 1.666666667.) In this sense, physical numbers differ from pure mathematical numbers.

DEFINITION

A significant figure[9] is any one of the digits 1, 2, 3, 4, 5, 6, 7, 8, 9, and 0. Note that zero is a significant figure *except* when it is used simply to fix the decimal point or to fill the places of unknown or discarded digits.

The number 234 has three significant figures, and the number 7305 has four significant figures, since the zero within the number is a legitimate significant digit. But leading zeroes before a decimal point are not significant.

[7]D. W. Hughes, "Measuring the Moon's Mass," from *Observatory*, Vol. 122, April 2002, p. 62.

[8]The averaging of a number of repeated measurements of the same quantity might seem to violate this statement, but it allows only increased confidence in the interval in which the averaged number will lie.

[9]There are many good websites on the Internet that deal with this concept. The following site has a self-test that you can use to check yourself: http://lectureonline.cl.msu.edu/~mmp/applist/sigfig/sig.htm.

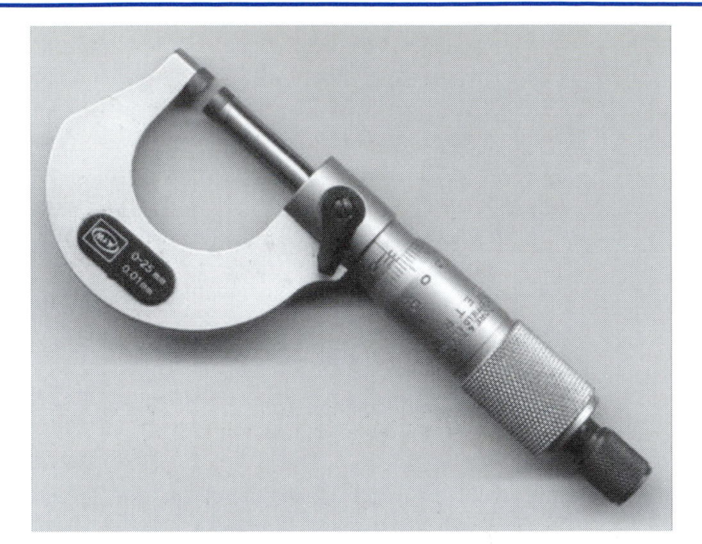

Figure 1

Machinist's Micrometer (Picture by permission of the Rochester Schools, New Hampshire)

Thus, the number 0.000452 has three significant figures (4, 5, and 2), the leading zeroes (including the first one before the decimal point) being place markers rather than significant figures. How about trailing zeroes? For example, the number 12,300 is indeed twelve thousand, three hundred, but we can't tell without additional information whether the trailing zeroes represent the **precision**[10] of the number or merely its magnitude. If the number 12,300 was precise only to ± 100, it has just three significant figures. If it were truly precise to ± 1, then all five figures are significant. In order to convey unequivocally which ending zeroes of a number are significant, it should be written as 1.2×10^4 if it has only two significant figures, as 1.23×10^4 if it has three, as 1.230×10^4 if it has four, and as 1.2300×10^4 if it has five.

The identification of the number of significant figures associated with a measurement comes only through knowledge of how the measurement was carried out. For example, if one measured the diameter of a shaft with a ruler, the result might be 3.5 inches (two significant figures), but if it were measured with a micrometer (Figure 1), it might be 3.512 inches (four significant figures).

Engineering calculations often deal with numbers having unequal numbers of significant figures. A number of logically defensible rules have been developed for various computations. These rules are actually the result of strict mathematical understanding of the propagation of errors due to

[10]The word "precision" has a specific meaning: it refers to how many times using independent measurements we can *reproduce* the number. If we throw many darts at a dartboard and they all cluster in the double three ring, they are precise but, if we were aiming at the bull's eye, they were not accurate!

arithmetical operations such as addition, subtraction, multiplication, and division. The rule for addition and subtraction is that:

RULE 1

The sum or difference of two values should contain no significant figures farther to the right of the decimal place than occurs in the least precise number in the operation.

For example, $113.2 + 1.43 = 114.63$, which must now be rounded to 114.6. The less precise number in this operation is 113.2 (having only one place to the right of the decimal point), so the final result can have no more than one place to the right of the decimal point. Similarly, $113.2 - 1.43 = 111.77$ must now be rounded to 111.8. This is vitally important when subtracting two numbers of similar magnitude, since their difference may be much less significant than the two numbers that were subtracted. For example, $113.212 - 113.0 = 0.2$ has only one significant figure even though the "measured" numbers each had four or more significant figures.

There is another rule for multiplication and division:

RULE 2

The rule for multiplication and division of figures is *The product or quotient should contain no more significant figures than are contained by the term with the least number of significant figures used in the operation.*

For example, $(113.2) \times (1.43) = 161.876$, which must now be rounded to 162, and $113.2/1.43 = 79.16$, which must now be rounded to 79.2 because 1.43 contains the least number of significant figures (i.e., three) in each case. Finally a rule for "rounding" is as follows:

RULE 3[11]

The rule for "rounding" numbers up or down is *When the discarded part of the number is 0, 1, 2, 3, or 4, the next remaining digit should not be changed. When the discarded part of the number is 5, 6, 7, 8, or 9, then the next remaining digit should be increased by one.*

For example, if we were to round 113.2 to three significant figures, it would be 113. If we were to round it further to two significant figures, it would be 110, and if we were to round it to one significant figure, it would be 100 with the trailing zeroes representing placeholders only. As another

[11]There is another round-off rule corresponding to Rule 3: The so-called "Bankers' Rule" was used before computers to check long columns of numbers. When the discarded part of the number is exactly 5 followed only by zeros (or nothing), then the previous digit should be rounded up if it is an odd number, but it remains unchanged if it is an even number. It was meant to average out any rounding bias in adding the columns.

example, 116.876 rounded to five significant figures is 116.88, which further rounded to four significant figures is 116.9, which further rounded to three significant figures is 117. As another example, 1.55 rounds to 1.6, but 1.54 rounds to 1.5.

SUMMARY

To summarize this chapter, the results of an engineering analysis must be correct in four ways. It must involve the appropriate variables. It must be expressed in the appropriate units. It must express the correct numerical value with the appropriate number of significant figures.

Newton's Second Law of Motion presents additional challenges in dimensional analysis, since so many possibilities are open to define its proportionality constant. In the MKS system, the proportionality constant is unity so one can write it as $F = ma$; in the English Engineering system, the proportionality constant g_c is chosen to have a value of $32.174\,\text{lbm ft/lbf s}^2$ and Newton's law is written as $F = ma/g_c$. Dependent quantities such as kinetic energy are likewise modified in the English Engineering system of units by dividing quantities that contain lbm by g_c.

EXERCISES

To help get you in the habit of applying these elements, in the following exercises you will be graded on use of all of the elements we have discussed in this chapter. In these problems, where necessary, assume that on the surface of the Earth, $g = 9.81\,\text{m/s}^2 = 32.2\,\text{ft/s}^2$ (i.e., each to three significant figures). Make sure you are reporting the solution to the proper number of significant figures.

1. If a U.S. gallon[12] has a volume of $0.134\,\text{ft}^3$ and a human mouth has a volume of $0.900\,\text{in}^3$, then how many mouthfuls of water are required to fill a 5.00 (U.S.) gallon can? (**A: 1.29×10^3 mouthfuls.**)

2. Identify whether you would perform the following unit conversions by definition, by conversion factors, by geometry, or by scientific law.

   (a) How many square miles in a square kilometer?
   (b) How many microfarads in a farad?
   (c) What is the weight on Earth in N of an object with a mass of 10 kg?
   (d) How many square miles on the surface of the Earth?

[12] Distinguish U.S. gallons from the old English measure of Imperial gallons; 1 Imp. gallon = 1.20 U.S. gallons.

3. The height of horses to the shoulder is still measured in the old unit of hands. There are 16 hands in a fathom and 6.0 feet in a fathom. How many feet high is a horse that is 13 hands tall? (**A: 4.9 ft.**)

4. An acre was originally defined as the amount of area that an oxen team could plow in a day.[13] Suppose a team could plow 0.4 hectare per day, where a hectare is $10^4\,m^2$. There are 1609 meters in a mile. How many acres are there in a square mile?

5. There are 39 inches in a meter. What is the area in the SI system of the skin of a spherical orange that is 4.0 inches in diameter? (**A: $3.3 \times 10^{-2}\,m^2$.**)

6. There are 39 inches in a meter. What is the volume in the Engineering English system of a spherical apple that is 10. cm in diameter?

7. Acceleration is sometimes measured in g's, where $1g = 9.8\,m/s^2$. How many g's correspond to the steady acceleration of a car doing "zero to sixty"[14] in 10.0 seconds? (**A: 0.27*g*.**)

8. What is your mass in kilograms divided by your weight in pounds? Do you have to step onto a scale to answer this question? How did you answer the question?

9. If power (measured in W, or watts) is defined as work (measured in J, or joules) performed per unit time (measured in s), and work is defined as force (measured in N or newtons) $\times$ distance (measured in m) and speed is defined as distance per unit time (measured in m/s), what is the power being exerted by a force of 1000. N on a car traveling at 30. m/s. (Assume force and speed are in the same direction, and treat all numbers as positive.) (**A: 3.0×10^4 W.**)

10. A rocket sled exerts 3.00×10^4 N of thrust and has a mass of 2.00×10^3 kg. What does it do "zero to sixty" in? How many g's (see problem 7) does it achieve?

11. A person pushes a crate on a frictionless surface with a force of 100. lbf. The crate accelerates at a rate of 3.0 feet per second^2. What is the mass of the crate in lbm? (**A: 1.07×10^3 lbm.**)

12. The force of gravity on the Moon is one-sixth as strong as the force of gravity on Earth. An apple weighs 1.0 N on Earth. What is (a) the mass of the apple on the moon, in lbm? (b) the weight of the apple on the Moon, in lbf? (Conversion factor: 1.00 kg = 2.20 lbm.)

[13] The furlong or "furrow-long" was the distance of 220 yards that the oxen could plow in a day times a width of one "chain" or 22 yards. Multiplied together they defined the area of an acre.

[14] In the language of the car enthusiast, a standard test to accelerate a car from a standing start to 60 mph is called "zero to sixty."

13. How many lbf does it take for a 4.0×10^3 lbm car to achieve 0 to 60 mph in 10. seconds? (**A: 1.1×10^3 lbf.**)

14. Suppose a planet exerted a gravitational force at its surface that was 0.6 the gravitational force exerted by Earth. What is g_c on that planet?

15. Suppose you were going to accelerate a 2000. kg car by the "Rube Goldberg" contraption shown in this picture. The fan (A) blows apples (C) off the tree (B) into the funnel and thus into the bag (D). The bag is pulled downward by the force of gravity (equal to the weight of the apples in the bag), and that force is transmitted via the pulley (E) to accelerate the car (F). About how many apples each weighing 1.00 N would have to fall into the bag in order to achieve 0 to 60. mph in 7.00 seconds? Assume the filled bag applies a constant force to the car, equal to the weight of the apples in the bag. (**A: 7.66×10^3 apples.**)

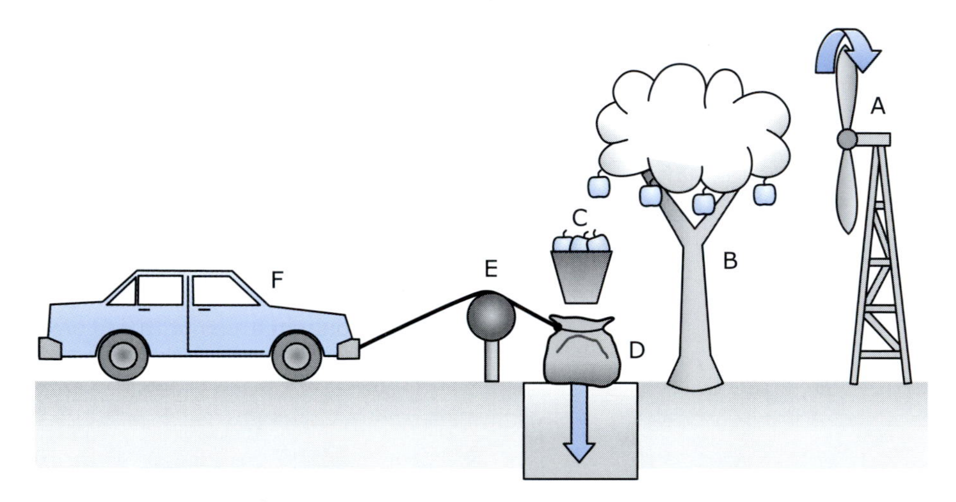

16. Calculate with the correct significant figures: (a) $100/(2.0 \times 10^2)$, (b) $1.0 \times 10^2/2.0 \times 10^2$. (**A: 0.5, 0.50.**)

17. Calculate with the correct significant figures: (a) 10/6, (b) 10.0/6, (c) 10/6.0, (d) 10./6.0, and (e) 10.0/6.00.

18. What is 2.68×10^8 minus 2.33×10^3 to the correct significant figures? (**A: 2.68×10^8.**)

19. A machinist has a sophisticated micrometer that can measure the diameter of a drill bit to 1/10,000 of an inch. What is the maximum number of significant figures that should be reported if the approximate diameter of the drill bit is:

    (a) 0.0001 inches
    (b) 0.1 inches
    (c) 1 inch

20. Round off to three significant places: 1.53, 15.345, 16.67, 102.04, −124.7, and 0.00123456.

21. Suppose you were going to design a front door and doorway to fit snugly enough to keep out the drafts, yet to be easy to open. (You are not showing off precision carpentry here, but merely designing a convenient ordinary door by standard methods.) The dimensions are to be given in inches. To how many significant figures would you specify the length and width of the door and doorway?[15] Assume a standard door is 30.0″ by 81.0″.

22. You are browsing the Internet and find some units conversion software that may be useful in this course. You would like to download the software on your PC at school and use it in this course. Do you?

    (a) Check with the Internet site to make sure this software is freeware for your use in this course.
    (b) Just download the software and use it because no one will know.
    (c) Download the software at home and bring it to school.
    (d) Never use software found on the Internet.

(**Suggested Answer:** Apply the Fundamental Canons: Engineers, in the fulfillment of their professional duties, shall:

1. Hold paramount the safety, health, and welfare of the public—**Does not apply.**

2. Perform services only in areas of their competence—**Does not apply.**

3. Issue public statements only in an objective and truthful manner—**Does not apply.**

4. Act for each employer or client as faithful agents or trustees—**Although in reality, the university is a vendor to the student, in the classroom situation, for educational purposes, the student plays the role of employee and the teacher plays the role of employer or client. Assuming the teacher is playing the role of employer or client, use of proprietary software can be harmful to that employer or client (the teacher or the teacher's employer, the university, could be sued by the software owner).**

5. Avoid deceptive acts—**This weighs against (c), which would be a deception aimed at circumventing the law.**

[15]The subject of tolerancing is important in mass manufacturing to ensure a proper fit with one part and another, since each part will not be exact their combined tolerance will determine how well, if at all, they fit together. It is the subject of significant statistical analysis. Applying the methods rigorously allowed the Japanese automotive industry to eclipse those of the rest of the world in terms of quality.

6. Conduct themselves honorably, responsibly, ethically, and lawfully so as to enhance the honor, reputation, and usefulness of the profession—**This weighs against options (b) and (c) and obliges you to check to see that the software is freeware; ignorance of the law is no excuse.**

Solution The applicable canons point in the same direction—do (a) or (d). Both are ethically acceptable. However, (d) represents an over-reaction in practical terms. You would be depriving yourself of a valuable tool.

23. On December 11, 1998, the Mars *Climate Orbiter* was launched on a 760 million mile journey to the Red Planet. On September 23, 1999, a final rocket firing was to put the spacecraft into orbit, but it disappeared. An investigation board concluded that NASA engineers failed to convert the rocket's thrust from pounds force to newtons (the unit used in the guidance software), causing the spacecraft to miss its intended 140–150 km altitude above Mars during orbit insertion, instead entering the Martian atmosphere at about 57 km. The spacecraft was then destroyed by atmospheric stresses and friction at this low altitude. As chief NASA engineer on this mission, how do you react to the national outcry for such a foolish mistake?

(a) Take all the blame yourself and resign?
(b) Find the person responsible, and fire, demote, or penalize that person?
(c) Make sure it doesn't happen again by conducting a software audit for specification compliance on all data transferred between working groups?
(d) Verify the consistent use of units throughout the spacecraft design and operations?

Chapter 3

© iStockphoto.com/Johnny Lye

Solving Problems and Spreadsheet Analyses

The Need to Know How to Solve Method

We suggest the **"need to know how to solve"** method for setting up and solving problems. It has the powerful benefit of enabling you to attack complicated problems in a systematic manner, thereby making success more likely. It may seem cumbersome in simple problems, but for those with an eye to their final class grades, it has an additional benefit of ensuring at least partial credit when a miscalculated number results in an incorrect answer! It also tracks an essential element of modern engineering practice by providing an "audit trail" by which past errors can be traced to their source easing subsequent corrections. The "need to know how to solve" method is a self-contained mnemonic device that should also help direct your thought processes.

Computers are essential for modern engineering; the engineer must communicate with them by learning a number of somewhat arcane computer languages. Fortunately, one of the more intuitive and powerful is **"spreadsheet analysis"**. This tool is simply another computer language that the engineer has to master. We will put its study at the end of this chapter so that the student can get practice as soon as possible, and we will use the technique throughout this book in the relevant exercise sections.

We have previously emphasized the need to find the proper variables, to express them in a consistent set of units, and to express the answer to the appropriate number of significant figures. But what guarantees do we have that the answer, even if correctly expressed in this way, is actually the correct solution to a problem or a useful result that accurately reflects or predicts the performance of an actual object or system in the real world? The correctness of an answer can be no better than the correctness of the methods and assumptions used to obtain it. Typically, the more systematic your method, the more likely you will be to avoid errors. In addition, the best way to ensure the correctness of a method is to submit it to evaluation and criticism. The use of an explicit method will leave an audit trail for your colleagues and customers (and for your future boss!).

In the "need know how to solve" method, the "need" is the variable for which you are solving. It is the very first thing you should write down. The "know" is the quantities that are known, either through explicit statement of the problem or through your background knowledge. Write these down next. They may be graphical sketches or schematic figures as well as numbers or principles. The "how" is the method you will use to solve the problem, typically expressed in an equation, although rough sketches or graphs may also be included. The "how" also includes the assumptions you make to solve the problem. Write it out in sentence or symbolic form before applying it. Only then, with *need*, *know*, and *how* explicitly laid out, should you proceed to *solve* the problem. Often it is also necessary to discuss the implications of your solution.

The pitfall this method guards against is the normal human temptation to look at a problem, *think* you know the answer, and simply write it down. Rather than gamble with this hit and miss method, the need know how to solve scheme gives a logical development that shows your thought processes.

Many of the problems you will solve in this book will involve specific equations. But do not be deceived into thinking that this method involves the blind plugging of numbers into equations if you do not understand where those equations came from. The essence of the method is to realize that once you have correctly defined the variable you **need,** you will realize that you **know** a lot more about the problem than you thought you did. With the aid of that knowledge, there is a method to **how to solve** the problem to an appropriate level of accuracy. Sometimes, this method will not require any equations at all.

In some sense, there is even a prior step before "need": ***Read the question and then reread the question.*** There is nothing more frustrating than solving the wrong problem!

It bears emphasizing: **The need know how to solve process is guaranteed to get a better grade even if the final answer is the same one that got an "F" when that was all you bothered to write down!** This parallels what a practicing engineer does to document his or her work; by leaving an "audit trail" prior mistakes can be spotted and rectified.

When applying the method, it may be especially helpful, as part of the "how" step, to sketch the situation described in the problem. Many people are discouraged from sketching by lack of artistic talent. Do not let this stop you. As the next example illustrates, even the crudest sketch can be illuminating.

EXAMPLE 1

A Texan wants to purchases the largest fenced-in square ranch she can afford. She has exactly \$320,000 available for the purchase. Fencing costs exactly \$10,000 a mile, and land costs exactly \$100,000 a square mile. How large a ranch, as measured by the length of one side of the square, can he buy?

Need: The length of a side of the largest square of land the Texan can buy.

Know: Fencing costs \$10,000 a mile, and land costs \$100,000 a square mile.

How: Let the unknown length = x miles. It may not be immediately obvious how to write an equation to find x. So sketch the ranch.

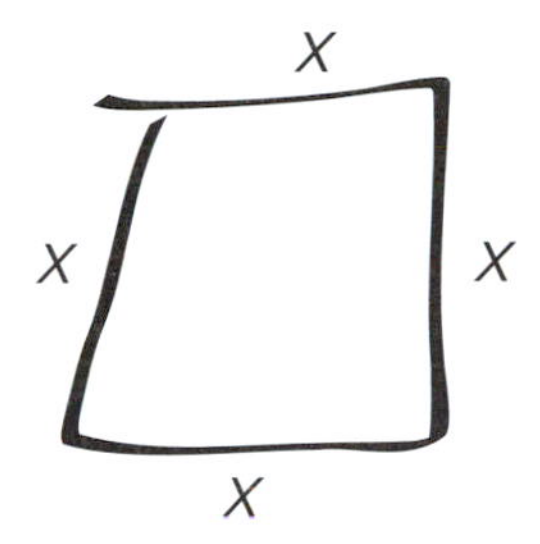

From the crude picture (and that is generally all you need), it is immediately obvious that the length of the fence surrounding the ranch is 4x, and the area of the ranch is x^2.
So the cost of the ranch is:

$$\text{Length of fence [miles]} \times 10{,}000\ [\$/\text{mile}] + \text{area of ranch [square miles]} \times 100{,}000\ [\$/\text{sq.mi}] = \$320{,}000 \text{ or}$$

$$4x\ [\text{miles}] \times 10{,}000\ [\$/\text{mile}] + x^2[\text{sq.mile}] \times 100{,}000\ [\$/\text{sq.mi}] = \$320{,}000$$

Solve:

$$(4x)(10{,}000) + x^2 \cdot (100{,}000) = 320{,}000$$

$\therefore\ 10x^2 + 4x - 32 = 0$, which is a simple quadratic equation whose solutions are: **x = +1.6 and –2.0 to two significant figures.**

So there are two solutions, $x = 1.6$ miles and $x = -2$ miles. Here one must apply a bit of knowledge so obvious that, although known, was not listed in the "known" section: that the length of the side of the ranch must be greater than zero (if for no other reason than to have room to put the cattle!). This yields the final answer. **The ranch is 1.6 miles on a side.** (It's always worth checking if this is correct: $1.6^2 \times \$100{,}000 + 4 \times 1.6 \times \$10{,}000 = \$3.2 \times 10^5$, which is arithmetically OK.)

EXAMPLE 2

Consider the following problem: How many barbershops are there in the city of Schenectady (population about 60,000 people)? Your first reaction may be "I haven't got a clue." (You may want to try the need to know how to solve method on this problem for yourself before looking at what follows.)

Need: The number of barbershops.

Know: There are about 60,000 people in the city of Schenectady, of whom about half are male. Assume the average male gets about 10 haircuts a year. A barber can probably do one haircut every half hour, or about 16 in the course of an eight-hour day. There are about three barbers in a typical barbershop.

How: The number of haircuts given by the barbers must be equal to the number of haircuts received by the customers. So if we calculate the number of haircuts per day received by all those 30,000 males, we can find the number of barbers needed to give those haircuts. Then we can calculate the number of shops needed to hold those barbers.

Solve: 30,000 males require about 10 [haircuts/male-year] × 30,000 [males] = 300,000 haircuts/year.
On a per day basis, this is about 300,000 [haircuts/year] × [one year/300 days] = 1000 haircuts/day.
This requires: 1000 [haircuts/day] × [1 barber-day/16 haircuts] = 62.5 barbers.
At three barbers per shop, this means 62.5 barbers × [1 shop/3 barbers] = 20.8 shops.

So the solution is **20 barbershops** (since surely not more than one of the digits is significant).

Looking in a recent yellow pages directory and counting the number of barbershops in Schenectady gives the result 23 barber shops. So we are within about 10 percent, which is a fortuitously good answer, given the roughness of our estimates and the many likely sources of error. Notice also in this problem how the method of carrying the units in [..] helps your analysis of the problem and directs your thinking. ■

Haircuts and barbershops and Texas ranches are not among the variables or units used by engineers. But *applying common sense* as well as equations is a crucial component of engineering analysis as demonstrated in the next example.

EXAMPLE 3

A 2.00 m steel wire is suspended from a hook in the ceiling by a mass of 10.0 kg that is tied to its lower end; the wire stretches by 15 mm. If this same mass is used to stretch a 4.00 m piece of the same steel wire, how much will it stretch?

Need: Stretch = ____ mm for a 4.0 m piece of wire.

Know: 10.0 kg mass will stretch a 2.0 m wire by 15 mm.

How: We need to deduce a possible law for extending a wire under load. Without experimentation we cannot know if our "theoretical law" is correct, but a mixture of common sense and dimensional analysis can yield a plausible relationship.

Solve: A longer wire should stretch further than a shorter one if otherwise equivalent. A larger mass presumably will also stretch the wire further.

A *plausible* model is thus the extension is proportional to the unstretched length of wire (all other things such as the stretching force and the wire's cross section being equal).

Then: $x \propto L$ and the ratio between the two cases is $\frac{x_2}{x_1} = \frac{L_2}{L_1}$

∴ The **new extension** $\mathbf{x_2} = 15 \times 4.0/2.0$ [mm][m/m] = **30.0 mm**.

The proportionality "law" derived in this manner is not guaranteed to be correct. To guarantee that it is physically correct, we need to either understand the underlying principles of wire stretching or have good experimental data relating wire stretching to the wire's length. In fact it is correct and is part of a physical law with the name "Hooke's Law." (See Chapter 11.)

In some problems, the need to know how to solve method is decidedly clumsy in execution; nevertheless, it is recommended that you use the basic method and learn how to modify it to suit the peculiarities of the problem at hand. You will soon find there is ambiguity among what you "know" and "how" and even to "solve." Never mind—the relevant information can go into one or the other baskets—just as long as it acts as a jog in the direction you need to solve the problem.

The five elements of engineering analysis introduced in this chapter will provide systematic ways of applying sense, both common and uncommon, to help meet engineering challenges.

Spreadsheet Analysis

Unfortunately, even though the need to know how to solve method will get you to an understanding of the answer, the subsequent mathematics may be too difficult to solve on a piece of paper, or perhaps you will need multiple cases, or graphical solutions, or other complications that prevent you from immediately writing down the answer you are seeking. Today, computers come to your aid; an experienced engineer on his/her laptop computer can obtain solutions to problems that were daunting a generation ago and might have required a room full of people hand cranking out small pieces of mathematical puzzles. In addition, spreadsheets do a superior job of displaying data—no small advantage if you want a busy person, perhaps your customer or your manager, to pay attention to the data.

Spreadsheets allow you to distinguish among three disparate concepts—***data, information,*** *and* ***knowledge.*** First, what are data?[1] What we call data are just a jumble of facts or numbers such 4, 3, 1, 1, 6, or an abstract set of colors such as red, violet, green, blue, orange, indigo, and yellow. They become "information" when you sort them out in some way such as 3, 1, 4, 1, 6, or red, orange, yellow, green, blue, indigo, and violet. They respectively become "knowledge" when the numerical sequence is interpreted as 3.1416 (i.e., π) and the color sequence is recognized as the visible spectrum arranged as dispersed by a prism (mnemonic "ROYGBIV").

Spreadsheets were invented at the Harvard Business School in 1979 by a student[2] who was apparently bored with repetitive hand calculations. He and a fellow entrepreneur made the first spreadsheet tool, "VisiCalc." It was originally intended simply to take the drudgery out of relatively mundane business ledgers. Not only do modern spreadsheets take the drudgery out of the computations, but they are also wonderful at displaying data in neat and tidy ways that can be real aids in understanding the displayed data.

The spreadsheet concept became one of the first so-called "killer-apps"—an application tool that no self-respecting personal computer could do without! Lotus 123 soon followed VisiCalc, and that was soon followed by Microsoft® Excel. In this book we will be explicitly using Excel, but the principles you learn here are readily transferred to IBM's Lotus 123 or to Corel's Quattro®. VisiCalc has faded from common use. Until VisiCalc, computers had tedious formatting requirements to display their results and typically as much time and energy was expended on getting results from the computer in an understandable way as was spent computing the results in the first place! What made spreadsheets interesting was the visual way they could handle large arrays of numbers; these numbers were identified by their position in these arrays by reading across their rows and down their columns. Interactions among these numbers were rapidly enhanced by providing sophisticated and complex mathematical functions that would manipulate the numbers in these arrays. Further, these arrays could be graphed in many ways so that an explicit crafted and individualized output was easily produced. Gone were most of the arcane skills in formatting that mainframe computers had demanded, and, since it was visual, silly mistakes instantly stood out like sore thumbs and could be expeditiously corrected. No wonder spreadsheets were one of the killer applications that drove the sales of PCs as indispensable tools for engineering, science, and for business.

[1]Singular form *datum*!
[2]http://www.bricklin.com/visicalc.htm.

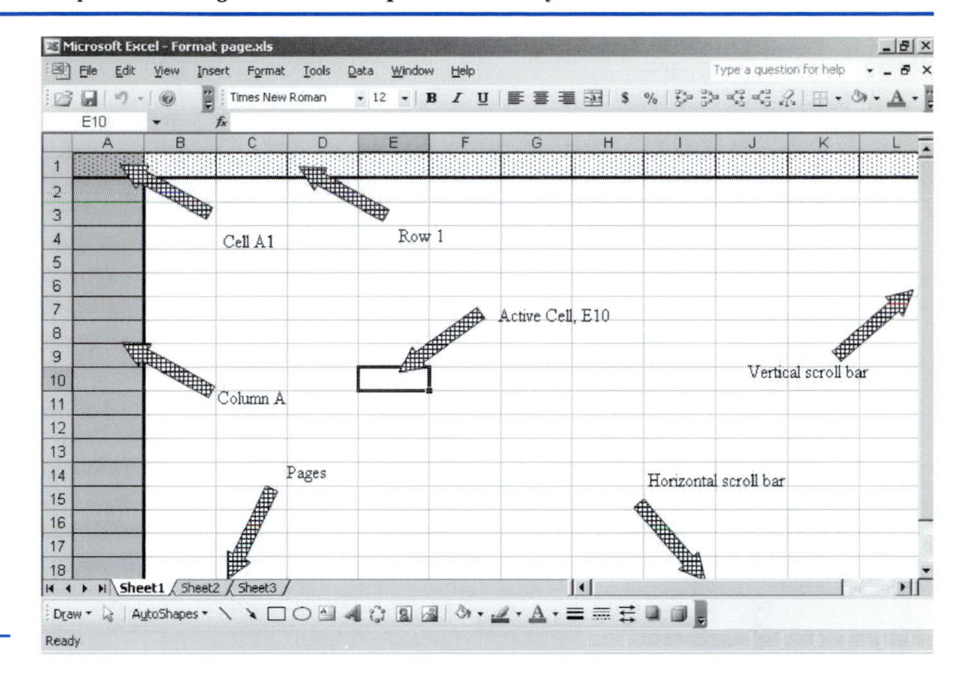

Figure 1

Typical Spreadsheet

Figure 1 is a reproduction of an Excel Spreadsheet with some guidelines as to its key parameters. It is navigated by the intersection of rows and columns. The first cell is thus A1. Active cells such as E10 in the example are boxed in a bold border. You can scroll across and down the spreadsheet using the horizontal and vertical scroll bars, respectively. Note there is even a provision to have multiple pages of spreadsheets that can be thought of as a 3D spreadsheet. Finally, the various icons and menu items that appear above and below this particular spreadsheet are idiosyncratic and can be changed at will to reflect your own personal preferences chosen among hundreds of different capabilities.

Cell Addressing Modes

The window that you view is just a small fraction of the whole spreadsheet; the remaining virtual spreadsheet has thousands of rows and hundreds of columns of available space as well as many interactive sheets per workbook, thus enabling "3D addressing." Movement within a window is by the scroll bars either across or down. Movement to virtual positions can also be effected by moving the cursor to the boundary you wish to extend.

Types of data that can be in a cell:

Alphanumeric—text
Headings
Labels

Numeric
Numbers: integers, floating point (meaning with decimals or exponents).
References to the contents of other cells (using the cell address).

Equations
Formulae such as = B1 + B4/H13 add the contents of cell B1 to the result of dividing the contents of cell B4 by the contents of cell H13.

Predefined functions
Sum
Average
Max
Min ... and many, many other functions, trigonometric, exponential, logical, statistical, and so on.

You will need to explicitly tell computer what data type is to be stored in a cell:

(a) Default is text or plain numbers
(b) " means a left-justified alphanumeric follows
(c) ' means left-justified text follows
(d) = means equation or numeric to follow
(e) + or – means numeric to follow[3]

EXAMPLE 4

FleetsR'US owns a fleet of rental cars; it wants to offer its customers an inclusive all-in-one-contract that takes account of the cost of the gasoline used. To estimate the cost and, therefore, pricing to their customers, they do a survey of the typical renter's journey and determine how many miles are driven by a selection of typical drivers. The raw data are given in this table.

Renter	City Miles	Suburban Miles	Highway Miles
Davis	16	28	79
Graham	10	31	112
Washington	4	7	158
Meyers	22	61	87
Richardson	12	56	198
Thomas	5	22	124
Williams	4	14	142

[3] But "+ name" means a specially defined macro or independent program.

Using a spreadsheet, give the total miles in each category of driving by each driver and the average miles driven in each category. Show the algorithm used in the spreadsheet.

Need: Driver mileage and average miles driven in each segment of renter's journey.

Know: Mileage data in table.

How: Total miles for each driver is the sum of the individual miles driven and the average mileage in each category of driving is the total miles divided by the number of drivers.

Solve: Use a spreadsheet to input the data table and the indicated arithmetical operations.

Table 1

Output of Spreadsheet Corresponding to Example 1

	A	B	C	D	E
1	**Miles Driven in Various Categories**				
2					
3					
4	**Renter**	**City Miles**	**Suburban Miles**	**Highway Miles**	**Total Miles**
5	Davis	16	28	79	123
6	Graham	10	31	112	153
7	Washington	4	7	158	169
8	Meyers	22	61	87	170
9	Richardson	12	56	198	266
10	Thomas	5	22	124	151
11	Williams	4	14	142	160
12					
13	Average	10.4	31.3	128.6	170.3

■

The sheet title heading conveniently appears in cell A1 (even though the actual input sprawls across adjacent cells). It helps in organizing the spreadsheet to entitle it descriptively so that it can be read by you (or others) later when the purpose of the spreadsheet has faded from immediate memory. In addition to the preceding spreadsheet, descriptive labels for some columns are given in row 4 as A4:E4. (Read the colon to mean it includes all the cells in the rectangular area prescribed by the limits of the addresses.) The labels may be simple such as these are, or they may be informative such as the units involved in the column (or row) that follows. Other descriptive labels that appear in this spreadsheet are A5:A11 and A13. All of these are **text** statements. The labels are the first steps in interpreting data as more than mere entries. Labels are there to codify what you want to arrange as information.

The numerical input data for this example are given in the block B5:D11. The results of some arithmetical operations appear as row E5:E11 and also as B13:E13. These data are the beginnings of information about the journeys. Note the variables are formatted to a reasonable number of significant figures. (The formatting structure appears under the headings "Format, Cells, Number" and it is highly recommended you experiment with fonts, borders, alignment, and number formats.) A block of cells can be highlighted before each operation and the formatting applied *en masse*.

Table 2 shows the mathematical operations behind Table 1 and what's in each cell. (Toggle these from the spreadsheet of Table 1 by using a *simultaneous* control key and a tilde (~). This is written in shorthand as ^~.)

Table 2 Expanded Spreadsheet from Table 1 (Using ^~)

	A	B	C	D	E
1	**Miles Driven**				
2					
3					
4	**Renter**	**City Miles**	**Suburban Miles**	**Highway Miles**	**Total Miles**
5	Davis	16	28	79	=SUM(B5:D5)
6	Graham	10	31	112	=SUM(B6:D6)
7	Washington	4	7	158	=SUM(B7:D7)
8	Meyers	22	61	87	=SUM(B8:D8)
9	Richardson	12	56	198	=SUM(B9:D9)
10	Thomas	5	22	124	=SUM(B10:D10)
11	Williams	4	14	142	=SUM(B11:D11)
12					
13	Average	=AVERAGE(B5:B11)	=AVERAGE(C5:C11)	=AVERAGE(D5:D11)	=AVERAGE(E5:E11)

Note the actual arithmetical operations now appear in cells E5:E11 and B13:E13. We declared them as such by inputting the mathematical operation after an "=" sign. The mathematical operations are the sum and the average, both with the specified ranges. Finally, notice in this view that the individual formatting for each cell has been removed and the actual input data as typed are revealed, while the long statement that appeared in cell A1 has now been truncated.

FleetsR'US now realizes that the gas mileage in each segment of the journeys is different; the length of the journey is less important than the gallons used in each segment. It also realizes that gas mileage is a moving target; vehicles can get both more and less fuel efficient (a large SUV vs. a compact hybrid vehicle, for example). It needs to keep its options open and to be able to update the spreadsheet when the market changes. Spreadsheets have a very powerful way of showing you how to do this. It introduces a powerful new tool: ***absolute and relative cell addressing.*** An absolute address is preceded by a "$" sign for either or both rows and columns; relative addressing is the default mode.

EXAMPLE 5

FleetsR'US wants to add the gas mileage to their spreadsheet in a flexible manner so it can be updated later if needed. It wants to now know the gallons used per trip and the average used by their typical drivers.
Here are the miles per gallon (mpg) by journey segment:

	City	Suburbs	Highway
mpg	12	18	26

Need: Average gallons per journey segment and average per driver.

Know: Mileage data from Example 1 plus mpg given above.

How: Divide mileage by mpg to get gallons used in each segment of the journey.

Solve: Here are several ways we can reprogram our spreadsheet. (We will just look at the spreadsheet mathematics using our ^~ method. Be assured that each cell *will* give the correct answer.)

Table 3 Constants and Absolute Addressing

	A	B	C	D	E
4	**Renter**	**City Gallons**	**Suburban Gallons**	**Highway Gallons**	**Total Gallons**
5	Davis	=16/12	=28/18	=79/26	=SUM(B5:D5)
6	Graham	=10/B17	=31/$C$17	=112/D$17	=SUM(B6:D6)
7	Washington	=4/B17	=7/$C$17	=158/D$17	=SUM(B7:D7)
8	Meyers	=22/B17	=61/$C$17	=87/D$17	=SUM(B8:D8)
9	Richardson	=12/B17	=56/$C$17	=198/$D17	=SUM(B9:D9)
10	Thomas	=5/B17	=22/$C$17	=124/$D17	=SUM(B10:D10)
11	Williams	=4/B17	=14/$C$17	=142/$D17	=SUM(B11:D11)
12					
13	Average	=AVERAGE(B5:B11)	=AVERAGE(C5:C11)	=AVERAGE(D5:D11)	=AVERAGE(E5:E11)
14					
15					
16		**City**	**Suburbs**	**Highway**	
17	**mpg**	12	18	26	

■

What, if anything, is wrong with these solutions? To start with, cells B5:D5 do exactly what we have been asked, but they are *inflexible* should we later want to enter different mpg, since we would have to hand input them to every affected cell (here just three, but possibly 10,000s or more).

Table 3 shows that we can use a "constant," such as the contents of cell B17, to divide each cell in column B; then, when we wish to again change, all we subsequently do is change the number in the cell B17, and it will be updated wherever it is used. This is a powerful improvement over entering the data in individual cells.

We can also change the weightings of the cells to reflect mpg in columns C and D. We'll use a slightly different method. In fact, we will fix the contents of the cell C17 in a tricky way and indicate that as $C$17. Such an address, for reasons seen shortly, is called "absolute" addressing. It will be used here to modify cells C6:C11; we can also use as a constant either D$17 or $D17 (called a "mixed" mode of relative and absolute addressing) to modify the cells in D6:D11. What these do is *fix* the reference to the row or number operated on by the "$" sign. What does this mean and why do it? The importance of these operations is best explained by copying and pasting a section of the spreadsheet as in the next example.

EXAMPLE 6

FleetsR'US wants to copy part of its spreadsheet to another area on the spreadsheet to make it clearer for someone to read. Assume the origin is no longer cell A1, but shift the whole spreadsheet to a new origin, H1. Analyze what you see and report whether the translocated spreadsheet is correct or not; if not, explain what is wrong.

Need: Copied spreadsheet.

Know: You can copy and paste across the spreadsheet form.

How: Use "Edit, Copy, Paste" commands.

Solve: Suppose you want to copy the table to another part of the spreadsheet. All you do is highlight the block of information A1:E13 that you want to move and go to "Edit," then "Copy."[4] Move your cursor to the corner cell you want to move to, say cell H1, and then go to "Edit," "Paste." Observe the effects after this is done in Table 4. ■

Notice we did not get all the expected results: Instead of modifying cells to the correct answers, cells I6:I11 and K6:K8 give the wrong answers. (Other divisions by zero also occur, since the spreadsheet's mathematical functions of sum and average also reference cells with one or more zero divisions.) What has happened? Use the toggle ^ ~ again and observe the results in Table 5.

Of course, the manually transplanted input cells I5:K5 are correct, since they call for no information outside of the particular cells, but the entries to I6:I11 are divided, not by B17 but by I17. In other words, when we translocated the cells, we also shifted B17 "relative" to the move (i.e., cell B17 is now displaced to I17 as is cell A1 is to H1). And cell I17

[4]There are alternate ways of copying and pasting using "smart tags"; however, the method suggested above is conceptually easier for a beginner in spreadsheet manipulations.

Table 4

Effect of Shifting the Spreadsheet of Table 3

	H	I	J	K	L
1	**Miles Driven in Various Categories**				
2					
3					
4	**Renter**	**City Gallons**	**Suburban Gallons**	**Highway Gallons**	**Total Gallon**
5	Davis	1.3	1.6	3.0	5.9
6	Graham	#DIV/0!	1.7	#DIV/0!	#DIV/0!
7	Washingt	#DIV/0!	0.4	#DIV/0!	#DIV/0!
8	Meyers	#DIV/0!	3.4	#DIV/0!	#DIV/0!
9	Richardso	#DIV/0!	3.1	7.6	#DIV/0!
10	Thomas	#DIV/0!	1.2	4.8	#DIV/0!
11	Williams	#DIV/0!	0.8	5.5	#DIV/0!
12					
13	Average	#DIV/0!	1.7	#DIV/0!	#DIV/0!

Table 5 Contents of Cells after Moving Location

	H	I	J	K	L
4	**Renter**	**City Gallons**	**Suburban Gallons**	**Highway Gallons**	**Total Gallons**
5	Davis	=16/12	=28/18	=79/26	=SUM(I5:K5)
6	Graham	=10/I17	=31/$C$17	=112/K$17	=SUM(I6:K6)
7	Washington	=4/I17	=7/$C$17	=158/K$17	=SUM(I7:K7)
8	Meyers	=22/I17	=61/$C$17	=87/K$17	=SUM(I8:K8)
9	Richardson	=12/I17	=56/$C$17	=198/$D17	=SUM(I9:K9)
10	Thomas	=5/I17	=22/$C$17	=124/$D17	=SUM(I10:K10)
11	Williams	=4/I17	=14/$C$17	=142/$D17	=SUM(I11:K11)
12					
13	Average	=AVERAGE(I5:I11)	=AVERAGE(J5:J11)	=AVERAGE(K5:K11)	=AVERAGE(L5:L11)

(not shown) contains a default empty cell value of zero—hence, the apparent disaster that the gallons used in city driving are now all infinite!

Notice that cells J6:J11 are multiplied by the contents of cell $C$17 as desired, since we used the "$" designator for both row and column. Cell $C$17 has not translocated relative to the move; its address is thus "absolute."

The sundry results of the translocated highway mileage column K is explained by the "mixed" mode of addressing used there. Cells K6:K8 now have been divided by K$17, since the column designator translocated from D to K (again, just as did column A to H). Since we fixed the row designator at $17, it did not translocate to a new row. But the cell K$17 contains default value of zero and hence the cells K6:K8, all of which contain this as a divisor, are each infinite.

However, cells K9:K11 gave the correct answer because they were divided by $D17. The $ fixed the column designator at column $D. We did not choose to change the rows when we copied the block of cells from A1 to H1 and the divisor $D17 still refers to the original mpg in cell $D17. We simply took advantage of our simple horizontal move from A1 to H1 in that it did not change the row locations. Had we also changed the row location—say, by pasting the block at A1:E13 to H2 instead of H1—the rest of highway gallons calculations would also have been in error due to an imputed mpg call to cell $D18.

Sometimes we want to use relative, absolute, or mixed modes. It depends on what we want to do. With experience, your ability to do this will improve. One nice feature of spreadsheets is that your mistakes are usually embarrassingly and immediately clear. If, for example, you forget to fix an absolute cell address and add, multiply, divide, exponentiate, and so on, by some unintended cell, the worksheet will complain as the target cell picks up whatever cell values were in the inadvertent cells.

EXAMPLE 7

Correct Example 6 using relative and absolute addressing modes.

Need: Corrected spreadsheet.

Know: How to use different cell addressing modes.

How: Make sure that no translocation cell addresses occur to undesired cells.

Solve: Had all the cells B5:D11 row by row been divided by their respective absolute constants, $B$17:$D$17, the final translocated result would have been as initially desired—see Table 6.

Table 6

Results of Spreadsheet That Used Absolute Cell Constants

	H	I	J	K	L
1	**Miles Driven in Various Categories**				
2					
3					
4	**Renter**	**City Gallons**	**Suburban Gallons**	**Highway Gallons**	**Total Gallons**
5	Davis	1.3	1.6	3.0	5.9
6	Graham	0.8	1.7	4.3	6.9
7	Washington	0.3	0.4	6.1	6.8
8	Meyers	1.8	3.4	3.3	8.6
9	Richardson	1.0	3.1	7.6	11.7
10	Thomas	0.4	1.2	4.8	6.4
11	Williams	0.3	0.8	5.5	6.6
12					
13	Average	0.9	1.7	4.9	7.6

Graphing in Spreadsheets

One of the big advances of spreadsheeting is that you can display your results in many ways that can visually convey a great deal of information. In a real sense most of your "knowledge" will be accessible only then, since the raw data, even if arranged as logical information in neat tables, may still be difficult to interpret.

We will confine ourselves for the moment to simple Cartesian graphing, but you will soon recognize the pattern to follow if you want to use other modes of display.

EXAMPLE 8

SpeedsR'US wants to sell after-market booster kits for cars. They measure a booster-modified vehicle that has the following characteristics on a level road:

Time, s	Mph Actual	Time, s	Mph Actual
0.0	0.0	5.0	78.6
1.0	18.6	6.0	86.9
2.0	32.1	7.0	96.9
3.0	46.5	8.0	102.9
4.0	58.2		

The car manufacturer says that the *un*modified car speed characteristic on a level road obeys the following equation: mph $=136.5(1 - e^{-0.158t})$ with t in seconds. Is the modified car faster to 100 mph than the standard car?

Need: To compare the tabular data to the equation given.

Know: A spreadsheet will provide multiple ways of looking at data.

How: Graph the tabular data and compare to the theoretical equation.

Solve: Set up the following spreadsheet:

Table 7

Setup for Graphing Function

	A	B	C	D
1	**Compare Measured Car Speed to Manufacturer's Specification**			
2				
3		Speed in mph = $136.5\left(1-e^{-0.158t}\right)$		
4				
5		with t in seconds		
6	**Time, s**	**Mph actual**	**Mph, theory**	
7	0	0	=136.5 *(1-EXP(-0.158*A7))	
8	=A7+1	18.6	=136.5 *(1-EXP(-0.158*A8))	
9	=A8+1	32.1	=136.5 *(1-EXP(-0.158*A9))	
10	=A9+1	46.5	=136.5 *(1-EXP(-0.158*A10))	
11	=A10+1	58.2	=136.5 *(1-EXP(-0.158*A11))	
12	=A11+1	78.6	=136.5 *(1-EXP(-0.158*A12))	
13	=A12+1	86.9	=136.5 *(1-EXP(-0.158*A13))	
14	=A13+1	96.9	=136.5 *(1-EXP(-0.158*A14))	
15	=A14+1	102.9	=136.5 *(1-EXP(-0.158*A15))	

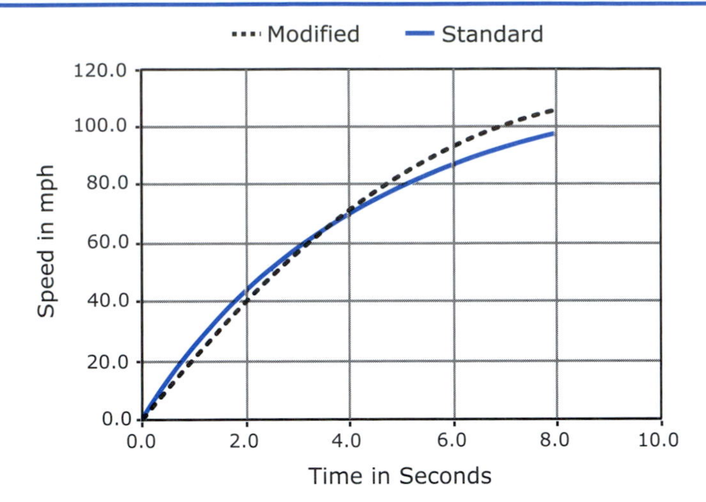

Figure 2

SpeedsR'Us Modification

Start with a spreadsheet title as in cell A1. Make it descriptive. The cell labels in A6:C6 should be helpful as to what the immediate columns under them mean. Of course, the numbers generated as A7:C15 are just raw data—not easy to interpret by merely looking at them. In columns A, B, and C are the data you want to plot—that is, the value of the variable "mph" given as input data and calculated by the manufacturer's equation.

You don't have to enter too much to get a lot of mileage.[5] In cells A7:B7, enter 0, and in C7, enter "136.5*(1 – exp(–0.158*A7)." The result in C7 is zero, since $1 - e^0 = 1 - 1$ is zero. Now in A8 write = A7 + 1. The value that appears in A8 will be 1, since A7 is zero. Now copy A8 to A9:A15, and then copy C7 to C8:C15, and you will have the manufacturer's function defined corresponding to each time of measurement.

You might want to graph your results, A7:A15 vs. B7:B15 and also A7:A15 vs. C7:C15. No sweat: Just highlight A7:C15 and then go to "Insert, Chart." Then tell the spreadsheet what kind of graph you want. You want an "xy ***(Scatter)***" plot. ("Scatter" is just a regular Cartesian graph!) You will see two sets of points corresponding to each column of speed data. Then hit "Next" and fill in your preferred titles. Optionally toggle off the "Legend" box. You can finish by pasting the graph on the current worksheet (i.e., the working page in the spreadsheet). If you want to, you can highlight either series of data and you can format them to discrete points or to a continuous line.

You should get the graph in Figure 2 for your troubles. Spreadsheets leave you plenty of flexibility if you use your imagination. On the basis of Figure 2, has the speed modification achieved its goal?

[5] As previously noted, you can short-circuit much of this discussion by taking advantage of "smart tags"; these are pulldowns on the cells that will automatically copy or increment adjacent cells. See the help menu for your particular spreadsheet to see if this usage is supported.

SUMMARY

A problem-solving method that guarantees you a path to the answer to engineering solutions is a valuable addition to an engineer's tool kit; we recommend the self-prompting mnemonic approach we call the **need to know how to solve** method. It allows for a systematic approach to engineering problems. We suggest you couple this method with the units method, [..], that we introduced in Chapter 2. Between these two methods you should be able to reason your way through even arcane problems that appear impossible at first glance. In addition, the systematic approach means you leave an auditing trail that can be used in your absence to check on, or to extend, or even to correct, your solution.

Many solutions to engineering problems require either or both complex analysis and repetitive solutions for many cases; still others require graphical representation. In other cases we may be confronted with jumbles of apparently uncorrelated **data;** spreadsheeting allows you to organize these data, sort them so you have some understanding or gross **information** extracted from the raw data and finally interpret the data to gain **knowledge** from the model you have developed. These situations are where **spreadsheet analysis** stands out. The heart of this method is to recognize that cells in a spreadsheet may be manipulated either in an **absolute** sense or in a **relative** sense. The former allows for facile repetition of algebraic and arithmetical operations using **constants,** and the latter allows one to easily extend tables of data, even while carrying their embedded mathematical operations.

Together, with an appreciation of the significance of engineering numbers and dimensional analysis, the contents of Chapters 2 and 3 are fundamental to what an engineer does and how successful he or she will be in pursuing an engineering career.

This lesson on spreadsheeting merely skims the surface of the very powerful tool of spreadsheet analysis. With time, both as a beginning engineer and then as a practicing engineer, you will discover many of the additional features of this tool. Used with skill and precision, this will become one of the most useful and generic skills that you will acquire. The sooner you make use of it, the sooner you will acquire these skills. They are urged upon you and you will find practice problems in virtually every chapter of this text that will be susceptible to spreadsheet analysis.

EXERCISES

Many of these exercises can be done using hand calculators; alternatively, you can also use a spreadsheet even for simple problems. If you choose a spreadsheet, then for *all* your spreadsheet exercises for this course, you should go to "file," then "print setup," then "sheet," and tick the boxes for "gridlines" and for "row and columns headings" so that your final cell locations will be printed when you submit the answers to these problems. *You will lose points otherwise*, since it is nearly impossible to grade answers without the row and column indicators!

You should use the **need to know how to solve** method in setting up all these problems. *You may lose points otherwise.*

1. The great physicist Enrico Fermi used to test the problem-solving ability of his students at the University of Chicago by giving them the following problem: How many piano tuners are there in the city of Chicago? (Assume the population of Chicago is 5 million people.) **(Plausible answers 50–250 tuners.[6])**

2. Of all the rectangles that have an area of one square meter, what are the dimensions (length and width) of the one that has the smallest perimeter? Solve by graphing on a spreadsheet.

3. Suppose you want to make a cylindrical can to hold 0.01 m^3 of soup. The sheet steel for the can costs \$0.01/m^2. It costs \$0.02/m to seal circular pieces to the top and bottom of the can and along the seam.

   What is the cost of the cylindrical can that is least expensive to make? (**A: 3 cents/can.**)

Exercises 4–6 involve the following situation:
Suppose that the weight of the gasoline in lbf in a car's gas tank equaled the weight of the car in lbf to the 2/3 power (If $G =$ gasoline weight, and $W =$ car weight, then $G = W^{2/3}$.) Assume further that gasoline weighs 8.0 lbf/gal, and gas mileage varies with weight according to the empirical formula:

$$\text{mpg} = (84{,}500\,\text{lbf mi/gal}) \times (1/\text{car's weight in lbf}) - 2.9\,\text{miles/gal}$$

4. What is the miles per gallon of a 3.00×10^3 lbf car?

5. What is the heaviest car that can achieve a range of 600. miles? (**A: 3.7×10^3 lbf.**)

6. Suppose the formula for weight of the gas was $G = W^b$, where b is can be varied in the range 0.50–0.75. Graph the range of a 3.69×10^3 lbm car as a function of b.

Exercises 7–10: These problems are concerned with bungee jumping as displayed in the figure.

[6]See http://www.grc.nasa.gov/WWW/K12/Numbers/Math/Mathematical_Thinking/fermis_piano_tuner.htmS.

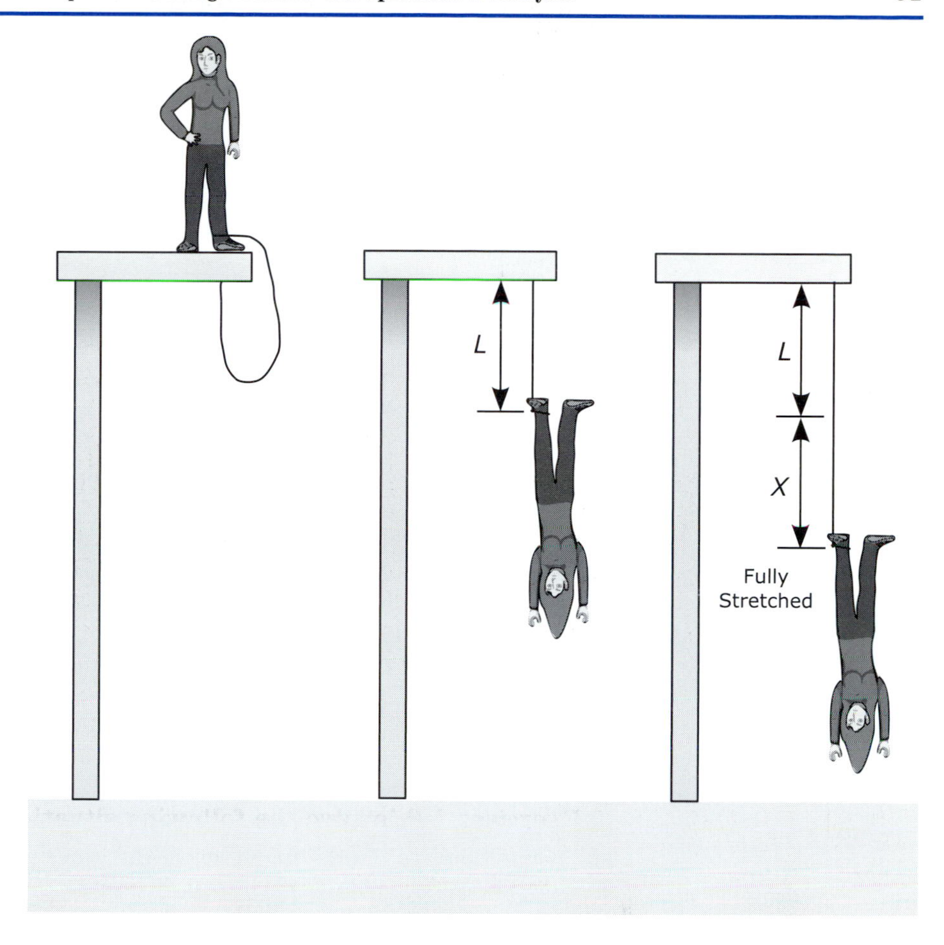

At full stretch, the elastic rope of original length L stretches to $L+x$. For a person whose weight is W lbf and a cord with a stiffness K lbf/ft, the extension x is given by this formula:

$$x = \frac{W}{K} + \sqrt{\frac{W^2}{K^2} + \frac{2W \times L}{K}}$$

and that can be written in "spreadsheetese" as $x = W/K + \text{sqrt}(W\verb|^|2/K\verb|^|2 + 2W * L/K)$.

7. If the height of the cliff is 150.0 ft, $K=6.25$ lbf/ft, $L=40.0$ ft, and the person's weight is 150.0 lbf, will the person be able to bungee jump safely? Support your answer by giving the final value for length $=L+x$. (**A: 114 ft <150 ft, OK.**)

8. Americans are getting heavier. What's the jumper's weight limit for a 40.0 ft unstretched bungee with stiffness of $K=6.25$ lbf/ft? Graph final length $L+x$ vs. W for weights from 100 lbf to 300 lbf in increments of 25 lbf. Print a warning if the jumper is too heavy for a 150 ft initial height. (**Hint:** Look up the application of the "IF" statement in your spreadsheet program.)

9. If the height of the parapet is 200.0 ft, and the weight of the person is 150.0 lbf, and the unstretched length $L = 45.0$ ft, find a value of K that enables this person to stop exactly five feet above the ground. (**A: 2.60 lbf/ft.**)

10. By copying and pasting your spreadsheet from problem 8, find and plot the values of L needed (in ft) vs. W, weight of jumper (in lbf) for successful bungee jumps (coming to a stop 5 ft above the ground) for $K = 6.25$ lbf/ft and from a cliff of height 150.0 ft above the ground. The graph should cover weights from 100 lbf to 300.0 lbf in increments of 25 lbf. (**Hint:** The function "Goal seek" under "Tools" is one way to solve this exercise.)

Exercises 11–13: The fixed costs per year of operating an automobile is approximately 20.% of the initial price of the car. Thus, the operating cost/mile $=$ 0.20/yr $\times$ (purchase price of automobile)/(miles driven per year) $+$ (price of gasoline/gallon) $\times$ (gallons used per mile). In the problems that follow, assume that the automobile is driven 2.0×10^4 miles per year. Assume gasoline costs $2.00/gallon.

11. Estimate the operating cost per mile of an automobile with a price of $15,000 that gets 30.0 miles per gallon. (**A: $0.22/mile.**)

12. If one were to double the price of the automobile in problem 11, what would its gas mileage have to be in order to cost the same to operate per mile as the automobile in problem 11?

13. Suppose that the purchase price of automobiles varies with weight according to the formula that weight in lbf $\times$ $8.00, and gas mileage varies according to mpg $=$ (84,500 mile-lbf/gal)/$W - 2.9$ miles/gal. Graph the cost per mile of operating a car as a function of the car's weight, in increments of 500.0 lbf from 2000.0 lbf to 5000.0 lbf. (**Partial A: 21 cents/mile for 2000 lbf car & 54 cents/mile for 5000 lbf car.**)

Exercises 14–16: In visiting stores, one finds the following prices for various things. Broccoli crowns cost $2.89 per pound. Soft drinks cost $2.00 per two-liter bottle. (A liter is 0.001 m^3.) A new automobile weighs 2.50×10^3 lbf and costs $\$1.50 \times 10^4$. A dozen oranges, each of which is 0.06 m in diameter, costs $2.05. A 1.5 lb package of chicken thighs costs $5.35. A dictionary weighs 5.00 pounds and costs $20. A refrigerator weighs 200.0 lbf and costs $900. Assume that one cubic meter of any solid object or liquid weighs 1.00×10^4 N.

14. For the objects listed above, make a table and graph of the cost of objects in dollars as a function of their weight in newtons. It is suggested for this graph to use a *line* (not "scatter") graph using line with markers displayed at each data value. (Get rid of the unwanted line using the "format series function.") The value of the line graph is that everything plotted is at the same horizontal displacement and not dependent on its value.

15. What (perhaps surprising) simple generalization about the cost of things might one make based on the table and graph of problem 14? (**Hint:** Does the comparative lack of spread in price surprise you?)

16. Name a product or group of products that does *not* fit the generalization you made in problem 15, and add and label its point to the graph in problem 14. To get a better perspective use a *log scale* for the y-axis, $/N, since the ordinate should be much larger than the coordinates of those points from problem 14.

17. An unnamed country has the following population of passenger cars on its roads as determined by 250 kg mass differences. You have to make these data clear to the undersecretary to that country's transport minister. Plot these data by two methods: (1) as a "pie chart" and (2) as a histogram to show the distribution in an effective manner.

	B	C
1	**Upper limit, kg**	**% all vehicles**
2	1000	12.1
3	1250	13.1
4	1500	15.4
5	1750	18.6
6	2000	14.8
7	2250	9.2
8	2500	7.5
9	2750	6.3
10	4000	3.0

Exercises 18–20 deal with "Hubbert's Peak."[7] This is a model of supply and demand for oil. It looks at the amount of available oil and its rate of consumption to draw conclusions about continuing the current course of our oil-based economy.

18. Suppose the world originally had three trillion ($3. \times 10^{12}$) barrels of oil and its exploration began in 1850. Suppose 10.0% of the remaining *undiscovered* oil has been found in every quarter century since 1850.

[7]M. King Hubbert was a geologist with Shell Oil who, in the 1950s, pointed out that the U.S. supply of oil was going to fall short of demand by the 1970s. His methods have since been applied to world oil production and, based on demand exceeding production, predict an ongoing oil supply crisis. See http://www.hubbertpeak.com/hubbert/.

Call the *discovered*, but not yet consumed, oil "*reserves*." Suppose oil consumption was 1.0×10^8 barrels in 1850, and further suppose oil consumption has grown by a factor of 5 in every quarter century since 1850.

When will the oil start to run out? (That is, when will the reserves become negative?) Give your answer to the nearest 25 years and provide a spreadsheet showing reserves and consumption as a factor. (**A: 2025.**)

19. Suppose the world originally had 10 trillion (10.0×10^{12}) barrels of oil. Use the data of problem 18 to again predict when the oil will start to run out.

20. Repeat problem 19, but instead of assuming the exponential growth in consumption continuing unabated by a factor of 5 in every quarter century since 1850, curtail growth since 2000 and assume consumption has stayed constant since then. Again predict when the oil will start to run out.

21. Your friend tells you that the need to know how to solve problem-solving method seems overly complicated. He/she just wants to find the answer to the problem in the quickest possible way—say, by finding some formula in the text and plugging numbers into it. What do you tell him/her?

(a) Go ahead and do whatever you want, then you'll flunk out, and I'll survive.
(b) Talk to the instructor and have him/her explain why this methodology works.
(c) Find someone who has used this method and ask to copy his/her homework.
(d) Explain why this technique will lead to a fail safe method of getting the correct answer.

22. You e-mail a classmate in this course for some information about a spreadsheet homework problem. In addition to answering your question, your classmate also attaches a spreadsheet solution to the homework. What do you do?

(a) Delete the spreadsheet without looking at it.
(b) Look at the spreadsheet to make sure he/she did it correctly.
(c) Copy the spreadsheet into your homework and change the formatting so that it doesn't look like the original.
(d) E-mail the spreadsheet to all your friends so that they can have the solution, too.

Chapter 4

Courtesy of Daimler-Chrysler

Energy: Kinds, Conversion, and Conservation

Using Energy

Is the world running out of energy? Where does the energy come from that provides the light by which you are reading these words? Does a car possess more energy when it is sitting in your driveway or 15 minutes later when it is traveling down the highway at 60 miles per hour? In this chapter, you will learn how to answer these questions, as well as how to address the more quantitative ones involving energy that engineers of all types encounter in their work.

Energy is *the capability to do work*. An important point is that energy comes in several discrete *kinds*. These kinds are capable of *conversion* from one kind to another. In the course of these conversions, energy is neither created nor destroyed, but rather *conserved*. That is, the total amount of energy in the universe remains constant. This is one of the most important principles you will have to master as an engineer. You will have to construct a useful system boundary known as a "control boundary" to monitor the flows of energy.

Using energy as a variable, engineers create models that help with a wide range of applications, such as deciding what fuel to use in an automobile, designing methods of protecting buildings from earthquakes or lightning strikes, or advising citizens as to the technical capability of a society to replace fossil fuels with such renewable resources as solar power, wind power, and biomass, or with such controversial but abundant energy sources as uranium. But in order to achieve these wide-ranging applications, engineers must be very narrow and specific in the way they use the concept of energy.

The way engineers use the word *energy* is examined in this chapter. The following key ideas are discussed. Energy is *the capability to do work*. Work is narrowly defined from Newton's Second Law of Motion. In the simplest mathematical terms, work is defined as:

$$\text{Work} = W = F \times d$$

in which a force F moves through a distance d. The concept is broader than it appears from this straightforward definition and can be applied to a number of situations apparently far removed from this simple statement, while still inherently dependent on it. Energy, E, has the same units as W. In an imaginary system one might convert energy to work and then recover an equivalent to the original energy.

An important point is that energy comes in several discrete kinds. These kinds are capable of conversion from one kind to another and, as such, are a core component of the work of several kinds of engineers. In the course of these conversions, energy is neither created nor destroyed but conserved. That is, the total amount of energy in the universe remains constant.[1]

Figure 1 represents one of the most important principles you will have to master as an engineer. The engineer works on something that we can

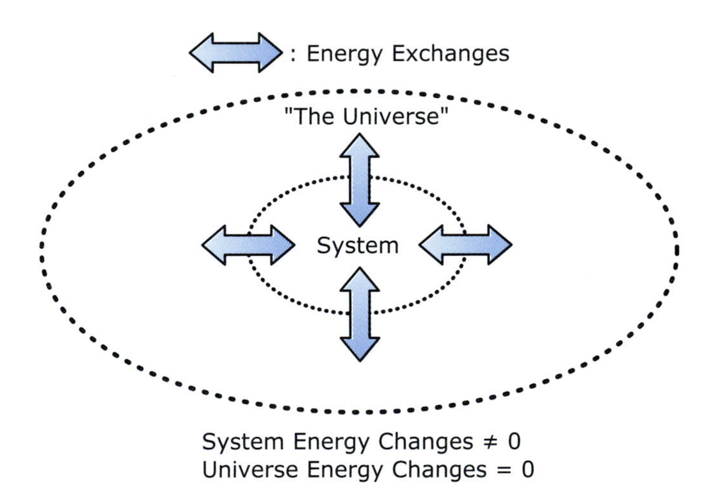

Figure 1

"Conservation of Energy"

[1]There is one caveat that cosmological physicists might add: the radiant energy in the universe leaks to beyond the observable edge of the expanding universe—something that is unlikely to disturb any engineers in the foreseeable future! For all practical purposes, the energy in the universe is constant.

generically call a "system." It might be an engine, a sailboat, a lawn mower, an electric kettle, a crane, or anything. It is that which is contained within the inner dotted boundary. It connects to the rest of the "universe" through whatever elements are interposed between the system and the universe. You will have to use engineering skills and know- how to construct a useful system boundary known as a **control boundary** such that you can meaningfully monitor the flows of energy (perhaps of several kinds) across it. When you do this, the **Law of Conservation of Energy** will enable you to determine the flows of energy necessary to maintain the universal balance.

When engineers use the term *energy conservation* they are not offering exhortations to turn off the lights, nor are they opposing the use of gas-guzzling sport utility vehicles. Rather, they are expressing a scientific principle that is fundamental to engineering and that will remain in effect regardless of the energy policies that people and nations choose to follow. If you remember only three words from this chapter remember this: "*Energy is conserved.*"

As a concrete example of what the engineers do in their analyses, one can divide the universe into at least two discrete regions: one of direct interest (such as a car's engine) and everything else. This is a very powerful tool in the hands of a capable engineer, since it allows the engineer to concentrate on the localized system.

Energy Is the Capability to Do Work

Understanding the variable "energy" begins by defining variables for the concepts of force and work. Force is a variable that may be thought of as a push or a pull, as measured with a spring balance or by the weight of an object in a gravitational field. As we have already seen, force has the units of newtons in the SI system and pounds force (lbf) in the Engineering English system. For example, at sea level, to lift a book of mass 1 kilogram requires a force of 9.8 newtons. In the Engineering English system, the force to lift this 1 kg book is 2.2 lbf (since 1 kg = 2.2 lbm).

Work as defined above is the product of force and the distance over which the force is applied, where the distance is measured in the same direction as the application of the force. The units of work are **joules** in the SI system, and **foot pounds force (ft lbf)** in the Engineering English system. A joule is exactly 1 N-m or equivalently 1 kg m^2/s^2. For example, to raise a book weighing one kilogram from the surface of the Earth to a level one meter above the surface of the Earth requires an amount of work that equals 9.8 [N] $\times$ 1 [m] = 9.8 [N-m] = 9.8 joules. In the Engineering English system we have to account for the pesky g_c to calculate the amount of work as $(F \times d)/g_c = 2.2 \times 32.17 \times 3.28/32.17$ [lbm] [ft/s^2] [ft] [lbf s^2/lbm ft] = 7.2 ft lbf (since 1 m is 3.28 ft).

Again, energy is the capability to do work. It may be stored in many objects, such as liquid gasoline, solid uranium, or a speeding bullet. Equally, the stored energy in a body can be released, such as water running through a water turbine, heat from burning gasoline, or heat from a piece of uranium. More precisely, any object that does an amount of work consisting of some number of joules will see its energy content decrease by that same number of joules. Any object that has an amount of work done upon it by its surroundings will see its energy similarly increase.

Often, an engineer wants to know how fast that work was done or how rapidly the amount of energy possessed by an object changed. This is determined using the variable **power,** which is defined as the time rate of doing work or, equivalently, as the time rate of change of energy. It is measured in watts in SI, where one watt is 1 joule per second. For example, if a person takes two seconds to lift a 1 kg book a height of 1 meter above the surface of the Earth, that person is expending (9.8 joules)/(2.0 seconds) = 4.9 watts of power during those two seconds.

In the Engineering English system, power is normally expressed in units of ft lbf/s or in horsepower,[2] where one horsepower is 550 ft lbf/s. It is indeed roughly the power exerted by a working horse and roughly 5–10 times the power exerted by a person doing sustained physical labor such as shoveling or carrying. A useful mnemonic connecting the SI and Engineering English systems is that a kilowatt and a horsepower are of the same order of magnitude. Some other examples of quantities of energy in everyday life are shown in Table 1.

In the past three centuries, a vast increase has occurred in the amount of power at the command of the ordinary person in the United States and

Table 1

Typical Energy Magnitudes

Number of joules	Approximate equivalent
100	~ Lighted match
10,000	~ Speeding bullet
5×10^5	~ Kinetic energy of a small car at 65 mph
10^6	~ A small meal (!!!); or kinetic energy of an SUV at 65 mph
3×10^7	~ Lightning stroke
10^8	~ One gallon of gasoline
10^{11}	~ One g of uranium in fission
4.2×10^{12}	~ One kton of TNT
60×10^{12}	~ Hiroshima-sized atomic bomb
10^{14}	~ Annihilation of one gram of any matter

[2]The Scottish engineer James Watt measured the rate of working that a good brewery horse could sustain as a way to sell his early steam engines.

other developed nations. In 1776, when Thomas Jefferson was writing the Declaration of Independence and James Watt was perfecting his steam engine, the average adult in the United States or in Britain could call on considerably less than one kilowatt of power for some fraction of a day. Today, the average person (adults and children included) in the United States and other industrialized nations typically has at his or her command more than 10 kW (or ~15 hp) every day of the year, every second, day and night. This includes several kilowatts of electric power per person to say nothing of over 100 kilowatts of automotive power for selected periods of the day. The other kilowatts per person serve us each day—for example, in the form of the trucks that haul goods for us, the oil or gas that heats us in winter, the airplanes that transport us and our overnight deliveries, and the machinery that manufactures all these energy-using artifacts. So each citizen of an industrialized nation continually has on call the power of hundreds of horses at his or her service for a full 24/7. At the same time, vast numbers of people elsewhere in the world are still in that eighteenth century less-than-one-horsepower state or, worse, a one-person-power state. Resolving this global power discrepancy in an environmentally acceptable manner is one of the great engineering and social challenges of the twenty-first century.

Let us now turn from these general concepts of work, energy, and power to the specific kind of energy that scientists and engineers have defined and to the phenomena in nature and technology that these kinds of energy help engineers to quantify and to use in engineering models.

Kinds of Energy

Energy comes in various forms. Energy due to motion is called kinetic energy and includes the first two major types of energy we will discuss, *translational kinetic energy*[3] and, less obviously, *thermal energy*.

Energy due to position is called potential energy and includes *gravitational potential energy, GPE*, and, less obviously, *chemical energy*[4] and *electromagnetic energy*.[5] There are other important types such as

[3]In some cases matter is not moving in a straight line but rather rotating around a central axis. In this case, the kinetic energy is called ***rotational kinetic energy***, or RKE. It is no different in principle from TKE. It does, however, require more complicated mathematics, since many pieces of matter are moving in different directions and at different speeds. Due to these mathematical complications, RKE will not be required knowledge in this book.

[4]Unburned fuel such as a can of gasoline clearly has potential energy locked inside of it. Obviously this stretches the simple definition that potential energy is mass × height × g (see section on GPE). Some part of this picture can be retained if you think of the electrons that surround the atoms of the fuel being in high energy states and endeavoring to reach lower or more stable states.

[5]Electromagnetic energy depends on the potential of electrons to do work; hence voltage is also sometimes called "potential."

Einstein's discovery that mass is a form of energy, as embodied in his famous equation $E = mc^2$. We will not deal with it in this course, but note that this insight is the basis of nuclear energy in which about 1% or 2% of matter in uranium is converted to thermal energy, then to electrical energy.

Translational kinetic energy (TKE) is the energy of mass in straight line motion. It is calculated by the formulae:

$$\text{TKE} = (½)mv^2(\text{SI}) \text{ or } (½)mv^2/g_c(\text{English})$$

where m is mass and v is speed. TKE is often assumed to be the only form of kinetic energy, and then it is simply called "kinetic energy" and then abbreviated "KE."

Note first that the formula results in the correct SI units of joules, since speed is m/s, and mass is in kilograms, so ½mv^2 has units [kg][m^2/s^2], which are the same units that make a joule. In Engineering English units, we have to divide by g_c to get the proper units of [ft lbf] from [lbm ft^2/s^2] [lbf s^2/lbm ft].

EXAMPLE 1

What is the translational kinetic energy of an automobile with a mass of 1.00×10^3 kg traveling at a velocity of 65 miles per hour (29 m/s)?

Need: TKE of vehicle.

Know: Mass is 1.00×10^3 kg, velocity is 29 m/s.

How: Apply the formula TKE = ½mv^2.

Solve: $\text{TKE} = ½ \times 1.00 \times 10^3\,[\text{kg}] \times 29[\text{m/s}]^2 = 423{,}410\,\text{kg m/s}^2 = 4.2341 \times 10^5\text{J}$.
Only two of the digits are significant (since the speed was stated only to two digits), so the answer is 4.2×10^5 J = $\mathbf{4.2 \times 10^2}$ **kJ.** ■

Anything that has mass and is moving in a line has TKE. Prominent examples of TKE in nature include the winds and the tides.

EXAMPLE 2

Estimate the total kinetic energy of the wind on Earth. As an introduction to this problem, consider that the wind is a movement of air that is produced by temperature differences in the atmosphere. Since hot air is less dense than cold air, air heated by the Sun at the equator rises until it reaches an altitude of about 6.0 miles (9.0 km) and then it spreads north and south. If the Earth did not rotate, this air would simply travel to the North and South Poles, cool down, and return to the equator along the surface of the Earth as wind. However, because the Earth rotates, the prevailing winds most of us see travel in a west-east rather than a south-north direction (in the northern hemisphere). In places where winds are strong and steady, it may make economic sense to install a windmill or wind turbine to capture that translational kinetic energy.

Need: Total TKE of the wind in joules.

Know: The atmosphere is about 9.0 kilometers thick (i.e., the height of Mt. Everest). The radius of the Earth is about 5.0 million meters. The surface area of a sphere is $4\pi R^2$, so the volume of an annular "shell" of thickness T around the Earth is about $4\pi R^2 T$, since the thickness T is very small compared to the radius R. Air has a density of about ½ kg per cubic meter (averaging from sea level to the top of Everest). This air is typically moving at about 10 m/sec.

How: Find the mass of air in that 9.0 km thick shell around the Earth, and apply the formula TKE $= ½ mv^2$.

Solve: Volume or air around the Earth is about $4\pi(5.0 \times 10^6)^2 \times 9{,}000$ $[\text{m}^2][\text{m}] = 2.8 \times 10^{18}\,\text{m}^3$. The mass of air around the Earth is therefore $½ \times 2.8 \times 10^{18}[\text{kg/m}^3][\text{m}^3] = 1.4 \times 10^{18}$ kg.
$\therefore \text{TKE} = ½ \times 1.4 \times 10^{18} \times 10^2[\text{kg}][\text{m/s}]^2 = 7 \times 10^{19}\,\text{J} = 70\,\text{EJ}$
(E ≡ Exa = 10^{18}; also note that only one digit is significant.) ■

This is a number sufficiently large as to be meaningless to most people. But to an engineer it should inspire such questions as: Where does this energy come from? Where does it go? In fact, the total Sun's energy reaching the Earth is about 1.4 kW for every m^2 of the Earth's surface. Of this, about 31% is reflected back into space. We thus receive about $1\,\text{kW/m}^2$ net solar radiation. Over the Earth this amounts to about 8×10^{16} kW.[6] Since we receive this on the average for 12 hours every day, our beneficent Sun delivers 3.4×10^{24} joules/day; apparently, our estimate of the winds accounts for only $7 \times 10^{19}/3.4 \times 10^{24} = 2\%$ of the daily received solar energy.

In addition, the Sun is the ultimate source of the energy in the fossil fuels that power virtually all of our automobiles and about two-thirds of our electric power stations, as well as the source of the biomass that, in the form of food, powers us. Moreover, it drives the cycles of flowing or falling water that power hydroelectric generating plants.

The role that the winds or other solar energy sources might play in meeting the needs of those billions of people who still live at an eighteenth-century energy standard is still debated. The major problems are that this source of energy is diffuse, usually unpredictable, daily, and seasonal.

Thermal energy, often referred to as heat, is a very special form of kinetic energy because it is the *random* motion of trillions and trillions of atoms and molecules that leads to the perception of temperature. Heat is simply the motion of things too small to see, an insight captured by

[6]And even this huge amount is only about 5×10^{-10} of the total energy generated by the Sun.

the nineteenth-century German physicist Rudolf Clausius when he defined thermal energy as "the kind of motion we call heat".

There is a macroscopic analogue to thermal energy that is familiar to many of you. (At least by reputation, if not personally experienced!) The "mosh pit" at a rock concert is an example of the jostling motion of many bodies expending kinetic energy but not actually moving anywhere. It is the molecular amplification of this picture with trillions and trillions of atoms or molecules bouncing together that produces the net effect of what we call thermal energy. Of course, it is impossible to measure the velocities for every atom or molecule in even a tiny sample of matter. So thermal energy is typically determined by statistical means, as well as experimentally with the aid of such analogue averaging devices as a thermometer. Temperature, as measured using a thermometer, is *not* the same thing as thermal energy (but it can be used to calculate thermal energy). The details of measuring and calculating thermal energy will not be covered in this book. The important thing to remember is that thermal energy cannot be left out of engineering calculations. When one of the other kinds of energy seems to "disappear" in a model or calculation, it eventually has been converted into thermal energy. Friction is a commonly observed manifestation of energy conversion.

Gravitational potential energy (GPE) is the energy acquired by an object by virtue of its position in a gravitational field—typically by being raised above the surface of the Earth. In SI units, it is calculated by the formula

$$GPE = mgh$$

in which h is the height above the ground. In Engineering English units, this definition is modified by g_c to ensure that it comes out in ft lbf units:

$$GPE = \frac{mgh}{g_c}$$

Note that gravitational potential energy only has meaning relative to some reference datum level. The choice of such a level is arbitrary but must be applied consistently throughout a problem or analysis.

EXAMPLE 3

Wile E. Coyote holds an anvil of mass 100.0 lbm at the edge of a cliff, directly above the Road Runner who is standing 1000.0 feet below. Relative to the position of the Road Runner, what is the gravitational potential energy of the anvil?

Need: GPE.

Know: Anvil has mass 100.0 lbm. The reference datum level, where the Road Runner is standing can be chosen as zero. The height h of anvil is 1000.0 feet referenced to Road Runner's datum. Because we are calculating in Engineering English units, we will have to use g_c.

How: In Engineering English units, $GPE = \dfrac{mgh}{g_c}$.

Solve: $\text{GPE} = \dfrac{100.[\text{lbm}]1000.[\text{ft}]32.17[\text{ft/s}^2]}{32.17[\text{lbm ft/lbf s}^2]} = 1.00 \times 10^5 \text{ ft lbf}$ correct to 3 sig. figs. (as is the least known variable, the mass of the anvil.) ■

Electromagnetic energy (often merely called "electricity") is a form of energy that is typically carried by electric charges[7] moving through wires, or electromagnetic waves (or particles) moving through space. It will be the subject of Chapter 7, so here we just touch on it briefly.

Like all forms of energy and power, **electromagnetic energy** can be measured in joules, and electromagnetic power can be measured in watts. Consider the electromagnetic power used by an electrical device that is connected by wires to an electrical battery to create an electrical *circuit*,[8] as shown in Figure 2.

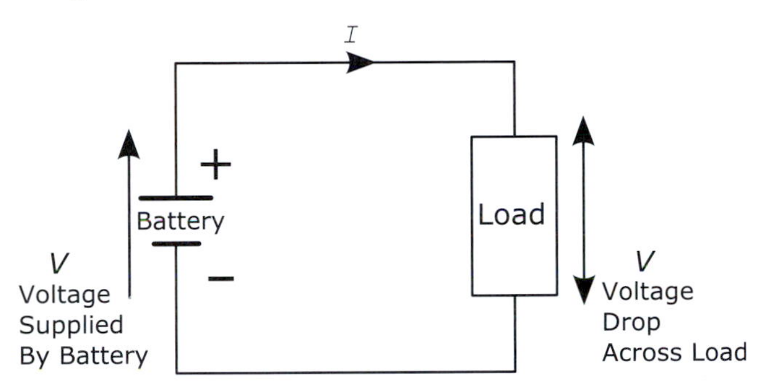

Figure 2

An Electrical Circuit

An electrical current, represented by the letter *I*, flows into and out of the device. Electrical current is measured in units called amperes (symbol A), which is simply a measure of the number of electrons passing through any cross section of the wire every second. An electrical potential, represented by the letter *V*, can be measured at any point on the circuit. Electrical potential is measured in units called volts (V). It might be thought of as an "electrical pressure," originating in the battery that keeps the current flowing through the wire. The voltage has its most positive value at the terminal on the battery marked plus (+), and drops throughout the circuit, reaching its minimum value at the battery terminal marked minus (−). By measuring the voltages at any two points along the wire and subtracting to determine the difference between those two measured voltages, one can determine the voltage drop between those two points on a circuit. One determines the voltage drop by subtracting the voltage at the point nearer

[7]Specifically negatively charged electrons.

[8]The word *circuit* has the idea of *circle* built in. Thus, an electrical circuit must start and finish in an unbroken loop. The physical reason is that the electrons as charge carriers are not consumed in the circuit.

the minus terminal of the battery from the voltage at the point nearer the plus terminal of the battery. Again, that voltage drop might be thought of as a pressure "pushing" the current from the point of higher voltage to the point of lower voltage.

In mechanical systems, power, the rate of doing work, can be computed as the product of force $\times$ velocity. From electrostatics, the *force* on a charge Q coulombs[9] in a voltage *gradient* is VQ/d (d is the distance over which the voltage changes from 0 to V)—hence, the *work* in moving the charge is $d \times QV/d = VQ$. If the charge is moved along the voltage gradient in time t, the *power* to move the charge is VQ/t. We define the rate of movement of charge as the electric current, $I = Q/t$. If Q in coulombs, t in seconds, then I is in amperes. Further, if V is in volts, then the power is in watts. Simply put, what you need to remember is that the electrical power is expressed by the equation:

$$\text{Power } = P = I \times V$$

EXAMPLE 4

A battery sustains a voltage drop of 3 V across a small lightbulb and sustains a current of 0.1 A through the lightbulb. What is the power of the lightbulb?

Need: Power of lightbulb in watts (W).

Know: Voltage across bulb is 3 V. Current through lightbulb $I = 0.1$ A.

How: Use the knowledge that $P = IV$—that is, power = voltage $\times$ current.

Solve: $P = 3$ [V] $\times$ 0.1 [A] = 0.3 W.

While the power into the lightbulb departs from the electric circuit, it does not disappear. Instead, some of it is radiated away from the bulb in the form of a wave of electromagnetic power (or, equivalently, in the form of massless particles called photons) traveling through space at the speed of light. Some of this electromagnetic energy is in the form of light; another part is in the form of radiant heat. Yet, more energy is lost in lower-grade heat by heating the local surroundings. Calculating the value of the power carried by these mechanisms will not be covered in this book. However, once again, that power can measured in watts.

Chemical energy is another form of potential energy in that it is determined by the relative distribution of electrons in the atoms that make up the structure of molecules. It is so important to our theme of Smart Cars that we will devote an entire chapter to chemical energy. Suffice it here to say that it too is most conveniently measured in joules.

[9]The charge on a single electron is 1.6×10^{-19} coulombs.

Conversion

Sunlight drives the winds. An anvil hoisted to a clifftop can be used to deliver kinetic energy to an unwary Road Runner below (though it is more likely to end up falling on the head of Wile E. Coyote himself!). Burning fossil fuel (a form of stored-up solar electromagnetic energy) results in the rotational kinetic energy (known there as "shaft work") of a turbine generator that is then converted back into electromagnetic energy to light up a lightbulb. All these occurrences suggest a second key fact about energy: Its various kinds can be converted from one to another.

Because all types of energy can be expressed in the same units, joules, this conversion can be expressed quantitatively in simple models. Even in Engineering English units, the number of conversion factors is relatively few.[10]

We mentioned earlier that the Sun delivers more than 10^{24} J of energy daily to the Earth in the form of the electromagnetic energy called light. What happens to that electromagnetic energy? A discussion provides the clues that can be made into a model. Part of the energy heats the atmosphere and drives the winds as previously noted as well as driving the water cycle. Part of the solar energy that reaches the Earth is converted into chemical energy via photosynthesis in trees, grass, and agricultural crops. Part of it simply heats the Earth and the ocean. Much of it is reflected away back into deep space. One thing the model must answer is the following question: If all that energy is arriving from the Sun every day, and much of it is converted to thermal energy, why isn't the Earth accumulating more and more thermal energy and continuously getting hotter? (Try to answer that question yourself before reading the next paragraph.)

The fact that the temperature of the Earth is staying roughly constant requires another element in our simple model. Roughly speaking, every day the Earth must reflect or radiate away into space the same amount of energy that it receives from the Sun. Should anything happen, either due to natural or human causes, to interfere with this energy conversion balance between absorption and radiation of solar energy, we will be in big trouble! The Earth might either cool off (as it did in the various ice ages) or heat up. (As it is alleged to be doing right now by many scientists who believe that global warming is taking place.) So understanding energy conversion is crucial to projecting the future of life on our planet.

We can perform a similar analysis of energy conversion on a technological system, such as an automobile. In this case, we can regard the initial

[10]One useful conversion factor in the Engineering English system shows the relative magnitude of mechanical and thermal units: 1 Btu is the heat sufficient to raise the temperature of 1 lbm of water by 1°F; it is equivalent to 778 ft lbf of energy. In other words, if you dropped 1 lbm of water through 778 ft and all of the original GPE were converted to thermal energy, you would heat that water only by 1°F. Another useful conversion factor is 1 Btu ~ 1 kJ (actually 1.055 kJ).

source of the energy as the chemical energy pumped into the vehicle at the gas station. In the automobile's engine, that chemical energy is converted to the translational kinetic energy of a piston in a cylinder. That translational kinetic energy is then converted into rotational energy in the transmission, axles, and wheels. That rotational kinetic energy in turn provides the translational kinetic energy of the automobile in its motion down the road. In the process, part of that energy—indeed, a substantial majority of the initial chemical energy—is transferred to the environment in the form of heat, either through the car's radiator and exhaust pipe or through road friction or air resistance.

EXAMPLE 5

A gallon of gasoline can provide about 1.30×10^5 kJ of chemical energy. Based on Example 1, *if all* the chemical energy of a gallon of gasoline could be converted into the translational kinetic energy of an automobile, how many gallons of gasoline would be equivalent to the vehicle's TKE if the automobile is traveling at 65 miles per hour on a level highway?

Need: Gallons of gasoline to propel vehicle at 65 mph (29.0 m/s).

Know: TKE of vehicle $= 4.2 \times 10^2$ kJ to 2 sig. figs. at 65 mph (as per our previous calculation in Example 1). Energy content of gasoline is 1.30×10^5 kJ/gallon.

How: Set TKE of vehicle equal to chemical energy in fuel. Let x = number of gallons needed to accomplish this.

Solve: $x \times 1.30 \times 10^5$ [gallons][kJ/gallon] $= 4.2 \times 10^2$ [kJ].
$\therefore$ **x = 0.0032 gallons.**

This is a tiny amount of gasoline. As we will see later, this is a misleadingly small amount of gasoline. In fact, most of the gasoline is *not* being converted into useful KE but is (in fact mostly *necessarily*) "wasted" elsewhere. The misleading nature of this calculation will become obvious as we proceed! The error is in the conditional statement "***if all*** the chemical energy of a gallon of gasoline could be converted into the translational kinetic energy of an automobile." The lesson is to beware of assumptions!

Conservation of Energy

It is possible in principle (though difficult in practice) to add up all the energy existing[11] in the universe at any moment and determine a grand

[11]In this case, we would also have to add in all the mass in the universe because cosmic events freely convert E into mc^2!

total. Scientists and engineers express this fact by saying that *energy is conserved*. This may sound like a theoretical claim of little practical use. However, engineers have developed a method of applying the fact that energy is conserved to practical problems while avoiding the inconvenience of trying to account for all the energy in the universe. This method is called "control boundary" analysis and is schematically illustrated in Figure 1.

It consists simply of isolating the particular object or system under consideration, and making a simple model of the way that object or system exchanges energy with the rest of the universe. Making that model begins with a conceptual sketch of the object or system. You draw a dotted line representing the control boundary around the sketch to contain the item under analysis. Then draw arrows across the boundary representing specific types of exchanges of energy between the object or system and the rest of the universe. By limiting the energy exchange in this way, the principle of conservation of energy can be "imported" into the system under consideration to determine the results of various energy conversion processes that occur within the dotted line. Losses or gains are simply one form or another of energy that crosses that boundary.

As a simple example, consider your classroom as a closed system and draw an imaginary control boundary around it. Assume this boundary is impervious to energy flows so that what's inside in the classroom stays there. Now imagine you have a 1 kg book on your desk that is one meter high. The book has $GPE = mgh = 1.0 \times 9.8 \times 1.0$ [kg][m/s^2][m] $= 9.8$ J with respect to the floor of your classroom. Suppose that book now falls to the floor, thus losing all of its GPE. Since the classroom boundaries are impervious to energy exchanges, where did that GPE go? Have we violated conservation of energy?

The fact is that GPE is still trapped within the room but not in the same form. What physically happens is (1) the GPE of the falling book is converted to TKE; (2) when the book hit the floor, it sets up some sound waves in the floor material and in the surrounding air; and (3) eventually when all the transients died out, the forms of energy in (1) and (2) end up heating the room and its contents by an amount corresponding exactly to the 9.8 J of energy.

The ***principle of conservation of energy*** says energy is never lost, merely transformed somewhere into another form. This is one statement of energy conservation also called the First Law of Thermodynamics. Ultimately all forms of energy degrade to heat; this is one statement of the Second Law of Thermodynamics. Control boundary analysis is a very useful way to account for energy flows.

EXAMPLE 6

Remove the impervious boundary in Figure 3 and replace it with a control boundary that allows energy flows only in the form of heat transfer from the room to the rest of the universe. There is thermal equilibrium when the book is on the desk (i.e., no net heat flowing).

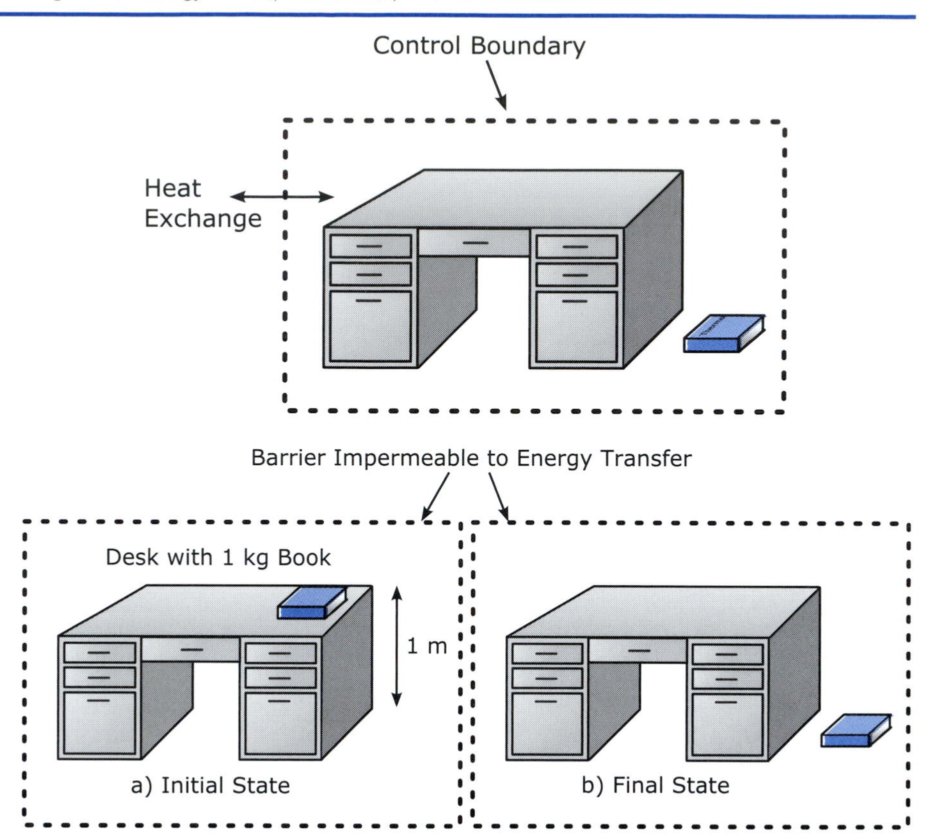

Figure 3

"Where Did the Book's Potential Energy Go?"

Suppose the initial energy with the classroom was 100.0 J when the room was isolated. What is the final energy of the classroom after the exchange of thermal energy?

Need: Energy of classroom, $Q_{\text{Final}} = \_\_$ J.

Know How: 9.8 J of potential energy was "lost" when a textbook fell to the floor. It cannot vanish, so what happened to it? It was eventually all converted to heat. When the book was on the table, the total energy in the room was 100.0 J.

Solve: After the textbook falls, 9.8 J of thermal energy was created from its original GPE. This eventually flows across the control boundary to the rest of the universe in the form of heat:
$\therefore$ initial total energy = final total energy or $100. = Q_{\text{Final}} + 9.8$

$\therefore$ $\boldsymbol{Q_{\text{Final}}} = 100.0 - 9.8 = \mathbf{90.\ J.}$ In other words, the classroom's energy falls by 9.8 J.

■

This kind of model captures enough of reality that it can be highly useful in engineering analysis and design, as Figure 4 shows. Figure 4

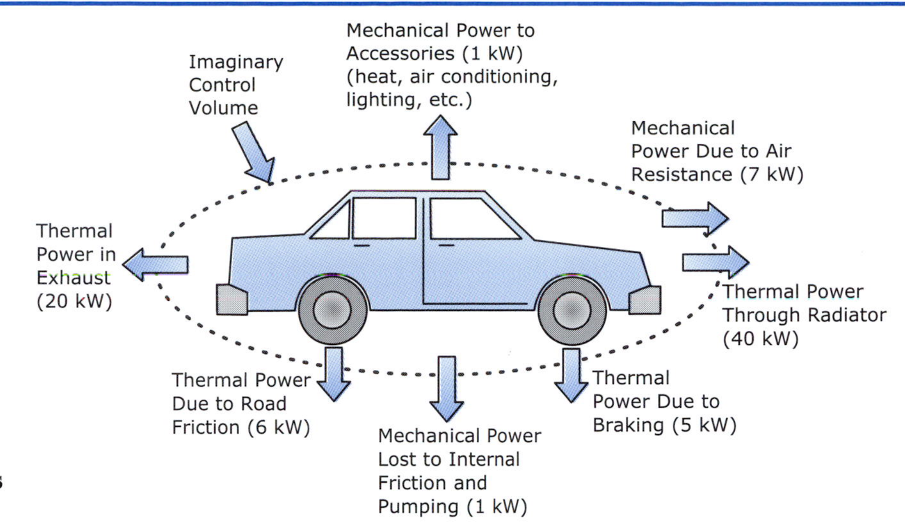

Figure 4

Control Boundary Analysis Model of Automobile

shows a control boundary analysis model as applied to an automobile. In words, the picture illustrates the following "energy accounting" of a typical automobile trip. As indicated by the arrows crossing the dotted line, an automobile traveling at 65 miles per hour (29 m/s) transfers about 40.0 kilowatts to the atmosphere in the form of thermal energy from the radiator, about 20.0 kilowatts to the atmosphere in the form of heat and chemical energy out the exhaust, about 7.0 kilowatts of mechanical power[12] to the atmosphere in overcoming air resistance, about 11 kilowatts in frictional work (mainly the result of elastic compression and expansion of the rubber in the tires and applying the brakes), and about 2.0 kilowatts of work for other purposes such as pumping and operating such accessories as heat, lights, and air conditioning. All of these contributions total about 80 kW.

EXAMPLE 7

If gasoline contains 1.30×10^5 kJ/gallon, how many gallons of gasoline must be used per second to provide the energy needed to sustain travel at 65 miles per hour on a level road? From this, estimate the fuel economy in mpg (miles per gallon).

Need: Amount of gasoline consumed per second in gallons.

Know: An automobile traveling at 65 miles per hour (29 m/s) transfers about 80.0 kW or kJ/s (or 110 hp) into various forms of energy.

How: Apply the principle of the conservation of energy. The energy lost by the car through the boundary of the control surface must come from somewhere. The only place it can come from within the car is by

[12]In the atmosphere, due to "friction" the displaced air will eventually slow, and the energy imparted to it will dissipate as an equivalent amount of thermal energy.

the decrease in the chemical energy of some of the gasoline. By the principle of conservation of energy, the decrease in chemical energy must equal the transfer out of the dotted lines of all the other types of energy.

Solve: Chemical energy needed per second = 80.0 kW or 80.0 kJ/s. Since gasoline contains about 1.30×10^5 kJ/gallon of chemical (i.e., potential) energy, this means $80.0/1.30 \times 10^5$ [kJ/s][gallon/kJ] = **6.15×10^{-4} gallons/s** of gasoline are consumed.

Is this a reasonable answer? Note that a car going at 65 miles per hour travels roughly one mile per minute or 65/3600.0 miles/s. So converting to miles per gallon as follows:

$$65/3600. \times 1/6.15 \times 10^{-4} \text{ [miles/s][s/gallon]} = 29 \text{ mpg}$$

This is a reasonable estimate given the roughness of our calculations.

Note a couple of implications of this example. First, one might, from an energy standpoint, view an automobile as a device for converting the chemical energy of fuel into thermal energy in the atmosphere. Second, operating an automobile requires a process for liberating 80,000 joules of energy every second from liquid fuel. This process, called combustion, will be the subject of the next chapter.

SUMMARY

The principle of conservation of energy can help us answer those questions that began this chapter. Is the world running out of energy? Drawing a dotted line around the Earth and applying control boundary analysis indicates that the amount of energy here on Earth is either remaining constant or if you believe that global warming is happening, increasing slightly. So the answer is no—the world certainly is not running of the total energy, although useful primary energy sources might be in decline.

So where does the energy come from that provides the light by which you are reading these words? This gets to the real point that people are making when they assert we are in an “energy crisis”. Though the amount of energy around us is constant, we are rapidly converting reserves of conveniently exploited chemical energy (which is actually a form of stored-up solar electromagnetic energy derived from past sunlight) into much less useful thermal energy. The light that you are reading this by probably came from a lightbulb powered by a fossil-fuel burning electric power plant. As you read these words, that light is being converted from electromagnetic energy into heat at room temperature—a form of energy of very little use for anything beyond keeping you warm.

Originally we asked, “Does a car possess more energy when it is sitting in your driveway or 15 minutes later when it is traveling down the highway at

65 miles per hour?" Our control boundary analysis model of the automobile tells us that the automobile is continually transferring energy from its gas tank to the atmosphere. So it possesses less energy on the highway than it did 15 minutes earlier in the driveway, and until you pull into a gas station and fill up the gas tank, it will steadily decrease in the energy content within its dotted control boundary.

To sum up, an understanding of the concept of energy is one of the indispensable analytical tools of an engineer. That understanding begins with the following key ideas. Energy is *the capability to do work*. Energy comes in many kinds. These kinds are capable of conversion from one kind to another. In the course of these conversions, energy is neither created nor destroyed but, rather, conserved. This is a principle that an astute engineer can exploit to advantage in the analysis of complex systems using the concept of a control boundary that separates the universe into that which we are studying and the rest of the universe. Using this concept we can directly calculate energy flows to and from the control boundary as an aid in analyzing its parts.

EXERCISES

Pay attention to the inferred number of significant figures in your answers! Conversion factors: 1.00 J = 0.738 ft-lbf; 1.00 kg = 2.20 lbm and g_c = 32.2 lbm ft/lbf s^2.

1. Determine the gravitational potential energy (GPE) of an 8.00 × 10^3 kg truck 30. m above the ground. (**A: 2.4 × 10^6 J** to two significant figures, since h is known only to two significant figures.)
2. A spring at ground level—that is, at height = 0 m—shoots a 0.80 kg ball upward with an initial kinetic energy of 245 J. Assume that all of the initial TKE is converted to GPE, how high will the ball rise (neglecting air resistance)?
3. Chunks of Earth orbital debris can have speeds of 2.3 × 10^4 miles per hour. Determine the translational kinetic energy (TKE) of a 2.0 × 10^3 lbm chunk of this material in SI units. (**A: 4.8 × 10^{10} J** to two significant figures.)
4. Repeat the calculation of Example 3 in Engineering English units. Check that the answers to problems 3 and 4 agree using the appropriate conversion factors at the top of this section.
5. An airplane with a mass of 1.50 × 10^4 kg is flying at a height of 1.35 × 10^3 m at a speed of 250.0 m/s. Which is larger—its translational kinetic energy or its gravitational potential energy? (Support your answer with numerical evidence.) (**A:** TKE = **4.69 × 10^8 J; GPE = 1.99 × 10^8 J**, ∴ TKE is **greater than** GPE.)

6. A vehicle of mass 1.50×10^4 kg is traveling on the ground with a TKE of 4.69×10^8 J. By means of a device that interacts with the surrounding air, it is able to convert 50% of the TKE into GPE. This energy conversion enables it to ascend vertically. To what height above the ground does it rise?
7. Aeronautical engineers have invented a device that achieves the conversion of kinetic to potential energy as described in problem 6. The device achieves this conversion with high efficiency. In other words, a high percentage of the translational kinetic energy of motion is converted into vertical "lift" with little lost to horizontal "drag." What is the device called? (**Hint:** This is not rocket science.)
8. A hypervelocity launcher is an electromagnetic gun capable of shooting a projectile at very high speed. A Sandia National Laboratory hypervelocity launcher shoots a 1.50 gram projectile that attains a speed of 14.0 km/s. How much electromagnetic energy must the gun convert into TKE to achieve this speed? Solve in SI. (**A: 1.5×10^2 kJ.**)
9. Solve Exercise 8 in Engineering English units. (Check your answer by converting the answer to Exercise 8 into Engineering English units.)
10. Micrometeoroids could strike the International Space Station with impact velocities of 19 km/s. What is the translational kinetic energy of a 1.0 gram micrometeoroid traveling at that speed? (**A: 1.8×10^5 J.**)
11. Suppose a spaceship is designed to withstand a micrometeoroid impact delivering a TKE of a million joules. Suppose that the most massive micrometeoroid it is likely to encounter in space has mass of 3g. What is the maximum velocity relative to the spaceship that the most massive micrometeorite can be traveling at for the spaceship to be able to withstand its impact?
12. A stiff 10.0 g ball is held directly above and in contact with a 600.0 g basketball and both are dropped from a height of 1.00 m. What is the *maximum* theoretical height to which the small ball can bounce?
13. An electric oven is heated by a circuit that consists of a heating element connected to a voltage source. The voltage source supplies a voltage of 110. V, which appears as a voltage drop of 110. V across the heating element. The resulting current through the heating element is 1.0 A. If the heating element is perfectly efficient at converting electric power into thermal power, what is the thermal power produced by the heating element? (**A: 1.1×10^2 W** to two significant figures.)
14. A truck starter motor must deliver 15 kW of power for a brief period. If the voltage of the motor is 12 V, what is the current through the starter motor while it is delivering that level of power?
15. A hybrid car is an automobile that achieves high fuel efficiency by using a combination of thermal energy and electrical energy for propulsion. One of the ways it achieves high fuel efficiency is by regenerative braking. That is, every time the car stops, the regenerative braking system converts part of the TKE of the car into electrical energy, which is stored in a battery. That stored energy can later be used to propel the car.

The remaining part of the TKE is lost as heat. Draw a control surface diagram showing the energy conversions that take place when the hybrid car stops. (**A:** See diagram.)

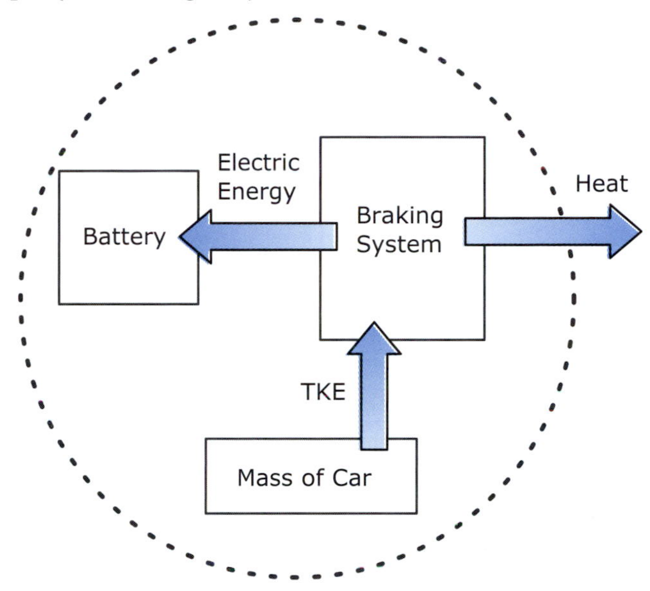

16. Suppose the car in problem 15 has a mass of 1000. kg and is traveling at 33.5 miles per hour. As it comes to a stop, the regenerative braking system operates with 75% efficiency. How much energy per stop can the regenerative braking system store in the battery? Illustrate with a control boundary showing the energy flows.
17. Suppose the car in problems 15 and 16 has stored 1.00×10^2 megajoules (MJ) of energy in its battery. Suppose the electric propulsion system of the car can convert 90% of that energy into mechanical power. Suppose the car requires 30. kw of mechanical power to travel at 33.5 miles per hour. How many miles can the car travel using the energy in its battery? (**A: 28 miles.**)
18. In order to maintain a speed v, a car must supply enough power to overcome air resistance. That required power goes up with increasing speed according to the formula:

$$\text{Power, } P = K \times v^3 \text{ kW}$$

where v is the speed measured in miles/hour and K is a constant of proportionality. Suppose it takes a measured 7.7 kW for a car to overcome air resistance alone at 30.0 mph.

(a) What is the value of K in its appropriate units?
(b) Using a spreadsheet, prepare a graph of power (kW on the y axis) as a function of speed (mph on the x axis) for speeds from 0 mph to 100 mph.

19. Review Exercises 7–10 in Chapter 3 concerning the dynamics (and consequent fate) of bungee jumpers.
    Draw a control surface around the jumper and cord. Show the various forms of energy possessed by the jumper and cord, along with arrows showing the directions of energy conversion inside and across the control surface: (a) when the jumper is standing on the cliff top, (b) when the jumper is halfway down and, (c) when the cord brings the jumper to a safe stop.
20. After working for a company for several years, you feel you have discovered a more efficient energy conversion method that would save your company millions of dollars annually. Since you made this discovery as part of your daily job you take your idea to your supervisor, but he/she claims it is impractical and refuses to consider it further. You still feel it has merit and want to proceed. What do you do?

    (a) You take your idea to another company to see if they will buy it.
    (b) You contact a patent lawyer to initiate a patent search on your idea.
    (c) You go over your bosses head and talk to his supervisor about your idea.
    (d) You complain to your company's human resources office about having poor supervision.

21. Your course instructor claims that energy is not really conserved. He/she uses the example of a spring that is compressed and then tied with a nylon string. When the compressed spring is put into a jar of acid, the spring dissolves and the energy it contained is lost. How do you react?

    (a) Ignore him/her and follow the established theories in the text.
    (b) Go to the department chairperson and complain that the instructor is incompetent.
    (c) Say nothing, but make detailed statements about the quality of the instructor on the course evaluation at the end of the term.
    (d) Respectfully suggest that the energy in his/her spring example really is conserved.

Chapter 5

Courtesy of Daimler-Chrysler

Chemical Energy of Fuels

Energy Conversion

How is the chemical energy of a fuel such as gasoline or natural gas converted into the kinetic energy of an automobile moving down the road or into the electrical energy from a turbine-generator set used to power a television or refrigerator? Each is a multistep process that effects the conversion from fuel to mechanical or to electrical energy. In this chapter, we will consider the first step in that process, conversion of chemical energy into thermal energy.

Combustion is the *oxidation* of a *fuel* generating heat, perhaps also accompanied by the emission of light. Oxidation is basically the chemical reaction of a substance with oxygen. Combustion may be either slow or rapid depending on the circumstances. The key points to understanding combustion are the concepts of **atoms, molecules,** and **chemical reactions;** a name for a very large number of atoms or molecules called the **kmol;** a "chemical algebra" method called **stoichiometry;** a measure of the amount of air present when combustion occurs, and the measure of this for an engine called the **air-to-fuel ratio,** and finally a way to determine the amount of heat produced by the combustion of a fuel, which is called the **heating value** of that fuel. One useful principle that is very helpful is the discovery of **Avogadro** that equal volumes of gases at the same conditions of temperature and pressure contain the same number of molecules.

Atoms, Molecules, and Chemical Reactions

An *atom* (from the Greek words meaning "cannot be sliced") is the smallest possible piece of a chemical element. Although now known to be sliceable, atoms are still used as the basic building blocks of matter in almost all engineering models. In combustion, as in all *chemical* reactions, *the number and type of participating atoms must remain constant.* A molecule is the smallest possible piece of a chemical *compound.* Molecules are made of atoms. In combustion, as in other chemical reactions, the number of molecules present typically does not remain constant, since molecules can be "sliced." This mathematics of molecules, where two plus one can equal two, is presented symbolically in chemical equations such as the following, which describes the combination of two hydrogen molecules with one oxygen molecule to form two molecules of water:

$$2H_2 + O_2 \rightarrow 2H_2O$$

This equation means that two molecules of hydrogen (symbol H_2, meaning each molecule contains two atoms of hydrogen, each symbol H) combine with one molecule of oxygen (symbol O_2 is a molecule that also contains two atoms of oxygen, each symbol O). The result of these molecules reacting is to form two molecules of water, each symbol H_2O, each molecule of which contains two atoms of hydrogen and one of oxygen. Note that there are four hydrogen and two oxygen atoms to the left of the arrow and also four hydrogen and two oxygen atoms to the right of the arrow.

The mol and the kmol

Since molecules are extremely small entities, it takes enormous numbers of them to provide useful amounts of energy for powering automobiles or performing any macroscopic task. So rather than counting molecules by ones or twos, they are counted in very large units called mols,[1] or even larger units called kmols[2] (thousands of mols).

> *The mol is defined to be the amount of substance containing as many "elementary entities" as there are atoms in exactly* 0.012 *kg of pure carbon*-12. *(The "kmol" is a factor* 10^3 *larger.)*

Just as a dozen eggs is a way of referring to exactly 12 eggs, a mol is a way of referring to 6.0221367×10^{23} molecules. This number of elementary entities is very large indeed and is referred to as "Avogadro's Number,"

[1]The "mol" is an abbreviation for "mole," just as the symbol for "degrees kelvin" is K. Without a prefix, mol/mole always means a g. mole that contains Avogadro's Number of elementary entities.

[2]Note that we can also define a kilogram mole (or kg mole, further abbreviated as kmol) as 1000 mols. Because of its virtual universal acceptance, the fundamental molar unit is the mol rather than the more logical SI-compatible kmol.

Table 1

Atomic Masses of Some Common Elements to three Significant Figures

Hydrogen, H	1.01*	Nitrogen, N	14.0
Oxygen, O	16.0	Helium, He	4.00
Carbon, C	12.0	Argon, Ar	40.0
Sulfur, S	32.1	Chlorine, Cl	35.5

**A more precise molecular mass for hydrogen atoms is 1.0079 to reflect that it contains a small amount of the double mass hydrogen isotope "deuterium" and a much smaller and a slightly variable amount of the radioactive triple mass hydrogen isotope, "tritium."* (For this book, writing the atomic mass of hydrogen as 1.00 is acceptable.)

symbol N_{Av}. Obviously in a *k*mol, the number of elementary entities is 6.0221367×10^{26}.

Elementary entities may be such things as atoms, molecules, ions, electrons, or other well-defined particles or groups of such particles. The mole unit is thus nothing but an alternate unit to counting individual elementary particles, and it will be useful in the analysis of chemical reactions. Continuing our dozen-egg analogy, the elementary entities might consist of five individual chicken eggs and seven individual turkey eggs. If so, notice that not every egg will have the same mass. The ***atomic masses***[3] of some common elements correct to three significant figures are given in Table 1. They are measured *relative* to the mass of carbon–12 (written C^{12} or C-12) being exactly $12.0^{\bullet}$ (the superscript $^{\bullet}$ meaning the zero reoccurs to infinite length).

Many gases are divalent (i.e., chemically combined as a paired set) such as hydrogen, oxygen, and nitrogen molecules, written H_2, O_2, and N_2, respectively (obviously their ***molecular*** masses are 2.00, 32.0, and 28.0, respectively). Thus, every kmol of water has a mass of approximately 18.0 kg, since the atomic mass of every hydrogen atom is (approximately) 1.00 kg/kmol, and the atomic mass of every oxygen atom is (approximately) 16.0 kg/kmol.

EXAMPLE 1

(a) How many mols of water are in 10 kg of water?
(b) How many kmols of water are in 10 kg of water?

Need: Number of mols, kmols in 10.0 kg of H_2O.

Know: Atomic masses of O and H are 16.0 and 1.00, respectively.

How: The number of moles $\boldsymbol{n}$ of a substance with a mass $\boldsymbol{m}$ that has a molecular mass $\boldsymbol{M}$ is given by $\boldsymbol{n} = \boldsymbol{m}/\boldsymbol{M}$.

[3]The term "g ***atom***" is sometimes used to refer to the mass of atoms in Avogadro's number of the particular ***atom*** in question so that 1 (k)g atom of oxygen atoms has a mass of 16.0 (k)g. However, the definition of kmol in terms of "elementary entities" is inclusive, and we will not have to use the "g atom" terminology.

Solve: The molecular mass of water (H_2O) is $M = 2(1.00) + 1(16.0) = 18.0$ kg/kmol = 18.0 g/mol[4]. Then for 10.0 kg of water:

(a) 10.0 kg = 10,000 g, then $n = m/M = 10{,}000$ [g]/[18.0 g/mol] = **556 mol.**

(b) $n = m/M = 10$ [kg]/[18.0 kg/kmol] = **0.556 kmol.** ■

EXAMPLE 2 Determine the effective molecular mass of air assuming it is composed of 79% nitrogen and 21% oxygen.

Need: Molar mass of air with 21% O_2 and 79% N_2. Thus, air is a mixture. A kmol of a mixture of "elementary entities" must still have Avogadro's number of elementary particles, be they oxygen or nitrogen molecules. Thus, we need the combined mass of just these two kinds of elementary entities in the correct ratio, each of which has a different mass.

Know: Molar mass of O_2 is 32.0 kg/kmol, and the molar mass of N_2 is 28.0 kg/kmol.

How: Proportion the masses of each constituent according to their concentration.

Solve: $M_{air} = \%N_2 \times M_{N_2} + \%O_2 \times Mo_2 = 0.79 \times 28.0 + 0.21 \times 32.0 =$ **28.8 kg/kmol.**

Note: we have defined a kmol of air (even though air *molecules* have no real existence) in the sense there are two components, but in so doing we have preserved the notion that every mole should have Avogadro's number of elementary entities. ■

Stoichiometry

The goal of many important engineering models is determining the energy that can be provided by the combustion of a particular kind of fuel. For example, Chapter 4 posed the question of how an automobile engine achieves by combustion the conversion of some 80 kJ every second of chemical energy into the same quantity of thermal and mechanical energy. The model of combustion presented in this chapter will help answer that question.

A first step in the model is writing the sort of symbolic chemical reaction described above. The reaction has the following general form:

$$\text{Fuel} + \text{Oxygen} \rightarrow \text{Reaction Products}$$

[4] And equally = 18.0 lb/lb mole = 18.0 tons/ton mole, etc.

For example, the equation for the formation of water $2H_2 + O_2 \rightarrow 2H_2O$ is a combustion reaction, with hydrogen as the fuel, and steam or water (H_2O) as the reaction product.

For most of the combustion reactions we will consider, the fuel will be a ***"hydrocarbon,"*** a mixture of compounds containing only the two chemical elements: hydrogen and carbon. For example, "natural gas" is mostly methane (CH_4), which has four hydrogen atoms bonded to every carbon atom. Gasoline is a mixture of more than a hundred different hydrocarbons, with chains of carbon and hydrogen atoms containing from 4 to 12 carbon atoms. For many purposes it can be conveniently *modeled* by considering it to consist only of molecules of isooctane, C_8H_{18}, with a *molar* hydrogen-to-carbon ratio[5] of $18/8 = 2.25$.

The energy history of the modern world might be summed up as an increase in the hydrogen-to-carbon ratio of the predominant fuel. Coal, the dominant fuel in the nineteenth century, has about one atom of hydrogen per atom of carbon. The twentieth century saw increasing use of petroleum-based fuels (with about two hydrogen atoms per carbon atom). At the turn of the twenty-first century, wealthy societies rely increasingly on natural gas (with about four hydrogen atoms per carbon atom). A possible goal for the twenty-first century is the "hydrogen economy," with an infinite hydrogen-to-carbon ratio (that is, no carbon in the fuel at all). Advocates of this continuing reduction of the carbon content of fuels point out that that combustion of carbon produces carbon dioxide, a major "greenhouse gas" implicated in global warming. Nuclear, solar, wind, and hydroelectric energy are possible contenders for primary noncarbon energy sources.

In general, in an engineering analysis we will know the fuel we want to "burn" and (most of) the reaction products that result. But we will not initially know the number of kilogram moles of oxygen needed to combine with the fuel, and we will not know the number of kilogram moles of each kind of reaction product that will result.

We can express this situation symbolically by assuming that we have one kilogram mole of fuel present and putting undetermined coefficients in front of the other chemicals to represent those unknowns.

For example, for the trivial example we have been considering, this would look as follows:

$$H_2 + aO_2 \rightarrow bH_2O$$

Determining the numerical value of those coefficients a and b is done by a "chemical algebra" called ***stoichiometry*** (Greek for "component measuring"). It relies on that key fact mentioned earlier: The number of each kind of atom remains constant. So one simply writes an equation for each type of atom that expresses this equality. Then one solves for the unknown

[5]By mass, the hydrogen to carbon ratio is approximately 0.158.

coefficients just as one solves any other set of simple linear algebraic equations. In the reaction shown above, for example, we have:

Hydrogen equation: 2 = 2 b (because there are two kmols of hydrogen atoms in the hydrogen molecules on the left, and 2b kmols hydrogen atoms in the b kmols of water molecules on the right.)

Oxygen equation: 2a = b (because there are 2a kmols of oxygen atoms in the oxygen molecule on the left and b kmols of oxygen atoms on the right.)

Solving those two equations with two unknowns yields b = 1 and a = ½.

Substituting this back into our original reaction results in the **stoichiometric equation:**

$$H_2 + 0.5O_2 \rightarrow H_2O$$

You will note that this equation looks slightly different from the one that was presented earlier ($2H_2 + O_2 \rightarrow 2H_2O$), but you can reassure yourself that it is actually the same in meaning as the earlier equation by dividing each of the coefficients of that equation by two. Note, too, that 0.5 kmol of O_2 makes perfectly good sense, whereas there is ambiguity[6] if we think we can equate ½ molecule of oxygen (i.e., $0.5O_2$) with one atom of O.

The process of finding the stoichiometric coefficients can be more difficult as the number of atoms in the participating molecules increases and yet more complicated when there are repeat molecules on both sides of the chemical reaction. Consider some common fossil fuels: natural gas (methane, CH_4), coal[7] $\sim CHO_{0.1}S_{0.05}N_{0.01}$ + ash, oil $\sim CH_2$, and many others. Most novices believe incorrectly that they can reliably find stochiometric coefficients by inspection. Experienced engineers will always use a systematic method. We will use a systematic method using tabular entries.

EXAMPLE 3

Common ether, better identified as diethyl ether, can be written as $H_3CH_2COCH_2CH_3$ but can be more simply written[8] as $C_4H_{10}O$. What are the stoichiometric coefficients to burn it completely?

$$C_4H_{10}O + aO_2 \rightarrow bCO_2 + cH_2O$$

[6]We will learn later that the *energetics* of the reaction is dependent on the *state* of the reactants and products; O_2 is markedly different in this respect from 2O just as CO_2 is different from C and O_2.

[7]Coal is very variable and has significantly different composition and properties, depending on its geologic age and its location.

[8]Chemical composition of "organic" chemicals can be summarized as "Hill formulae" in which we first write the number of carbon atoms, then the number of hydrogen atoms and then the rest of the atoms in alphabetical order. What this loses in chemical structure, it makes up in simplicity for retrieval of data pertaining to a complex molecule.

Table 2

Tabular Method of Solving Stoichiometric Combustion Problems

Atoms	LHS*	RHS*	Solution
C	4	b	$b = 4$
H	10	2c	$c = 5$
O	1 + 2a	2b + c	$a = b + c/2 - ½ = 6$

**LHS = left-hand side of the equation, and RHS = right-hand side.*

Need: Stoichiometric cfts, a, b, & c.

Know: You could certainly find a, b, and c by trial and error.

How: The calculation can be systematized into a simple tabular method of finding three equations in the three unknowns a, b, & c, as per Table 2.

Solve:

$$\therefore\ C_4H_{10}O + 6O_2 \rightarrow 4CO_2 + 5H_2O$$

Notice you have three unknowns (a, b, & c) and three equations—one for C, one for H, and one for O. Hence, you can solve it uniquely by equating the LHS with the RHS for each element. ■

In this book we exclusively use this "tabular" method of determining stoichiometric coefficients. One secret is to make sure that simple balances, such as for C and for H in this illustration, each involving only one unknown, should be solved *before* those that involve more than one variable, such as for O in this instance.

Make a quick scan of your solution when complete to confirm that you have a "mass balance"—that is, that you have neither created nor destroyed atoms. In this case you can quickly spot that there are 4 C atoms (or kmols, etc.) on both sides of the equation, 10 H atoms, and 13 O atoms. If there were other elements present, you can use one additional row in the table to balance each element (one equation per unknown).

The Air-to-Fuel Ratio

Although rockets in space chemically react fuel with an oxygen source that they have to haul on takeoff, most combustion reactions on Earth react fuel with the oxygen in the air. In almost all their combustion models, engineers treat air as a mixture that has 3.76 kmols of nitrogen molecules (N_2) for every kmol of oxygen molecules (O_2), ignoring the other molecules such as carbon dioxide, water vapor, and argon that are present in much smaller quantities. This approximate composition of air works out to be 79% N_2 and

21% O_2 expressed equally as volume or mole fractions. (They are the same if Avogadro's law is obeyed.)[9]

Many models also incorporate the fact that in typical combustion systems 100–200 parts per million (ppm) of the nitrogen in the air react with oxygen in the air to form nitrogen oxides that are an important cause of acid rain. However, in this introduction to engineering, we will ignore this quantitatively small but important reaction and treat all the nitrogen in products as in its divalent (and inert) form N_2. Thus, the N_2 in air will be considered totally unreactive in this simple treatment of combustion. Thus, too, the N_2 in the reactant air remains unchanged and leaves quantitatively as N_2 as a product.

In order to design efficient combustion systems, it is important to know the amount of air needed to burn each gallon or kilogram of fuel. This amount of air is determined by the **air-to-fuel ratio.** To determine it, we begin by including those 3.76 kmols of nitrogen that accompany each kmol of oxygen in the stoichiometric equation. This is done simply by multiplying the stoichiometric coefficient for oxygen by 3.76 and using that as the stoichiometric coefficient for nitrogen. Since, in our model, the nitrogen does not take part in the combustion reaction, it *usually*[10] has the same coefficient on both sides of the reaction. Thus:

$$H_2 + [\tfrac{1}{2}O_2 + \tfrac{1}{2} \times 3.76N_2] \rightarrow H_2O + \tfrac{1}{2} \times 3.76N_2$$

or

$$H_2 + \tfrac{1}{2}[O_2 + 3.76N_2] \rightarrow H_2O + 1.88N_2$$

The term $O_2 + 3.76N_2$ (in the square brackets) is *not* one mol of air even though only 1 mol of it, the oxygen, is actually participating in the reaction; in fact, it is 4.76 mols of an oxygen + nitrogen mixture. It can be thought of as being 4.76 mols of air in the sense that 1.00 mol of air, as is any mol, will always contain Avogadro's number's worth of elementary entities irrespective of their identity as an oxygen or a nitrogen molecule, and thus, 4.76 "mols of air" contain $4.76 \times N_{Av}$ elementary particles.

The above stoichiometric equation makes it possible to determine the air-to-fuel ratio in two different ways: as a ratio of numbers of molecules (called the *molecular* or *molar* air-to-fuel ratio) and as a ratio of their masses (called the *mass* air-to-fuel ratio). We will use the nomenclature $(A/F)_{molar}$ and $(A/F)_{mass}$ to distinguish between these two dimensionless ratios. Since these ratios have no units, they are best written as $(A/F)_{molar} =$ [kmol

[9] After an extensive period of experimentation, the Italian chemist Count Amado Avogadro (1776–1856) proposed in 1811 that equal volumes of different gases at the same temperature and pressure contained equal numbers of molecules. It was not generally accepted by the scientific community until after Avogadro's death. The law is equivalent to claiming that each component in the gaseous mixture behaves as an *ideal gas*, a constraint that is best assumed at low pressures and high temperatures for most gases.

[10] The exception being the fuel-bound nitrogen case—e.g., burning ammonia gas according to $NH_3 + 1.5[O_2 + 3.76N_2] \rightarrow 3H_2O + \mathbf{6.64}N_2$ (and not $\mathbf{5.64}N_2$).

of air/kmol of fuel] and $(A/F)_{mass}$ = [kg of air/kg of fuel], respectively, to emphasize what is being stated.

EXAMPLE 4

Determine the molar and mass stoichiometric air-to-fuel ratios for the combustion of hydrogen.

Need: $(A/F)_{molar}$ = ___ moles of air per mol of hydrogen and $(A/F)_{mass}$ = ____ mass of air per mass of hydrogen.

Know: Stoichiometric equation: $H_2 + 0.5[O_2 + 3.76N_2] \rightarrow H_2O + 1.88N_2$.

How: Use the stoichiometric equation to ratio mols; then multiply by relative masses of atoms from Table 1.

Solve: $(A/F)_{molar} = \dfrac{0.5(1+3.76)}{1} = \mathbf{2.38\ [kmol\ air/kmol\ H_2]}$

To express the *mass* ratio, simply assign the molecular masses to each component—that is, $\mathbf{(A/F)_{mass}} = 0.5 \times (1 \times 32.0 + 3.76 \times 28.0)/(1 \times 2.00)$
{[kmol O_2] [kg O_2/kmol O_2] + [kmol N_2][kg N_2/kmol N_2]}/[kmol H_2] [kg H_2/kmol H_2] = **34.3 [kg air/kg H_2]** in which we have identified the mass of air term as that in braces "{ }." ■

Since the unit method is clumsy in this example, it is pointed out that the concept of a molar air mass, M_{Air}, of 28.8 kg/kmol gets around this minor complication in a simplified step.

Hence, $(A/F)_{mass} = 2.38 \times 28.8/(1 \times 2.00)$ [kmol air][kg air/kmol air]/[kmol H_2][kg H_2/kmol H_2] = 34.3 [kg air/kg H_2] (as above).

The fuel-to-air ratios are simply the inverses of the air-to-fuel ratios:

$$(F/A)_{molar} = 1/(A/F)_{molar} = 1/2.38 = 0.42\ [kmol\ H_2/kmol\ air]$$

$$(F/A)_{mass} = 1/(A/F)_{mass} = 1/34.2 = 0.029\ [kg\ H_2/kg\ air]$$

The molar and the mass types of air-to-fuel ratios have different uses in engineering analysis. For example, the ***molar*** air-to-fuel ratio is useful to a design engineer determining the ***volume*** of air involved[11] in a combustion system and therefore its dimensions of air intakes and passageways. The ***mass*** air-to-fuel ratio is useful to the engineer who wants to calculate how much fuel to provide, perhaps in estimating a vehicle's fuel economy. An environmental engineer may also calculate the masses of different types of pollutants that might result from a combustion process.

The stoichiometric fuel-to-air ratio is invariant for a given fuel composition, while *actual* fuel-to-air ratios can be varied by an engineer deciding just how much air or fuel to add. The term *equivalence ratio*, symbol

[11]Recall Avogadro's law that moles and volumes are proportional to each other.

φ (Greek phi), is defined as **(*F/A* actual)/(*F/A* stoichiometric),** and it is useful, since it quickly tells you the overall conditions of the burn. Stoichiometric combustion means that the equivalence ratio is 1.0. In the jargon of the combustion engineer, excess fuel is described as *rich* and excess air as *lean*—for example, equivalence ratios of 1.05 and 0.95, respectively.

The Heating Value of Hydrocarbon Fuels

Chemistry can also answer the question "How much energy can be obtained by the combustion of one kilogram of fuel under stoichiometric conditions?" That in turn is a key part of answering the question "How many miles per gallon of fuel is it possible for an automobile to travel?" Let us briefly consider the first of those two questions.

In this section we will discuss only a simplified method and leave the general case for the end of this chapter. To determine the amount of energy that can be obtained by the combustion of one kilogram of fuel under stoichiometric conditions, the type of fuel being used must be specified. Assume that fuel is composed only of a molecule with x atoms of carbon and y atoms of hydrogen, and its combustion is thus described by the following stoichiometric equation,

$$C_xH_y + (x + (y/4))O_2 \rightarrow xCO_2 + (y/2)H_2O$$

We can think of this as taking place in three imaginary steps: The first is the decomposition of the fuel to its constituent atoms, and the second and third are their respective combustion to the stable oxides CO_2 and H_2O. Hence, the conceptual model says that:

$$C_xH_y \rightarrow xC + yH \ \ldots \text{ is followed by:}$$

$$xC + xO_2 \rightarrow xCO_2 \ \ldots \text{ and}$$

$$yH + (y/4)O_2 = (y/2)H_2O$$

The energy released must be that released by burning x moles of carbon atoms and y moles of hydrogen atoms *less* the energy in step 1 to dissociate the original fuel into its elements xC and yH.

Some fuels have different molecular structure than others—for example, "aromatic" fuels such as benzene (C_6H_6) form hexagonal molecules and are structurally unlike the more linear "aliphatic" fuels such as isooctane (C_8H_{18}) and cetane ($C_{16}H_{34}$). We are mostly interested in common transport fuels such as gasoline (as modeled by isooctane) and by diesel fuel (as modeled by cetane).

We will initially simplify our study of the energy content of hydrocarbon fuels by *asserting* our typical hydrocarbon fuel has a heating value of 45,500 kJ/kg irrespective of precise composition. This says the energy to dissociate

Table 3

Heating Values of Hydrocarbon Fuels

Fuel	Heating Value, kJ/kg
Methane, CH_4	55,650
Propane, C_3H_8	46,390
Isobutane, C_4H_{10}	45,660
Gasoline, "C_8H_{18}"	45,560
Diesel fuel, "$C_{16}H_{34}$"	43,980
Benzene, "C_6H_6"	42,350
Toluene, "$C_6H_5CH_3$"	42,960

each step of these hydrocarbons is about equal. This is obviously not true of the smallest hydrocarbon member, CH_4, or indeed if the fuels also contain significant oxygenates (such as alcohols), but it is sufficiently accurate for our current purposes and can be easily corrected.

The amount of oxygen consumed by our simple hydrocarbon, C_xH_y, is:

$$\text{Mass of } O_2 = \frac{32.0(x + y/4)}{(12.0x + y)} \text{ kg/kg fuel}$$

The amount of carbon dioxide produced is:

$$\text{Mass of } CO_2 = \frac{44.0x}{(12.0x + y)} \text{ kg/kg fuel}$$

The amount of water produced is:

$$\text{Mass of } H_2O = \frac{9.0y}{(12.0x + y)} \text{ kg/kg fuel}$$

Thus, for typical gasoline-like compounds $x = 8$, $y = 18$, one kilogram of a compound of hydrogen and carbon reacts with 3.5 kilograms of oxygen to release 45,500 kilojoules of energy and to produce 3.1 kg of carbon dioxide and about 1.4 kg of water. This knowledge enables one to a crude first approximation to determine how many miles it is possible for an automobile to travel on one gallon of fuel. That calculation is one of the exercises of this chapter.

The Heating Value of Fuels: The General Case

The bottom line of our stoichiometric chemistry is answering the question "How much energy can be obtained by the combustion of one kilogram of *any* fuel under stoichiometric conditions?" So far in this book, we have asserted an approximation for *hydrocarbons* that the heating value HV = 45,500 kJ/kg. However, nonhydrocarbon molecules do *not* obey this simple rule, as shown in Table 4.

Why are these fuels different from the hydrocarbons? What is essentially different is that the energy consumed in making these fuels from their

Table 4

Some Nonhydrocarbon Fuels

Fuel	Heating Value, kJ/kg
Ethyl alcohol, C_2H_5OH	27,904
Hydrogen, H_2	120,000
Carbon monoxide, CO	10,100
Carbon, C	32,800

constituent atoms is not equal to the corresponding energy in making hydrocarbons. In the strict language of thermodynamics, the *heat of formation* of a compound is the heat of reaction when that compound is formed from its elements at the same temperature and pressure. In the case of elements such as H_2 and plain carbon, their "energy of formation" is taken as zero.

Table 5

Heats of Formation of Various Fuels

Substance	ΔH_f kJ/kg
Carbon, C (s)	0
Nitrogen, N_2 (g)	0
Oxygen, O_2 (g)	0
Hydrogen, H_2 (g)	0
Carbon Monoxide, CO (g)	−3,946
Carbon Dioxide, CO_2 (g)	−8,942
Water, H_2O (g)	−13,423
Methane, CH_4 (g)	−4,667
Acetylene, C_2H_2 (g)	+8,720
Hexane, C_6H_{14} (g)	−1,945
Ethanol, C_2H_5OH (*l*)	−5,771
Benzene, C_6H_6(*l*)	629

(g) indicates the gas or vapor state, (l) is a liquid, and (s) indicates a solid.

The numbers in Table 5 reflect the exothermicity (heat out) or endothermicity (heat in) of the (often imaginary[12]) act of putting these molecules together starting with their elements (e.g., C and ½ O_2 in the case of CO, and simply H_2 in the case of H_2[13], etc.). As long as the energies are all calculated on the same basis (such as at the same temperature and pressure as each other), these heats of formation can be comingled.

[12]Usually one has to devise an indirect chemical path to make these molecules from their constituents.

[13]This is why its energy of formation is zero.

The "heat of combustion" or the "heating value" of the fuel[14] is defined as:

Heating Value = Energy in fuel and oxidizer − Energy in the products of combustion

The energy in the fuel and its combustion products are their respective heats of formation, so we can use Table 5 of the heats of formation in conjunction with the stoichiometric method we used earlier. In order to determine a heating value of a fuel, we need to know (1) the amount and type of fuel used and (2) the composition of the combustion products. To do this we need to determine the chemical reaction for the combustion process. This is done by calculating the stoichiometric coefficients of each chemical in the reaction on both sides of the reaction equation using the tabular method. For example, in the combustion of ethanol in pure oxygen, the reaction equation is:

$$C_2H_5OH + 3O_2 \rightarrow 2CO_2 + 3H_3O + \text{"Heat"}$$

To systematically carry out this process of determining the heating value of fuels, yet another tabular method is suggested for this book; indeed, it should also be useful for practicing engineers, since it is a visual and systematic way to avoid algebraic (and particularly sign) errors (see Table 6). Using our definition, Heating Value = Energy in fuel and oxidizer − Energy in the products of combustion,

$$HV = -5{,}771 - (-17{,}106 - 15{,}759)$$

$$\text{or } \mathbf{HV = +27{,}100\ kJ/kg\ of\ fuel}$$

This method is exact, since we are no longer "hiding" the heat of formation of different fuels so that nonhydrocarbons are dealt with on the same basis as we can use for simple hydrocarbons.

Table 6

Heating Value Method

	LHS	LHS	RHS	RHS
Molar quantities	$C_2H_5OH(l)$	$O_2(g)$	$CO_2(g)$	$H_2O(g)$
Mass, kg	46.0	3 × 32.0	2 × 44.0	3 × 18.0
Mass m, kg/kg of fuel	1.00	2.087	1.913	1.174
ΔH_f, kJ/kg	−5,771	0.000	−8,942	−13,423
m × ΔH_f kJ/kg of fuel	−5,771	0.000	−17,106	−15,759

[14]There are two common conventions in thermodynamics in which the heat of combustion is positive in the "engineering" convention and negative in the "scientific" convention. Both systems are logically consistent.

The original method to determine the heating value of a hydrocarbon fuel was to assume it is fixed at 45,500 kJ/kg; our updated more general method takes into account the energy of formation of the fuel. In effect, our original approximation is valid because the energy to form H_2C-CH_2 bonds does not vary much in simple hydrocarbons; thus, the energy of formation for these bonds per mass of CH_2 also does not vary much either. However, for other fuels where the energy of formation is quite different, the more complicated calculations are then required.

SUMMARY

Determining useful engineering parameters, such as the miles-per-gallon achievable by an automobile, requires calculating the amount of thermal energy (heat) available from the chemical energy contained in a fuel. This calculation requires the mastery of several concepts from the field of chemistry. Those concepts include **atoms, molecules,** and **chemical reactions;** a name for a very large number of atoms or molecules called the **kmol;** a chemical algebra method called **stoichiometry** for keeping track of the numbers of atoms that occur in those reactions; a measure of the amount of air needed for combustion called the **air-to-fuel ratio;** and finally a way to understand the **heating value** of that fuel.

The term **equivalence ratio,** φ, is defined as **(*F/A* actual)/(*F/A* stoichiometric)** is used to quickly assess combustion conditions: a "fuel-rich" process operates at $\varphi > 1$ and a "fuel-lean" one at $\varphi < 1$.

EXERCISES

1. How many kmol are contained in 3.0 kg of ammonia NH_3? (**A: 0.18 kmols** to two significant figures.)
2. How many kmol are contained in 1.0 kg of nitroglycerine $C_3H_5(NO_3)_3$?
3. What is the mass of 5.0 kmol of carbon dioxide CO_2? (**A: 2.2 $\times$ 10^2 kg.**)
4. What is the mass of 1.00 kmol vitamin B1 disulfide $C_{24}H_{34}N_8O_4S_2$?
5. The effective molecular mass of air is defined as the mass of a kmol of elementary particles of which 78.09% are nitrogen molecules, 20.95% are oxygen molecules, 0.933% are argon atoms, and 0.027% are carbon dioxide molecules. What is the effective molecular mass of air? (Watch your significant figures!) What other factor could affect the effective molecular mass of air?
6. A gallon of gasoline has a mass of about 3.0 kg. Further, a kg of gasoline has an energy content of about 45,500 kJ/kg. If an experimental

automobile requires just 10. kW of power to overcome air resistance at steady speed of 30. miles an hour, and *if* there are no other losses, what would the gas mileage of the car be in miles per gallon? (**A: 110 mpg** to only two significant figures as problem is stated.)

7. Determine the value of the stoichiometric coefficients for the combustion of an oil (assumed molecular formula CH_2) in oxygen: $CH_2 + aO_2 = bCO_2 + cH_2O$. Confirm your answer is correct! (**A:** $a = 1.5$, $b = 1$, $c = 1$.)

8. Determine the value of the stoichiometric coefficients for the combustion of coal in oxygen given by the stoichiometric equation:

$$CHN_{0.001}O_{0.1}S_{0.05} + aO_2 = bCO_2 + cH_2O + dN_2 + eSO_2$$

9. Determine the value of the stoichiometric coefficients for the combustion of natural gas in air: $CH_4 + a(O_2 + 3.76N_2) = bCO_2 + cH_2O + dN_2$

10. Using the stoichiometric coefficients you found in problem 9, determine the molar air-to-fuel ratio $(A/F)_{molar}$ for the combustion of natural gas in air. (**A:** $\mathbf{(A/F)_{molar} = 9.52}$ kmols of air/kmols of fuel.)

11. Using the results of problems 9 and 10, determine the *mass* air-to-fuel ratio $(A/F)_{mass}$ for the combustion of natural gas in air. (**A: 17.2 kg of air/kg of fuel.**)

12. Determine the mass air-to-fuel ratio $(A/F)_{mass}$ for the combustion of an oil (represented by CH_2) in air.

13. Determine the mass air-to-fuel ratio $(A/F)_{mass}$ for the combustion of coal in air represented by the equation: $CHN_{0.01}O_{0.1}S_{0.05} + a(O_2 + 3.76N_2) = bCO_2 + cH_2O + dN_2 + eSO_2$.

14. Determine the mass air-to-fuel ratio $(A/F)_{mass}$ for the combustion of isooctane C_8H_{18} in air. Its stoichiometric equation is $C_8H_{18} + 12.5(O_2 + 3.76N_2) = 8CO_2 + 9H_2O + 47.0N_2$.

Exercises 15–20 use engineering considerations to give insight on the "global warming" issue.

15. Assume one kg of isooctane produces 45,500 kJ of energy. How much carbon dioxide is released in obtaining a kJ of energy from the combustion of isooctane in air? (**A:** $\mathbf{6.79 \times 10^{-5}}$ **kg** $\mathbf{CO_2}$**/kJ.**)

16. Assume a kg of hydrogen produces 1.20×10^5 kJ of thermal energy. How much carbon dioxide is released in obtaining a kJ of energy from the combustion of hydrogen in air?

17. The amount of carbon dioxide levels in the atmosphere has been increasing for several decades. Many scientists believe that the increase

in carbon dioxide levels could lead to "global warming" and might trigger abrupt climate changes. You are approached by a future U.S. president for suggestions on what to do about this problem. Based on your answers to problems 15 and 16, what might be the *simplistic* approach to reducing the rate of increase of the amount of carbon dioxide in the atmosphere due to combustion in automobiles and trucks? (**A: Replace isooctane with hydrogen.**)

18. Give three reasons why your suggested solution in problem 17 to global warming has *not* already been adopted.
19. A monomeric formula for wood cellulose is $C_6H_{12}O_6$ (it repeats a hexagonal structure based on this formula many times). The energy content by burning wood is approximately 1.0×10^4 kJ/kg. Determine the amount of carbon dioxide released in obtaining a kJ of energy by the combustion of wood in air **(A: 1.5×10^{-4} kg of CO_2 released/kJ of energy produced.)**
20. Repeat problem 19 for the combustion of coal. For simplicity assume coal is just carbon, C, with a HV of 32,800 kJ/kg.
21. Given the *heat of formation* of liquid methanol, $CH_3OH(l)$, is $-238{,}000$ kJ/kmol, what is its heat of combustion in kJ/kg?
22. In 1800 the main fuel used in the United States was wood. (Assume wood is cellulose whose representative repeating formula is $C_6H_{12}O_6$.) In 1900, the main fuel used was coal (assume coal in this example can be approximated by pure carbon, C). In 2000, the main fuel used is oil (assume it can be represented by isooctane). Assume that in 2100 the main fuel will be hydrogen (H_2). Use a spreadsheet to prepare a graph of carbon dioxide released per kJ of energy produced (y-axis) as a function of year (x-axis) from 1850 to 2050.
23. You have worked for a petroleum company producing automotive fuels for several years, and in your spare time at home you have developed a new fuel composition that has a higher heating value than ordinary gasoline. When you began work at this company, you signed a "confidentiality agreement" that gave them all of your intellectual property. What are your obligations to your employer?

    (a) It is your work on your time, so you have no obligations to your current employer.
    (b) Everything you used to develop your new fuel you learned on the job at your company, so your work really belongs to them.
    (c) Your company's confidentially agreement requires you to provide them with all your work.
    (d) Give your idea to a friend and let him/her pursue it while keeping you as a silent partner.

24. As a production engineer for a large chemical company, you need to find a new supplier for a specific commodity. Since this contract is substantial, the salespeople you meet with are naturally trying to

influence your purchasing decision. Which of the items below are ethical in your opinion?[15]

(a) Your meeting with a salesperson extends over lunch, and he/she pays for the lunch.
(b) In casual conversation at the sales meeting you express an interest in baseball. After the meeting, a salesperson sends you free tickets to your favorite team's game.
(c) After your meeting, a salesperson sends a case of wine to your home with a note thanking you for the "useful" meeting.
(d) As a result of the sales meeting, you are invited on an all expense paid trip to China to visit the salesperson's manufacturing facility.

Suggested answer: Fundamental Canons: Engineers, in the fulfillment of their professional duties, shall:

1. Hold paramount the safety, health, and welfare of the public. **Does not apply.**
2. Perform services only in areas of their competence. **Does not apply.**
3. Issue public statements only in an objective and truthful manner. **Does not apply.**
4. Act for each employer or client as faithful agents or trustees. **Accepting a gift larger than a specified level violates this canon. The specified level set by your company determines which, if any, of the options violates the canon. If your company has not set such a level, you can accept any gift without violating this canon. However, if such a level is set, you are obliged to know what it is. Ignorance is no excuse.**
5. Avoid deceptive acts. **Does not apply.**
6. Conduct themselves honorably, responsibly, ethically, and lawfully so as to enhance the honor, reputation, and usefulness of the profession. **Here the ethical standard of enhancing honor, reputation, and usefulness requires avoiding even the appearance of impropriety. Accepting a substantial gift appears improper, even if does not actually influence you. This is true even if your company's explicit standard permits you to accept the gift. So canon 5 suggests doing only (a), not (b), (c), or (d).**
   Solution: No tension here. Do only (a).

[15]How do you distinguish between a token gift and a bribe? Where do you draw the line?

25. You are attending a national engineering conference as a representative of your company. A supplier to your company has a booth at the conference and is passing out small electronic calculators to everyone who comes to their booth. The calculators are valued at about $25, and you are offered one. What do you do?

(a) Accept the gift and, since it has small value, there is no need to report it.
(b) Accept the gift and report it to your supervisor.
(c) Decline the gift, explaining that you work for one of their customers.
(d) Ask someone else to get one of the calculators for you.

Chapter 6

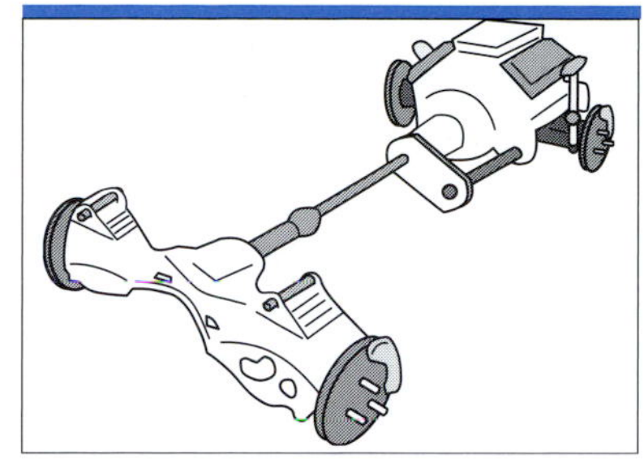

Automotive Driveline Illustration

The Automotive Drive Train

The Otto Cycle

How does the chemical energy of fuel become the kinetic energy of an automobile traveling along a road at 65 miles per hour? This chapter will trace this energy conversion process through the **automotive drive train,** which consists of an automobile's engine, crankshaft, transmission, axle, and wheels. We will detail the operation of **the Otto cycle,** the engineering name for the vast majority of today's automotive engines, and model the **power output** of that cycle. We will then consider the use of that power output to achieve **motion.** This consideration of motion will introduce the important concept of **gear ratios.** We will conclude by considering **some proposed improvements to the Otto cycle,** involving either making it better or replacing it with a different type of propulsion source.

Operation of the Otto Cycle

Central to the drive train of the modern automobile is the type of engine developed in the 1860s by a French engineer named Alphonse Eugene Beau de Rochas and then independently reinvented by a German engineer named Nikolaus Otto in 1876. The Otto cycle, as this invention is now universally called, has been victorious so far over rival ideas for powering automobiles ranging from the electric-and-steam powered propulsion systems of the early twentieth century to the gas turbine and rotary Wänkel engine of the 1960s and 1970s and recently reintroduced by Mazda. In the process, the Otto cycle has been improved through inventions ranging from the float-feed carburetor in the 1890s to the computer-controlled fuel-injection systems of today. This chapter will provide engineering models for understanding this ubiquitous and essential element of modern civilization.

The success of the Otto cycle rests on its combination of power, simplicity, and efficiency. That in turn results from its carrying out all the processes needed to convert chemical energy into mechanical energy within a cylinder. These processes include the four *strokes* of the Otto cycle: intake, compression, expansion, and exhaust. Since the process of combustion, which occurs between the compression and expansion strokes, is also within the cylinder, the Otto cycle belongs to the class of engines called **internal combustion engines.** This contrasts with the previously dominant prime mover, the steam engine, an *external* combustion engine. The steam engine requires, in addition to the cylinder, separate structures, such as a firebox, a boiler, and, in most cases, a condenser.

The essence of energy conversion in the Otto cycle is a series of controlled combustion processes within the cylinder, about 10–50 per second. Each combustion event converts about 1–2 kJ of chemical energy into heat, depending on the size of the cylinder. The heat and the extra volume of gases generated during the combustion processes expand the gaseous combustion products to push down a piston, thereby converting a portion of the heat energy into translational kinetic energy. Use of a **crank mechanism** converts that translational kinetic energy into the rotational kinetic energy of a **crankshaft,** providing the type of energy needed to turn the wheels of an automobile. A conceptual model of a four-cylinder Otto engine is shown in Figure 1.

Understanding the operation of this engine requires looking more closely at the cylinder and what happens inside it. The piston is cylindrical in shape

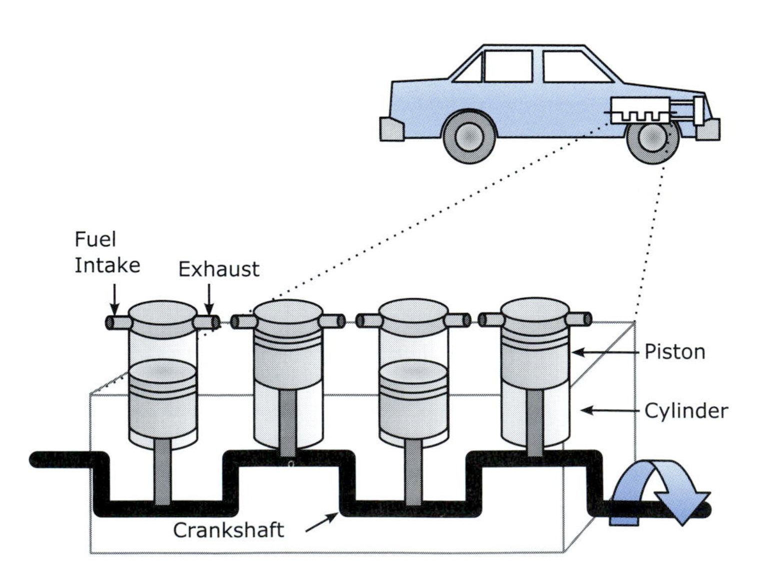

Figure 1

Conceptual Model of an Automobile Engine

and is a very tight fit within the cylinder. For the purposes of an engineering model, the cylinder and the piston moving within it are described by their common diameter, by the **stroke** (or distance traveled by the top of the piston from its highest to lowest point, "top and bottom dead center" respectively), by the **displacement,** which is the volume swept out by the piston when going through its stroke, and by the **clearance volume,** which is the volume within the cylinder above the highest point reached by the piston (see Figure 2).

At the top of the clearance volume are two ports that can be opened and closed by valves for the purpose of letting fuel and air into the cylinder and letting combustion products out. Within this cylinder the four

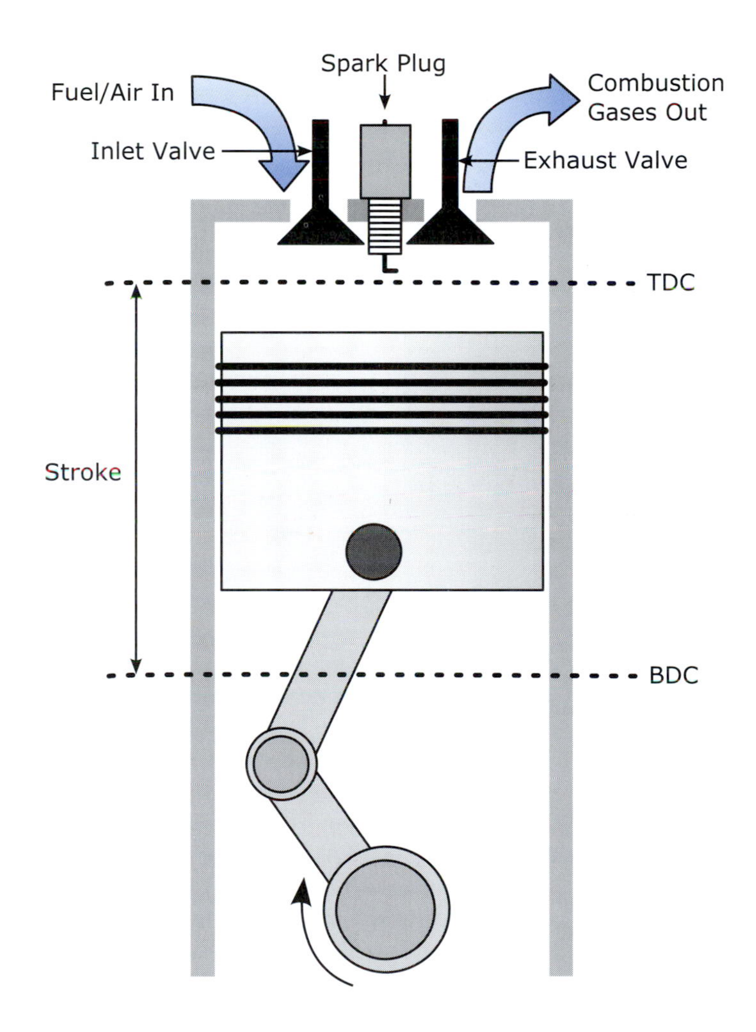

Figure 2

Stroke of an Engine Cylinder: TDC/BDC = Top/ Bottom Dead Center

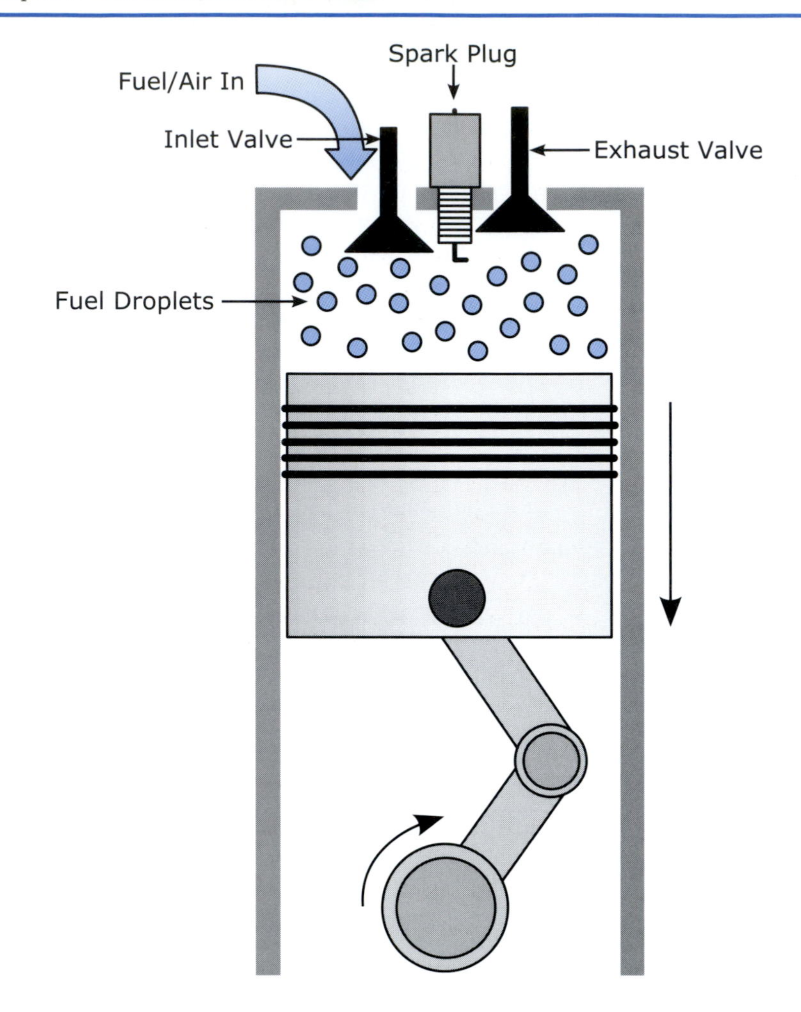

Figure 3

Intake Stroke

strokes of the Otto cycle occur sequentially, once per every two crankshaft revolutions.[1]

Figure 3 illustrates the **intake** stroke. In the first stroke, the intake valve is opened, the exhaust valve is closed, and the downward movement of the piston draws a mixture of air and fuel into the cylinder. Ideally, this is very close to a stoichiometric mixture as defined in the previous chapter.[2] While one of the virtues of the Otto cycle is its ability to run on a wide range of fuels, for the purposes of this model we will assume

[1]An Otto cycle animation can be found at http://techni.tachemie.uni-leipzig.de/otto/index_e.html.
[2]In practice, an Otto engine is normally run a few percent "lean." This means that there is up to 5% excess air to ensure that combustion is complete and fewer pollutants are emitted from the tailpipe.

the fuel is gasoline, which we can model by the compound isooctane, here denoted by the Hill formula C_8H_{18}. The stoichiometric equation also determines the mass stoichiometric **air-to-fuel ratio *A/F*,** also previously defined, and generally expressed as kilograms of air per kilogram of fuel.

In the second stroke, compression, the piston moves upward, compressing the fuel-air mixture from the full cylinder displacement down to clearance volume. This increases the pressure of the fuel-air mixture from nominally atmospheric pressure to a value greater than the product of (atmospheric pressure) × (compression ratio). The additional pressure is due to the fact that the compression process is rapid and thus heats the gases during the compression stroke.

Figure 4 illustrates the **compression** stroke. Following completion of the compression stroke, both valves are closed. A spark is produced in the spark plug at the correct instant. That spark ignites the fuel, and combustion is initiated.[3] As we have seen, combustion is a chemical reaction that converts the mixture of fuel and oxygen into a mixture of carbon dioxide and water, while leaving almost all of the nitrogen in its initial form and converting the chemical energy in the fuel-air mixture into heat.

During combustion the **power** stroke occurs. Both valves remain closed. The piston is driven downward by the expanding gases, delivering its mechanical energy to cause rotation of the crankshaft. (See Figure 5).

After the piston reaches its lowest point, the intake valve remains closed, and the exhaust valve is opened. The piston is then moved upward by rotation of the crankshaft, pushing the combustion gases out the exhaust port. The **exhaust** stroke is shown in Figure 6.

When the piston reaches its highest point of the exhaust stroke, the four strokes of the cycle are complete, and a new cycle begins as the piston proceeds down and more air/fuel enters through the inlet valve (Figure 3). Notice the crankshaft has rotated through two complete revolutions and the piston has traveled a total of four sweeps of the cylinder (up and down each being a single sweep or stroke)—hence the name four-stroke engine. For these four sweeps there are two revolutions of the crankshaft, as shown in Figure 7.

It is hard to get the dynamic picture of what is occurring from static pictures. There are several online animations that are useful and graphic. With animated graphics, it is easy to visualize that the crankshaft has to rotate twice to complete one power stroke of the engine.

[3] A successful engine and fuel design is such that the combustion process is not explosive per se but burns in a controlled manner during the "power stroke." Predetonation explosions or "knock" will destroy an engine.

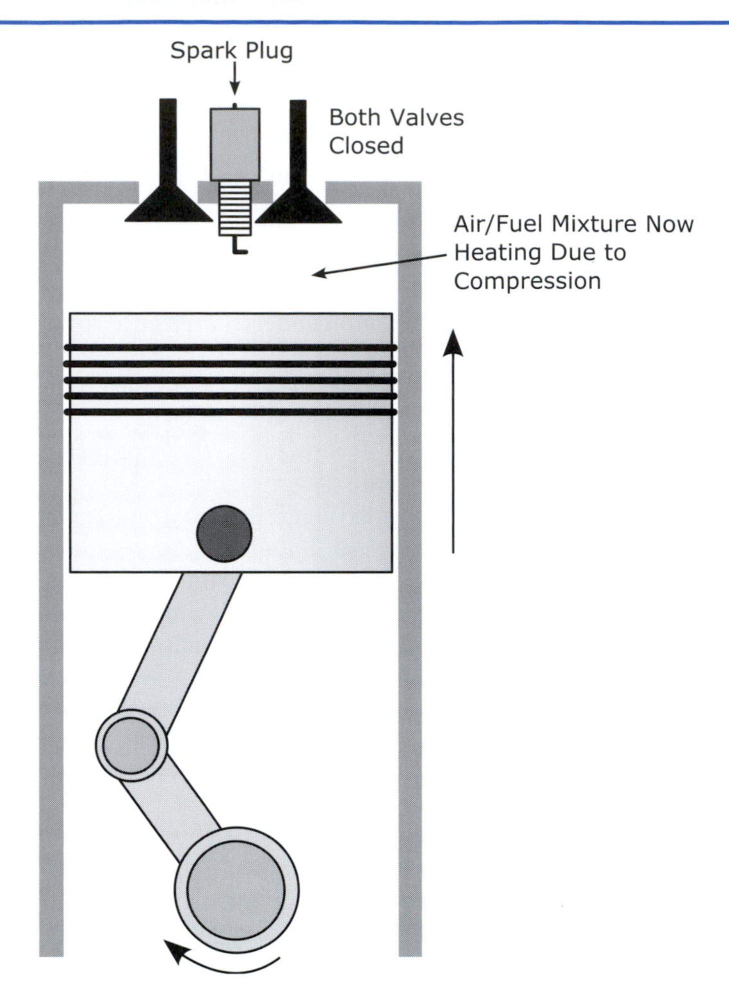

Figure 4

Compression Stroke

Modeling the Power Output of the Otto Cycle

To study the performance of the Otto cycle, the events that occur within it can be expressed in a mathematical model. Engineers summarize this model using the following equation:

$$\text{Ideal power} = \frac{\eta_{\text{Otto}} \times \rho_{\text{air}} \times \text{D} \times \text{N} \times \text{HV}}{120 \times A/F}$$

where ρ_{air} = Density of air at cylinder conditions

η_{Otto} = Otto cycle thermal to mechanical energy, conversion efficiency

D = Cylinder displacement (*volume* swept by piston in cylinder)

N = Crankshaft speed in revolutions per ***minute,*** RPM

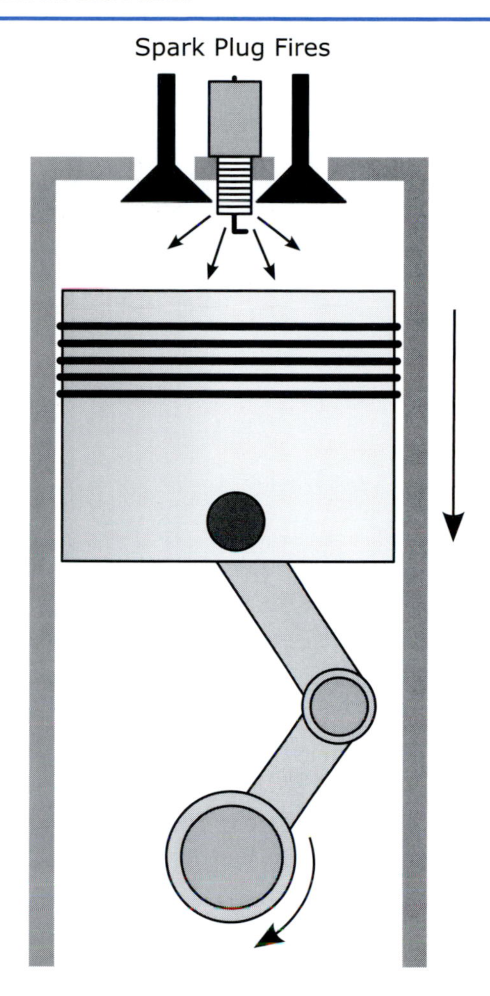

Figure 5

Power Stroke

A/F = Air-to-fuel mass ratio (usually close to the stoichiometric value)
CV = Valve clearance volume
CR = Compression ratio, $(D + CV)/CV$ (this will determine η_{Otto})
HV = Heating value of the fuel in consistent units (normally kJ/kg)

These variables must all be expressed *in a consistent set of units:* In fact, it often helps clarify the terms in such an equation by carrying a set of consistent units in its development (see following). The above equation looks somewhat frightening, but it is simply a matter of bookkeeping and is easily derived starting with the idea of representing the energy release in the Otto cycle as a series of relatively small "deflagrations." (A deflagration is a release of energy occurring at a controlled rate in which a "flame-front" advances in a progressive manner starting at the spark ignition point.) A certain number of such deflagrations occur every second, and each deflagration converts a certain amount of energy from chemical energy to heat energy. That conversion is carried out with a certain efficiency.

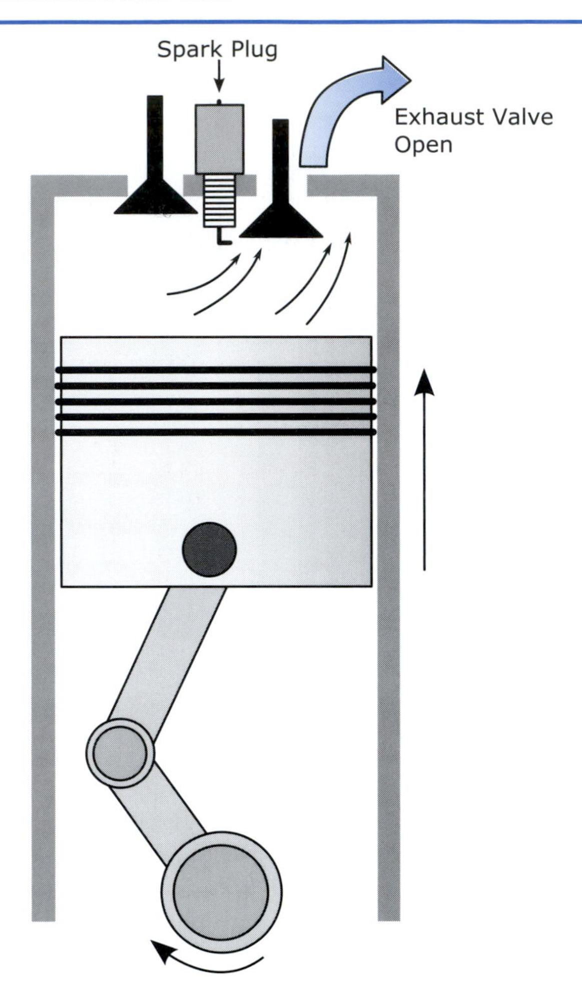

Figure 6

Exhaust Stroke

Breaking down into their elements each of these concepts—frequency of deflagrations, energy release, and efficiency—yields a quantitative model for operation of the Otto cycle. This model enables the engineer to derive that formidable looking equation from simple principles in the manner described here.

Derivation of Otto Cycle Power Output Equation

Step 1: How much air is drawn into the cylinder? The total cylinder volume is $D + CV$ expressed in m^3. However, the actual volume of air drawn into the cylinder is D as the piston cycles from top dead center to the bottom dead center. The mass of air drawn into the cylinder is thus $[\rho_{air} \times D]$ [kg air/m^3][m^3] = [kg air].

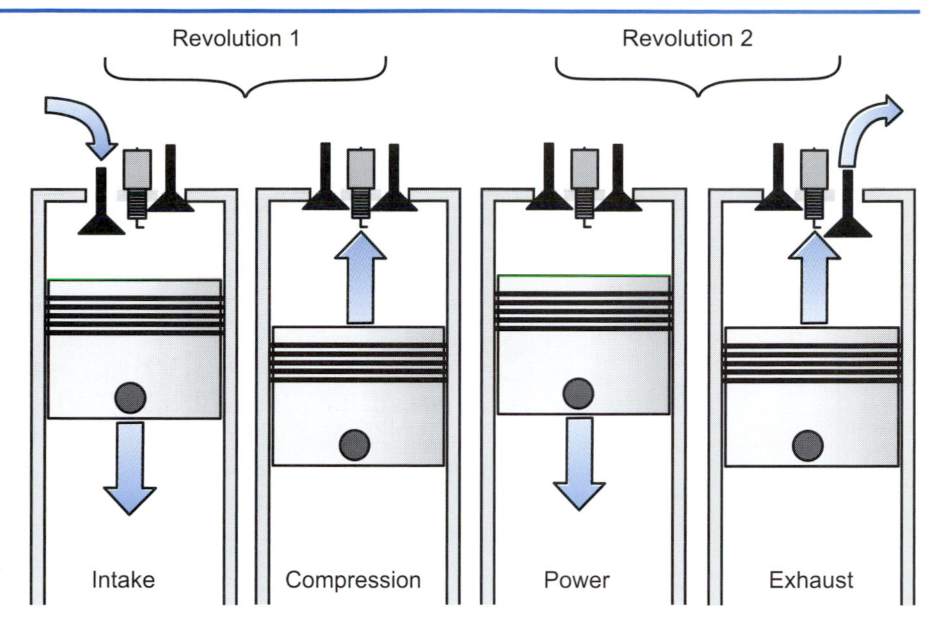

Figure 7

Summary of the Four-Stroke Engine Cycle

Step 2: How much fuel? $F = [\rho_{air} \times D] \times F/A$ [kg air] [kg fuel/kg air] = [kg fuel]

Step 3: How much thermal energy is released by this amount of fuel? $[\rho_{Air} \times D \times F/A] \times HV$ [kg fuel] [kJ (heat)/kg fuel] = [kJ(heat)]. Of course, this is for just one compression stroke, so we should remind ourselves that this is really $[\rho_{air} \times D \times F/A] \times HV$ [kJ (heat)/power stroke].

Step 4: Of this, only a portion is converted into mechanical work. Hence, mechanical work $= \eta_{Otto} \times [\rho_{air} \times D \times F/A] \times HV$ [kJ (heat)/power stroke] [kJ (work)/kJ (heat)] = [kJ (work)/compression stroke].

Step 5: How many power strokes are there? As we have seen, there is just one per every two engine revolutions, and if N is the number of revolutions/**minute**,[4] the number of power strokes/s $= N/(2 \times 60)$ [power strokes/rev] [rev/min] [min/s] = [power strokes/s].

Step 6: Hence, the rate of working is $\eta_{Otto} \times [\rho_{air} \times D \times F/A] \times HV \times N/(2 \times 60)$ [kJ (work)/power stroke] [power strokes/s] = [kJ (work)/s] = [kW].

Step 7: Finally, we can cast this in the suggested form:

$$\text{Ideal power} = \frac{\eta_{Otto} \times \rho_{air} \times D \times N \times HV_{fuel}}{120 \times A/F}$$

[4]"Minutes," a non-SI unit, will introduce a numerical factor of 60 somewhere in the equation provided the rest of the variables contain seconds—so look for a numerical constant to appear in the final result. (The other factor of "2" is there because there are *two* rotations of the crank per power stroke.)

In SI units this will give the answer in watts; in English units the result will usually be expressed in "horsepower" abbreviated "HP." This unit of power goes back to James Watt, an eighteenth-century Scottish engineer. 1 HP = 0.746 kW, so that it is of comparable magnitude to (what we hope is now) the more familiar kW. The final expression may be termed an "ideal" Otto cycle analysis, since it does not account for some real phenomena that limit the delivered power of an Otto engine.

The Otto cycle efficiency term in the model, η_{Otto}, can be derived from the field of ***thermodynamics.*** The result is $\eta_{Otto} = 1 - CR^{-0.40}$, a result directly related to the increase in absolute temperature during the compression of the air/fuel charge; it is a subject for more advanced courses. However, the point of the equation is clear: The higher the compression ratio, the higher the efficiency. Even if carried out with no frictional or thermal losses, an ideal Otto cycle will see its efficiency limited by its compression ratio. The ideal efficiency will always be less than one, although it will increase toward 100% with an increase in the compression ratio.

There are other mechanical losses associated with friction in the internals of the engineering. We can define a mechanical efficiency as:

$$\eta_{Mech} = \frac{\text{Power delivered by engine}}{\text{Power generated by engine}}$$

Unfortunately in real engines this term may only be only 50% so that one half of the mechanical power developed in the cylinders may not make it through to the final output shaft. (It will ultimately appear as parasitic heat generated somewhere in the engine.) One can define an overall Otto engine efficiency that includes this effect by:

$$\eta = \eta_{Mech} \times \eta_{Otto}$$

$$\therefore \text{ Power out} = \frac{\eta_{Mech} \times \eta_{Otto} \times \rho_{air} \times D \times N \times HV}{120 \times A/F}$$

The value of η_{Mech} may be empirical (meaning determined by experiment), or it may have some theoretical underpinnings. In either case, it suggests that there is still room for considerable improvement in overall gasoline engine efficiency to challenge another generation of engineers.

This simple model enables the engineer to draw conclusions about the effects of changing various engine parameters. For example, the model predicts that, *other things being equal*, increasing the displacement,[5]

[5]For a single cylinder, $D = \pi \times \text{radius}^2 \times \text{stroke} = \pi/4 \times \text{diameter}^2 \times \text{stroke}$. (In engineering, "diameter" is preferred to "radius," since that is what a measuring instrument would determine.) In multiple cylinder engines one can multiply the above result by the number of cylinders or simply use the total engine displacement, something the manufacturer usually provides (as in a "2.4 liter engine," etc.).

compression ratio (as it occurs in the Otto efficiency) and the revolution rate of the engine will increase its power, while increasing the air-to-fuel ratio will decrease its power. In this simple model, only by increasing the compression ratio can the ***efficiency*** of the engine be increased.

Like all simple models, this one has limitations. It would suggest, for example, that increasing the speed of revolution of the crankshaft (which really means carrying out more deflagrations of fuel per second) would increase the power without limit. This increase indeed occurs at revolution rates up to a few thousand per minute. But it ceases to be true at higher crankshaft revolution rates. At those higher rates, it becomes impossible to move matter (even including low-density matter such as air) or heat around fast enough to allow energy conversion to occur efficiently. Instead, as rotation rate is increased, power output first reaches a maximum, and then decreases. So to mathematically model performance of an Otto cycle for a wider range of rotational speeds, engineers add other correction terms, sometimes empirical terms guided by experiments but often based on solid theoretical underpinnings that show how power output rises and then falls. This additional term will be a subject for one of your exercises.

Motion

Another application of the model is to help understand why stepping on the accelerator increases engine power output. Which of the terms in the equation is affected? Stepping on the accelerator turns a plate positioned in the air intake to the engine in a way that enables more air to be sucked into the cylinder with each intake stroke (Figure 8).

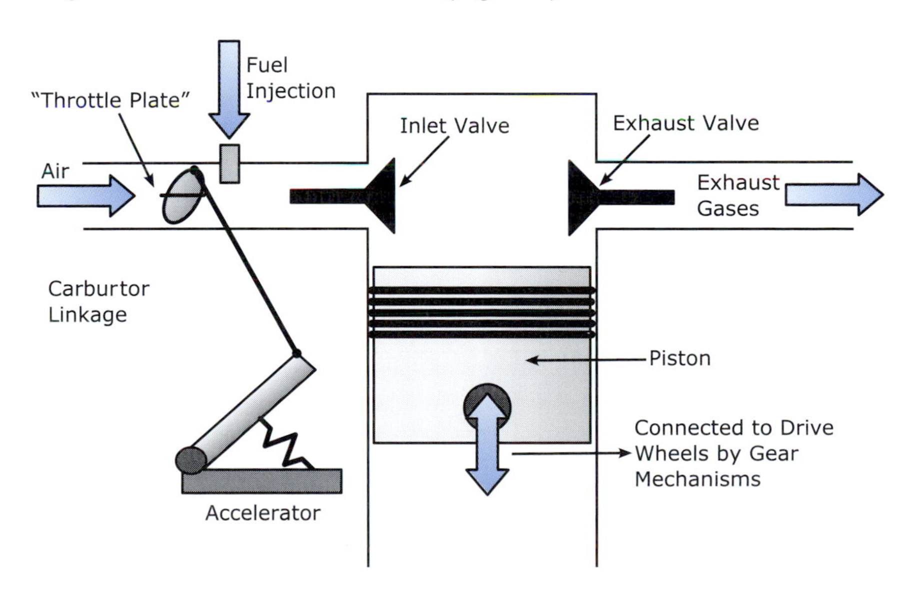

Figure 8

A/F System Schematic

In our model equation, the air density entering the engine maximizes at that of the surrounding atmosphere. Once air is sucked into the engine by the intake stroke, its pressure is partially controlled by the throttle plate. Opening the throttle plate more means an effective increase in ρ_{air}, the air density. Assuming the air-to-fuel ratio remains constant (modern cars have sophisticated controls that can ensure this is true), the fuel will increase. As a result, each deflagration within the cylinder will result in a greater power output.

The up and down movement of the piston is converted by a crank mechanism to the rotating movement of a crankshaft. That rotating movement is then transmitted to the rotation of the wheels by a technology called a **transmission.** The turning of the wheels on the road then becomes the forward motion of the automobile. This conversion of rotation of the wheels into linear motion of an automobile can be modeled by ordinary geometry. Assume a wheel has a radius R, or equivalently diameter D. Each time the wheel goes through one revolution, it moves the car forward a distance corresponding to the circumference of the wheel, which is $2\pi R$ or πD. So the forward speed of the automobile is simply the number of times the wheel rotates per second times the circumference of the wheel.

Gearing and Gear Ratios

In addition to the conventional concept of speed, engineers must also deal with the concept of **rotational speed.** It is defined in two ways, the more familiar being N, the revolutions *per minute* of a wheel. There is a corresponding "scientific" unit of rotational speed in terms of circular measure or radians/**s**. Its symbol is the Greek lowercase letter, omega (ω). There are 2π radians in a complete circle. Hence,

$$N = 60 \times \omega/2\pi \text{ [s/minute] [radians/s] [revolution/radian]} = \text{RPM}$$

conversely,

$$\omega = 2\pi N/60 \text{ radians/s}$$

Angular speed ω is directly related to ***linear speed*** v. Since each revolution covers $2\pi R$ in forward distance per revolution, moving at N RPM is a tangential speed of $v = 2\pi RN/60$ in m/s; this is ωR in m/s.

Hence, $v = \omega R$.

There is a problem in linking up the automobile models we have discussed so far. We have seen Otto engines that can sustain rotation speeds of approximately 1000–7000 RPM (as high as 12,000 RPM in very high-performance vehicles), but we have vehicles that are moving at linear speeds of, say, 115 km/h (32. m/s). Suppose the tire outer diameter is 0.80 m (radius of 0.40 m). What is the wheel's rotational speed?

In our current case, the wheels are rotating at a circular speed corresponding to the formula $\omega R = v = 32.$ m/s. Hence, $\omega = 32./0.40 = 80.$ (radians)/s or $80 \times 60/2\pi = 764$ RPM. Somehow the rotational speed of the engine is harnessed to the rotational needs of the wheels. How can these two different speeds of rotation be reconciled?

It is done by a mechanism called a transmission. A manual transmission is made of several gears.[6] A gear is simply a wheel with a toothed circumference (normally on the outside edge; see Figure 9).

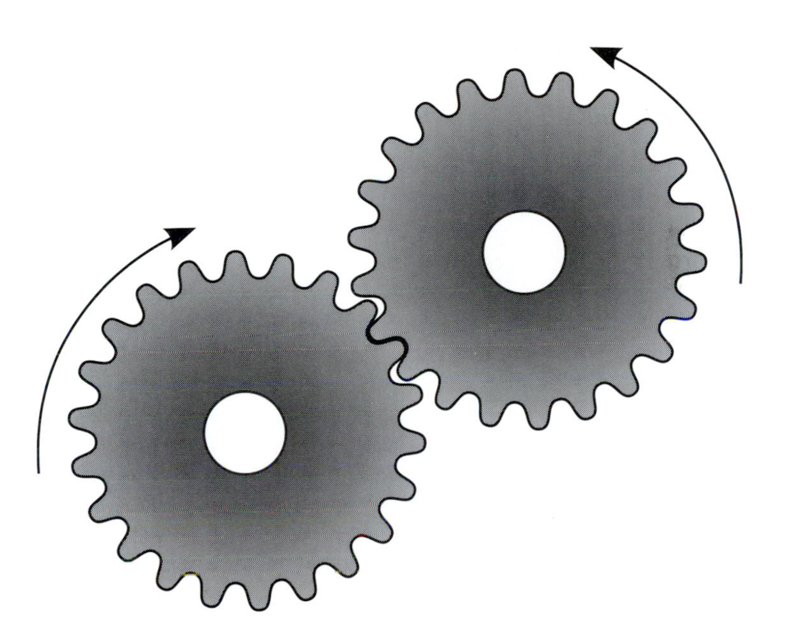

Figure 9

Intermeshing Gears

A gear *set* is a collection of gears of different sizes, with each tooth on any gear having exactly the same profile as every other tooth (and each gap between the teeth being just sized to mesh). The teeth enable one gear to drive another—that is, to transmit rotation, from one gear to the other. ***Note:*** A simple gear pair as per Figure 9 ***reverses*** the rotational direction of the driven gear from that of the driving gear. You need at least three gears in a set of simple gears as per Figure 10 to transmit in the same direction as the original direction.

There exists a simple relation between the rates of revolution of two gears when one is driving the other. The essential feature is that they must have an identical tangential or linear speed at their point of contact (or, at best, the gears will slip, and, at worst, the gear teeth will break and fly!).

[6]In this book, all the gears are toothed and intermeshing; other types of gear such as ratchet wheels and "worm" gears are excluded from the present discussion.

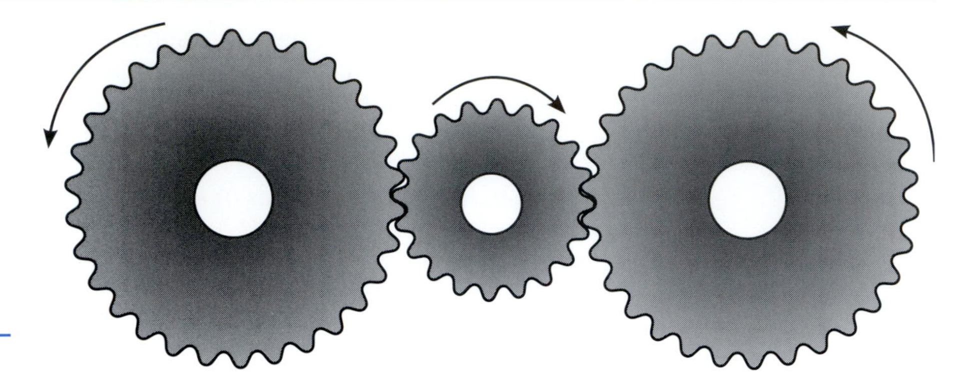

Figure 10

Nonreversing gear set

If gear 1 has radius r_1 and angular speed ω_1, and gear 2 has radius r_2 and angular speed ω_2, then the common tangential velocity is $\omega_1 r_1 = \omega_2 r_2$, so that $(\omega_2/\omega_1) = (r_1/r_2)$. This relationship applies to all no-slip systems such as smooth friction wheels (as in bicycle tires rubbing together) and equally to toothed gears. The relationship can be described in terms of gear radii as noted above or equally in terms of the number of teeth, since the number of teeth must scale directly with the gear radius, given that the meshing teeth must be of equal size.

In the context of gears, it is more usual to think in terms of *N*, the RPM rather than in terms of angular velocity, ω, in (radians) per second so that the **velocity ratio** for a simple gear train is:

$$VR \equiv \frac{\text{Output rotation}}{\text{Input rotation}}$$

$$\therefore VR = \frac{N_2}{N_1} = \frac{r_1}{r_2} = \frac{d_1}{d_2} = \frac{t_1}{t_2}$$

in which "*d*" stands for diameter and "t" stands for the number of teeth per gear. In other words, this relation, known as the velocity ratio[7] or *VR*, tells us that to increase the RPM (make the driven gear turn faster than the driving gear), choose a driven gear that is smaller than the driving gear. To decrease the RPM (make the driven gear turn slower than the driving gear), choose a driven gear that is larger than the driving gear.

EXAMPLE 1

An engine crankshaft is turning at 2000 RPM and is connected to a gear of radius 3.0 cm. That gear in turn is driving a gear of radius 20.0 cm. What is the RPM of the driven gear?

Need: RPM of driven gear, N_2.

[7]Gear ratio is defined as the inverse of velocity ratio, **GR** = 1/VR = input rotation/output rotation.

Know: Speed of driving gear $N_1 = 2000.$ RPM. Radius of driving gear (r_1) is 3.0 cm. Radius of driven gear (r_2) is 20.0 cm. A sketch as per the inset is recommended.

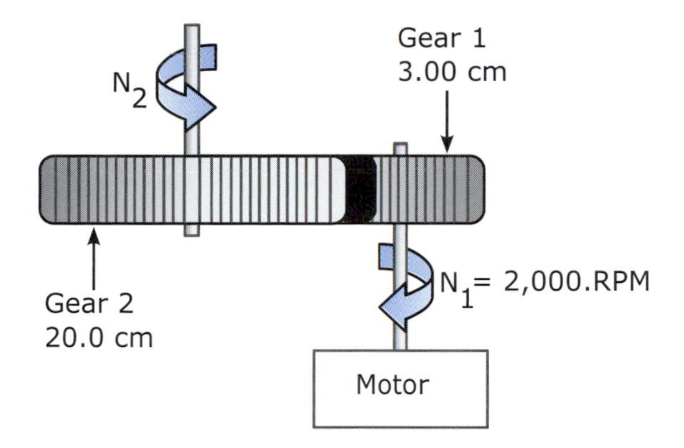

How: Use the VR equation $(N_2/N_1) = (r_1/r_2)$

Solve: $\therefore N_2$=2000[RPM]×3.00/20.0[cm/cm]=**3.00×10² RPM.** ■

Note that this example has solved the problem posed above: how to reconcile a high-speed engine crankshaft with the lower speed demanded by the wheels. Figure 11 is a highly simplified semiconceptual diagram of how this is achieved in an automobile.

Compound gear sets—sets of multiple interacting gears on separate shafts, as per Figure 12—can be very easily treated using the gear ratio concept introduced above. Consider this compound gear set: What is the velocity ratio for the full set of compound gears? Note that some of the gears are *driven* and some are *drivers;* in addition, some of the gears are

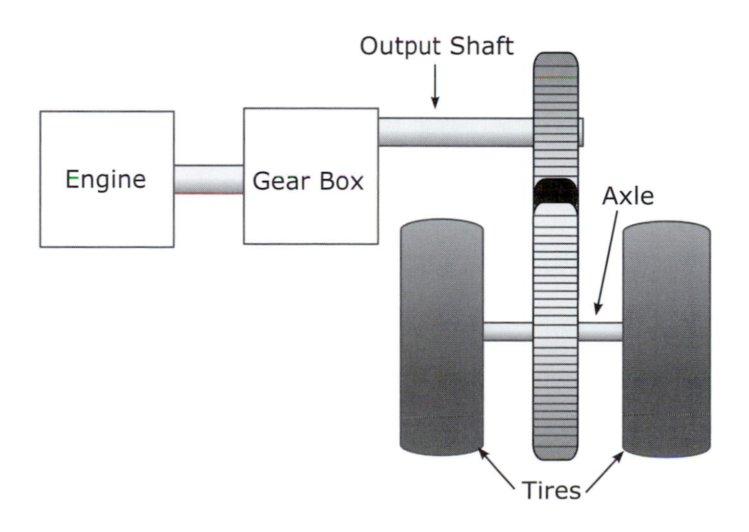

Figure 11

Simple Gear Train

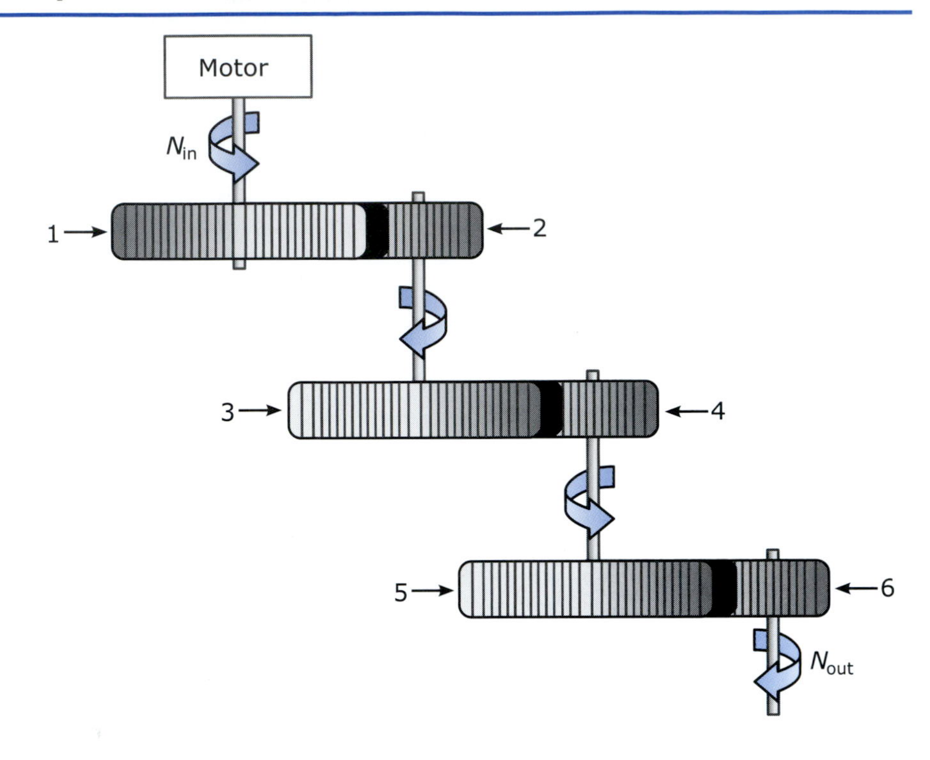

Figure 12

Compound Gear Train

connected by internal shafts so that these turn at a common speed. The analysis is straightforward:

$$N_1/N_{in} = 1 \text{ (same shaft)}$$
$$N_2/N_1 = r_1/r_2$$
$$N_3/N_2 = 1 \text{ (same shaft)}$$
$$N_4/N_3 = r_3/r_4$$
$$N_5/N_4 = 1 \text{ (same shaft)}$$
$$N_{out}/N_5 = r_5/r_6$$

$$\therefore \text{overall VR} \equiv N_{out}/N_{in} = N_1/N_{in} \cdot N_2/N_1 \cdot N_3/N_2 \cdot N_4/N_3 \cdot N_5/N_4 \cdot N_{out}/N_5$$
$$= 1 \cdot r_1/r_2 \cdot 1 \cdot r_3/r_4 \cdot 1 \cdot r_5/r_6$$
$$= (r_1 \cdot r_3 \cdot r_5)/(r_2 \cdot r_4 \cdot r_6)$$

These relationships make it very easy to analyze compound gear trains.

$$\therefore VR = \frac{\text{Product of diameter of driving gears}}{\text{Product of diameter of driven gears}}$$

$$\text{or } VR = \frac{\text{Product of number of gear teeth of driving gears}}{\text{Product of number of gear teeth of driven gears}}$$

EXAMPLE 2

A 70.0 RPM motor is connected to a 100-tooth gear that couples in turn to an 80-tooth gear that directly drives a 50-tooth gear. The 50-tooth gear drives a 200-tooth gear. If the latter is connected by a shaft to a final drive, what is its RPM? The setup is essentially the same as the previous example; we'll go straight to **solve** for this reason.
Again, use a sketch to help visualize the problem.

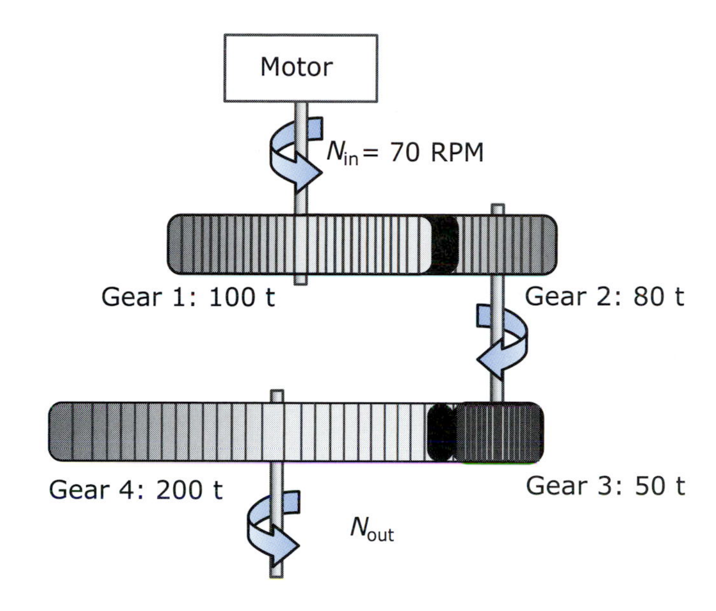

$$\text{Overall VR}{=}N_{\text{out}}/N_{\text{in}} = \frac{\text{Product of number of gear teeth of driving gears}}{\text{Product of number of gear teeth of driven gears}}$$

$$\therefore\ N_{out}/N_{in} = (t_1 \times t_3)/(t_2 \times t_4) = (100 \times 50)/(80 \times 200) = 0.313$$

$$\therefore\ N_{\mathbf{out}} = 0.313 \times 70.0 = 21.9 = \mathbf{22.0RPM}$$ ■

Another use of gears is to change not only the rotation speed, but also the twisting force, or **torque,** of the axle. We will merely note that you can't increase both the torque and the angular velocity (RPM) simultaneously using gears. That would violate conservation of energy—you would be getting something for nothing. In fact, the torque (T) transmitted is inverse to speed:

$$\frac{T_1}{T_2} = \frac{N_2}{N_1} = \frac{t_1}{t_2}$$

However, you can use gears to achieve a desired combination of torque and RPM. For example, you can first use one set of gears to provide the wheels with high torque and low RPM for initial acceleration (first gear). Then you can shift to another set of gears providing the wheels with lower torque and higher RPM as the car speeds up (second gear). Then you can shift to a third gear combination that offers low torque and high RPM for cruising along a level highway at 65 miles per hour. In a modern automobile there may be four, five, or even six forward gears.

The set of gears that achieves these different gear ratios at the appropriate times for each is called the transmission. Today most American transmissions are automatic[8] being based on a fluid coupling rather than simply mechanical gears, so drivers no longer experience gear ratios directly as they did in the days of the standard or "stick" shift,[9] when expertise with the left foot on the clutch (which temporarily disengages the gear train between engine and wheels) was as crucial a part of driving as was expertise with the right foot upon the accelerator.

Improving on the Otto Cycle

The Otto cycle's supremacy through the twentieth century has not convinced all engineers that it will remain the power source of the twenty-first century. At the beginning of this century, three competitors stand particularly prominent in offering a challenge to the Otto cycle due either to higher fuel efficiency, lower pollution emissions, or both.

The least radical of the challengers is the compression-ignition direct-injection (**CIDI**) engine, usually called the **diesel** engine after its inventor Rudolf Diesel, a German engineer active at the turn of the twentieth century. It is essentially an Otto cycle *without* a spark plug, using heat generated in the compression stroke to ignite the fuel, the fuel being injected into the cylinder *after* the compression stroke. It is used on locomotives and most large trucks today but on only a small fraction of automobiles. The CIDI or Diesel offers a higher efficiency than the Otto cycle, due to its higher compression ratio and higher internal temperatures. However, for use on typical automobiles, it requires further development in the areas of noise, vibration, cold-weather operation, and emissions control.

Another challenge is offered by the **hybrid** propulsion system. It uses a conventional Otto cycle or CIDI cycle as its prime mover, combined with an electrical generator and energy storage system. The internal combustion engine is run at its most efficient performance level or turned off altogether, and the increased power needed to achieve acceleration is

[8]To learn how automatic transmissions work visit the website http://www.edmunds.com/ownership/techcenter/articles/43836/article.html.

[9]Of course, many modern bicycles have from ~15 to 24 gears from which the rider can choose (and appreciate) their beneficial effects.

drawn from the energy storage system (typically an electric battery, but conceivably from another type of storage system such as a flywheel, or from an ultracapacitor).[10] The advantages are better fuel economy and lower emissions (since running an engine at constant speed makes possible slightly lean or substoichiometric combustion). A disadvantage is the higher weight and cost due to inclusion of the energy storage system. Some hybrid automobiles use the concept of regenerative braking to further increase efficiency. This concept begins from the fact that conventional braking systems reduce the translational kinetic energy of an automobile when stopping by converting that energy to heat. Regenerative systems instead capture that kinetic energy in a more useful form by using it to run an electrical generator to charge a battery or conceivably by using it to deliver rotational kinetic energy to an energy storage flywheel.

A more radical challenge to the Otto cycle is the **fuel cell.** It is a system for converting chemical energy *directly* to electrical energy, which then can drive an electric motor to turn the wheels. Because this conversion does not require the release of heat, it can in principle offer higher efficiency than the Otto cycle, while eliminating the production of such pollutants as nitrogen oxides that are produced at "hot spots" in the combustion process. To achieve these advantages to the fullest extent, engineers have proposed running fuel cells on hydrogen fuel, as was proved practical in fuel cells developed for the space program in the 1960s. This would indeed result in a clean-running, efficient automobile propulsion system with vehicle range comparable to that of today's cars. However, hydrogen is not today readily available as a fuel and is difficult to store in a compact form. So use of hydrogen fuel cells would require either a new fuel infrastructure of storage facilities and hydrogen "gas stations," or a component called a fuel reformer that would enable each automobile to convert gasoline or another hydrocarbon fuel into hydrogen. This in turn introduces significant system complications, as well as major cost and pollution issues.

The least radical approach to more efficient vehicles that might extend the life of the Otto engine is to simply reduce the vehicle's weight.

Today, compact American cars weigh about 2000 lbf and larger cars 4000 lbf (and SUV's as much as 7500 lbf). Figure 13 shows clearly that the vehicle weight is a principal determiner of fuel consumption—the heavier the vehicle the larger the fuel penalty. Capitalizing on weight reduction, Mercedes has introduced the "Smart Car" in Europe weighing about 1600 lbf and getting more than 50 mpg (Figure 14).

Unlike previous small European cars, the Smart Car has features suc as A/C, power options, and air bags designed to appeal to an affluent consumer. Not to be outdone, VW (Figure 15) has an in-line two-seat

[10] A hybrid of an electrical capacitor and a battery; it can deliver a limited amount of energy as a burst of power.

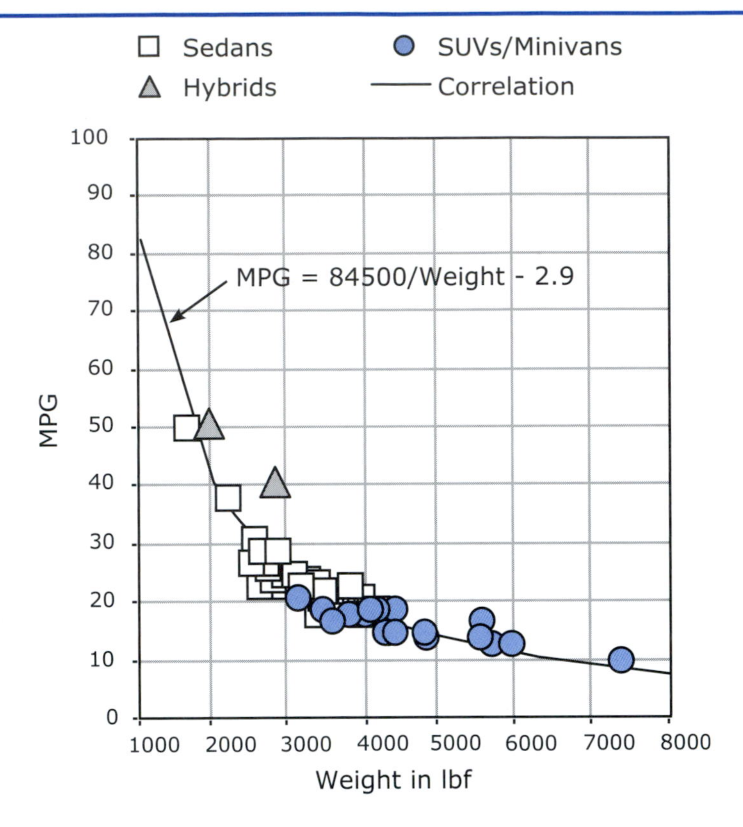

Figure 13

Weight and Mileage (*Consumer Reports*, April 2001)

Figure 14

Mercedes "Smart Car"

experimental 8.5 HP, 290 kg (639 lbm) vehicle that is capable of an astounding nearly 300 mpg.[11]

Figure 15

Experimental 300 mpg VW

Meanwhile, for several years, the Japanese companies Honda and Toyota have introduced commercial models of hybrid vehicles capable of achieving nearly 60 miles per gallon. In competition, some American companies have recently entered the hybrid market. Whether one of these options, or some other option, will replace or extend the Otto cycle remains one of the major engineering issues of the coming century.

SUMMARY

The futures of the automobile and the Otto cycle are not merely engineering issues but social, economic, and political issues, as well. The topics touched on range from global product competition to the geopolitics of relying on the Middle East for imported oil to the environmental impacts of increased atmospheric levels of carbon dioxide. Engineering models of energy conversion, such as the ones described in this chapter, cannot by themselves decide such issues. They remain, however, essential guides for providing realistic assessment of technical possibilities as society considers alternative energy futures. So engineers' abilities to understand the **operation of the Otto cycle, model the power output** of that cycle, use that power output to achieve motion, apply the power as needed through the use of **gears** as evaluated through **velocity ratios and gear ratios,** and assess proposed

[11]http://www.canadiandriver.com/news/020419-1.htm.

improvements to the Otto engine contribute not only to their technical expertise but also to their capability to participate in an important public debate.

EXERCISES

1. An Otto cycle engine has a compression ratio of 8.0. What is the engine's thermal efficiency? (**A: 0.56 or 56%.**)
2. Suppose you want to design an Otto Cycle with a thermal efficiency of 45%. What compression ratio must you use?
3. An Otto cycle engine has a total displacement of 4000. cm^3 (0.0040 m^3). Its compression ratio is 9.0. It is fueled with gasoline having a specific heating value of 45,500 kJ/kg. Its mass air-to-fuel ratio is 16.0. Assume the density of air is 1.00 kg/m^3. Determine its ideal power output at $N = 4.00 \times 10^3$ RPM, in kW and in HP. (**A: 222 kW, 298 HP** (kg air)/(kg fuel))
4. For the engine in problem 3, suppose you wanted to increase the power output at 4.00×10^3 RPM by 10%. One way to do this would be by changing only the compression ratio. What would the new compression ratio be?
5. For the engine in problem 3, suppose you wanted to increase the power output at 4.00×10^3 RPM by 50% while retaining a compression ratio of 9.0. One way to do this would be by changing only the displacement. What would the new displacement be (in cm^3)? (**A: 6000. cm^3**. This is a large engine for a car; it's easier and cheaper to achieve this performance by increasing ρ_{air} in the equation for engine power, since engine power is proportional to ρ_{air}).

   If you increase air density by compression from its nominal value of ~1.00 kg/m^3 (an approximate value at one atmosphere) to ~1.50 kg/m^3 by adding a turbo (compressor) to bring the air pressure up to 1.50 atmospheres (the volume of air falls by Boyles' law proportional to the increase in pressure—ergo, the density of a fixed mass of gas increases in the same proportion).

   Many cars do have some form of turbo charger to get additional power from a smaller engine.

Exercises 6–8 involve the world's fastest truck record holder, a 1940 Ford fire truck. Its top speed is 404 mph! It was powered by two Rolls Royce 601 Viper jet engines with afterburners, developing an equivalent to 1.20×10^4 HP! It consumes 2.0×10^2 lbm of fuel per second. (See http://www.jetfiretruck.com/).

6. Suppose the fuel used by the fire truck contains 45,500 kJ/kg of chemical energy. What is the overall efficiency (ratio of mechanical power out to chemical power in) of the truck's engines?

7. Suppose you wanted to design an Otto cycle engine with the same power output as the two jet engines powering the fire truck. Suppose your Otto cycle engine was able to achieve the far higher efficiency of $\eta_{mech} = 1/2$ and $\eta_{Otto} = 2/3$. Suppose you operated the engine at 4.00×10^3 RPM and $(A/F)_{Mass} = 16.0$ (kg air)/(kg fuel). Suppose your fuel was isooctane, with $HV_{fuel} = 45{,}500$ kJ/kW. How large would the displacement of the Otto cycle engine have to be in order to achieve the fire truck's power output of 1.20×10^4 hp? (**A: 0.28 m^3.**)

8. To get an idea of how realistic it would be to build an Otto cycle engine with the displacement you calculated in problem 7, assume that your engine had cylinders each the size of a two-liter soft drink bottle (a very large cylinder for an Otto cycle engine designed for automobile use!). How many cylinders would be required to produce the desired power?

Exercises 9–13 refer to a 3.5 horsepower lawn mower engine at a nominal speed of 2.50×10^3 RPM. A typical lawn mower engine, and the inside of its cylinder head, is shown above. If such an engine is available to you in class, you can disassemble it and measure the bore, stroke, and clearance volume of the engine.

9. If available, based on your measurements of bore (cylinder diameter), stroke (TDC to BDC), and clearance volume, *CV*, what is the displacement, *D*, of the lawnmower engine? (Use experimental numbers if available; otherwise, take these numbers as typical: stroke $= 0.100$ m, bore $= 0.050$ m, and CV $= 2.5 \times 10^{-5}$ m^3.) (**A: 196 cm^3.**)

10. Based on your measurements and calculations (exercise 9), what is the compression ratio, *CR*, of the lawnmower engine?

11. Based on your measurements and calculations so far, what is the Otto efficiency of the lawnmower engine? (**A: 0.58 = 58%.**)

12. Based on your measurements so far, if the mechanical efficiency of the lawnmower engine is 0.50, what is what is the power output of the lawnmower engine, in kW, when it is operating at 2.50×10^3 RPM with a mass air-to-fuel ratio of 15 (kg air)/(kg fuel) with a fuel that gives 45,500 kJ/kg of energy?

13. The power output of the lawnmower engine is not directly proportional to the number of RPMs, as the simple formula used so far suggests. It actually reaches a peak at about N = 2500 RPM and declines for higher RPM. This behavior can be approximated by the formula: $\eta_{Mech} = \exp(-1.26 \times 10^{-7} \times N^2)$ (which can be written in spreadsheet form as: η_{Mech} = EXP((−1.26E−7)*N^2) if N is in RPM units, and which varies from 1.0 at $N = 0$ RPM to 0.0 as N sufficiently increases).
Prepare a graph of power in kW versus RPM from 0 to 8000 in increments of 500; comment on the shape of the graph.

Exercises 14–19 concern a gear set consisting of up to the three gears shown below. In each case the larger (36 or 42 tooth) and smaller (12 tooth) portions of each gear are fastened together so that they rotate at the same rate.

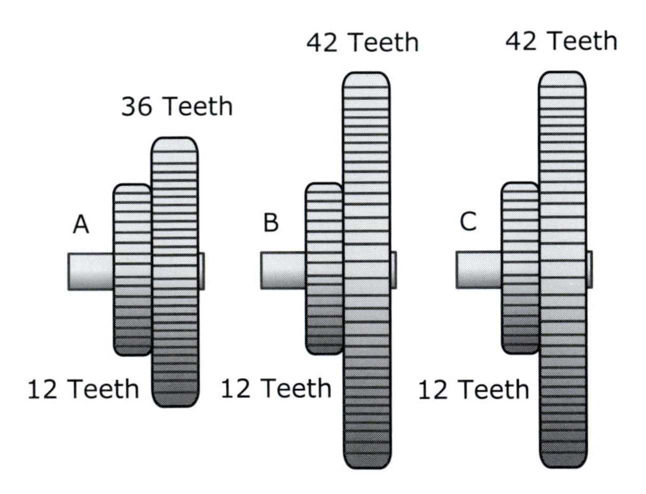

14. Suppose an 1800.0 RPM motor is directly connected to gear A. The 12-tooth portion of gear A drives the 42-tooth portion of gear B. What is the rotational speed in RPM of the shaft of gear B? (**A: 514 RPM.**)

15. If the drive gear in problem 14 is rotating *clockwise*, what is the direction of rotation of the driven gear?

16. Suppose gear A was driven by a motor at 1800. RPM. Suggest an arrangement of gears from the given gear set A, B, and C, to achieve an output speed of 18,900 RPM.

17. Suppose the gear set is to be used with a motor rotating at a constant speed of 1000. RPM. What are the lowest and highest output speeds that can be developed with the given three gears A, B, and C? (**A: Lowest speed cluster = 81.6 RPM; highest speed cluster = 12,250 RPM.**)

18. A motor drives a shaft at 1000.0 RPM with a torque of 35 newton-meters. A gear on that shaft has 42 teeth. That gear drives a second gear with 12 teeth. That second gear is connected to an output shaft. What is the torque produced by the output shaft? (**A: output torque = 10.0 Nm.**)

19. Consider only two of the gears in the gear set, the 12–36 and 12–42 tooth gears. For any given input speed, how many different output speeds are possible? Illustrate by categorizing the number of configurations that (1) increase or decrease the final shaft speed, (2) increase or decrease the final torque, and (3) leave the final speed and torque unchanged compared to the drive motor.

20. As a young engineer in the transmission division of a large automotive company, you are given an engineering assignment that originated in the engine division. Your supervisor is experienced and has a good idea of what the engineer in the engine division wants. Your solution is not what your supervisor expected, and he/she asks you to change your report. What do you do?

    a. Fully discuss the issues with your supervisor, and if you sincerely feel that you are right, then do not change your report.
    b. As long as your supervisor assumes responsibility for your report, then change it as requested.
    c. Since you are not very experienced, change your report so long as the changes are still correct engineering practice.
    d. Report your supervisor's behavior to the human resources office as harassment.

21. As an engineering student you are offered a lucrative summer job at a large tire manufacturing company. On your first day of work you are asked to sign an agreement assigning to the company all patent rights for your summer and the succeeding year. What do you do?

    (a) Sign the agreement, since your summer work will probably not produce anything patentable.
    (b) Ask that the agreement be altered to omit patentable work in areas outside those included in the summer job (e.g., thesis work).
    (c) Sign the agreement and hope the company does not hold you to it.
    (d) Refuse to sign and look for another job.

Chapter 7

© iStockphoto.com/Robert Kyllo

Electrical Circuits

Introduction

One of the most convenient ways to make energy useful is by converting it to electromagnetic energy, usually called electricity. In Chapter 4, electromagnetic power was briefly introduced as the result of multiplying the instantaneous current by the voltage drop that produced it. This chapter expands that introduction by giving a more detailed, but still highly simplified, discussion of electromagnetic energy and some of its uses.

Though electricity is everywhere around us at all times, the easiest way of getting electricity to go where it is needed is by means of an arrangement called an electrical circuit. In this chapter you will learn what **electric circuits** are, and you will apply a very simple model for analyzing their operation. The model has as its basic variables **charge, current, voltage,** and **resistance.** This highly simplified treatment relies on a model consisting of just two laws: **Ohm's Law** and what we call the **"Power Law."** Using this simple model, one can analyze the operation of two classes of direct current circuits called **series** circuits and **parallel** circuits. An entire field of engineering–electrical engineering–is devoted to the design and analysis of electrical circuits to meet a wide range of purposes, from melting metals to keeping food frozen, from projecting visual images to powering cell phones, from taking elevators to the top of skyscrapers, to moving submarines through the ocean depths. Rather than attempt to describe the whole range of electrical engineering applications, we will focus on one case study: The development of ever-faster **switches.** Different kinds of switches can be implemented in many ways, using such properties of electricity as the ability of a current to produce **magnetism,** and the use of electric charge to control the flow of current.

Electrical Circuits

A circuit is a *closed* loop of wire connecting various electrical components such as batteries, lightbulbs, switches, and motors. The word *circuit* should bring to mind the idea of a *circle.* We first recognize a natural phenomenon of **electric charge,** measured in a unit called **coulombs,** which flow through the electrically conductive wires as water flows through pipes. This is the essence of the simple model we will use for understanding electrical circuits. And just as water is actually made up of very small particles called water molecules, each of which has the same mass, charge is actually carried by very small particles called **electrons,** each of which has the same very tiny electrical charge of -1.60×10^{-19} coulombs (notice the minus sign). Because it is inconvenient to measure amounts of water in numbers of molecules, instead we use gallons, liters, or whatever, which are a huge number (about 10^{27}) of molecules. It is similarly inconvenient to measure charge in number of electrons, so we measure charge in coulombs. The coulomb is a huge number (about 10^{20}) of electrons. A car battery might hold 1 million or more coulombs of charge, while a C cell used in a flashlight might hold 10,000 coulombs.

Water would not be useful to us if it just sat in a reservoir, and electric charge would not be useful to us if it just sat in a battery. Think of a Roman aqueduct system in which water from a high reservoir is allowed to flow down to the city; it progressively loses its original potential energy as it flows. Electrons similarly flow from a high potential to a lower one. We call the latter potential **"voltage"** rather than gravitational potential, but the concept is equivalent. Thus electric charge is *pumped* by a battery or generator through the wires for use in electrical components that we value such as lights and air conditioning, and then it returns to the battery by means of wires for reuse.

Just as the flow of water from a hose can be separated into two of more exiting streams totaling the amount of arriving water, if trillions of electrons flow every second into an electrically conductive branch junction, you can be sure that the total number of electrons leaving the junction is exactly equal to the number of arriving electrons, a principle called *conservation of charge.* The flow of charge is a variable called **current** and is measured in **amperes,** where *one ampere is one coulomb per second* (and correspondingly equal to the flow of many trillions of electrons every second).

The hollow interior of a water pipe permits the water flow. Wires are not hollow, but they do allow the flow of electricity in the same way that a hollow interior of a pipe allows the flow of water when subjected to upstream pressure. We call materials used in wires **conductors,** which include metals. Other materials do not allow the significant flow of electricity through them. These materials are called **insulators.** They include such nonmetals as ceramics and plastics. Other materials, such as the

semiconductors, have properties somewhere between those of conductors and insulators. We distinguish among conductors, semiconductors, and insulators by a property known as **resistance.** Electric current, usually in the form of electrons, flows because of the voltage that is applied to them. The electrons in metal conductors flow from the negative to the positive direction in potential. However, traditionally (and unfortunately), the current is considered to flow from the positive pole of the battery to the negative one.[1] It was discovered well after this convention had been adopted that the current carriers in metals are electrons and flow in the opposite direction (see Figure 1).

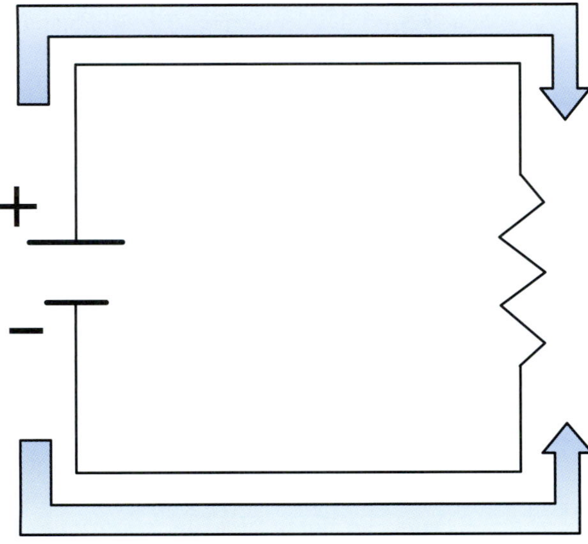

Figure 1

Blame Ben Franklin!

The motion of water through a pipe may be caused by a pump, which exerts a pressure on the water, forcing it to move through the pipe. In the same way, a battery, in addition to holding electrical charge, serves as a sort of electrical pump to force electricity to move through a wire. This electrical "pressure" is the variable we call voltage, measured in a unit called **volts.** Just as a hand pump might be used to apply a pressure of 15 pounds per square inch to move water through a pipe, a car battery might be used to apply 12 volts to move current through a wire.

Resistance is just the ratio of the given voltage drop to current. High resistance means that a large voltage drop is required to achieve a given current. Low resistance means that only a small voltage drop is required to achieve a given current. Resistive devices are called **resistors** (and often

[1]This convention was first suggested by Benjamin Franklin in the mid-eighteenth century.

colloquially just resistances) and may be made from coils of wire alloy, pieces of carbon, or other materials. In fact, all materials except superconductors[2] exhibit resistance to current flow. Some metals, such as pure silver and pure copper, have a very low resistance to electron flow. Other metals, such as aluminum and gold, have somewhat higher resistance but are still used as conductors in some applications. We usually use copper for connections, since we do not want much resistance there. Metal alloys and semiconductors offer appreciable resistance to electron flow, and we use those where we wish to exploit that property.

To summarize: In our model of an electric circuit, the voltage (i.e., the electrical "pressure") pushes a moving current of charge though a conductor. This charge (i.e., electrons) never disappears but simply circles around the circuit again and again, doing its various functions each time around the circuit. Once or more times each circuit, the electrons are pumped to high potential energy by the voltage source and then lose this energy as they perform their functions.

Symbolically, the simplest possible electrical circuit consists of a battery, a wire, and an electrical component located somewhere along the wire, as shown in Figure 2, which also introduces you to the visual symbol for a battery, with the longer line indicating the "plus" side of the battery and the shorter line representing the "minus" side of the battery.

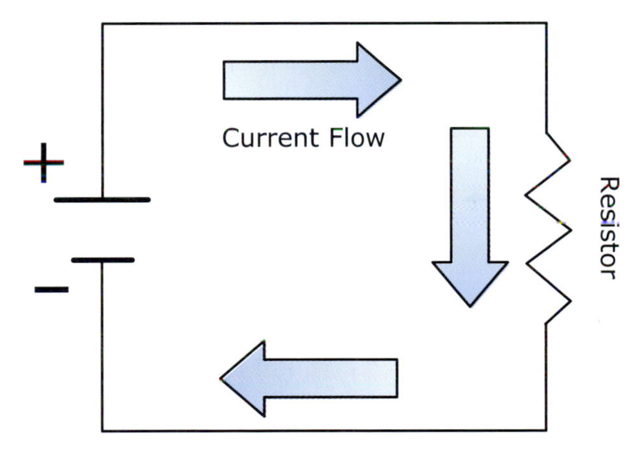

Figure 2

Current Flows in Wires

The zigzag line represents a resistor, and the straight lines are the wires connecting it to the battery. As previously stated, the current is considered to flow from the positive pole of the battery to the negative one. Again, the actual flow of current (electrons in wires) is in the opposite direction.

Since a circuit goes in a circle and ends up where it started, the voltage must also return to its initial value after a trip around the circuit. Since the

[2]These materials allow the *unrestricted* flow of electrons.

voltage is raised across the battery, it must be lowered back somewhere else in the circuit. In our simplest model, we will assume that none of this voltage "drop" occurs in the wires, but rather all of it occurs in the component or components in the circuit, although we will soon relax that restriction.

Resistance, Ohm's Law, and the "Power Law"

As water flows though a pipe, it experiences friction by coming into contact with the walls of the pipe, and this tends to slow down the flow. Pressure drop is needed to keep water moving though the pipe, and the magnitude of the change of the pressure is determined by the friction between the water and the walls.

In the same way, as charge flows through a wire, it experiences interactions that tend to reduce the voltage that is "pushing" the charge. We will initially assume that these interactions do not occur in the connecting wires but are confined to the components that are hooked together by the wires. The interactions that reduce the voltage are described by resistance. It is measured in units called **ohms** and abbreviated with the uppercase Greek letter omega, Ω. The voltage change (that is, an electrical "pressure change") needed to keep the charges moving through the wire is determined by the resistance. This relation between current, voltage, and resistance constitutes the first basic law of our model of electrical circuits, **Ohm's Law.** It states that the resistance, which is defined as the ratio of voltage drop to current, remains *constant* for all applied voltage drops. Mathematically stated,

$$R = V/I$$

If V is in volts and I is in amperes, R is defined in the units of ohms:

$$[\Omega] = [V]/[A].$$

For example, consider a resistor of resistance 2.00 Ω that is made of a material that exactly obeys Ohm's Law. If a voltage drop of 1.000 V is applied across the resistor, its current will be exactly 0.500 A. If a voltage drop of 1000.0 V is developed, the current will be exactly 500. A.

Few if any real materials obey Ohm's Law in this exact manner. The law is, however, a very useful approximation for most practical electric circuit. We now have a way of using numbers in our model of a circuit.

EXAMPLE 1 Find the current flowing in amperes through the wire in a circuit consisting of a 12. volt battery, a wire, and a 10. Ω resistor.

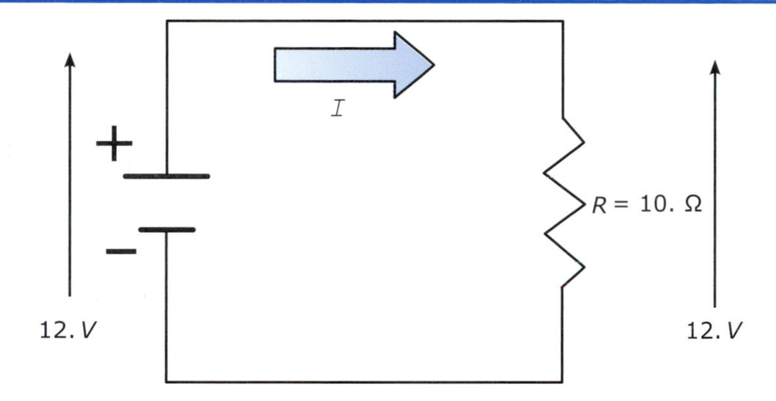

Need: Current flowing through the wire in amperes.

Know: Voltage provided by battery is 12. volts, resistance is 10.0 ohms.

How: Sketch the circuit and apply Ohm's Law.

Solve: The voltage *drop* across the resistor thus is 12.0 volts and, as given by Ohm's law:
$V = I \times R$. $\therefore \boldsymbol{I} = V/R = 12.0/10.0 = \mathbf{1.2\ A}$ (two significant figures). ■

Ohm's law implies a second important property about electric circuits. Recall that for the circuit as a whole, the total drop in voltage in the components other than the battery must be equal in magnitude but opposite in sign to the increase in voltage caused by the battery. (Again, we assume this entire voltage *drop* occurs across the resistive components, rather than in the wires.) Ohm's Law then enables us to find this voltage drop including, if necessary, its implied sign.

In continuing our water analogy, hydraulic resistance to water flow may be thought of as flow restrictions in the water pipe. Think of water valves half shut, extra long hoses, and so on. These contribute hydraulic resistance to the water pipe and reduce water flow. However, opening the pipe say from ½ ″ diameter to ¾″ (such as in a larger garden hose) reduces its hydraulic resistance. Since water is forced through pipes under pressure, it provides energy that might be used, for example, to power a water wheel to run your washing machine. To attempt this, it would be useful to know the relation between water pressure, amount of water flow, and power output from the water wheel. In the same way, electric charge can provide energy to run a wide variety of electrical appliances, from lightbulbs to computers. To attempt this, it is useful to know the relation between voltage, current, and power.

That relation is the second basic leg of our model—the **"Power Law"**—which was derived in Chapter 4 but bears repeating. We have seen that electrons with some charge—say, Q coulombs—"fall" through an electric potential—say, ΔV—in a resistor. This is analogous to a mass m falling through a gravitational potential energy change $mg\Delta h$. The equivalent electric work done is $Q\Delta V$. The power produced is thus $Q/t \times \Delta V$ in which Q/t is the *rate* of charge flow, which we call current. The "Power Law" thus simply states that electric power is given by:

$$P = I \times V \text{ or Power} = \text{Current} \times \text{voltage}$$

Note that power is not a new variable, but it is the same old one that we used earlier in talking about energy. We have inserted quotes around "Power Law" to emphasize that it is not an independent law but perfectly derivable from first principles using the Newtonian definition of power as rate of working. Power is still measured in watts, where a watt is a joule per second. This fact allows us to relate our model of electrical circuits to our earlier energy models. The analogy between water flow and electricity is thus reasonably complete (see Table 1).

Table 1

Water Flow in a Pipe vs. Electricity Flow in Wire

Water Flow in Hose	Electricity Flow in Wire
Pressure	Voltage
Flow rate	Current
Hydraulic resistance	Electrical resistance
Power in water stream = pressure × flow rate	Power in electrical circuit = voltage × current

Evidence of the basic "Power Law" relation is provided by a second important property of a resistor in addition to its ability to limit the flow of electricity. That property becomes evident if you force enough charge through the resistance at a high enough voltage and observe that the resistor may then give light (unless it melts first!). This is the basis of a very useful invention, the lightbulb (technically the incandescent lamp), invented not, as popularly believed, by Thomas Edison in 1879, but by Humphrey Davy in England and August De La Rive[3] in France in about 1810.

Series and Parallel Circuits

There are two basic types of electric circuits. In a **series circuit,** shown in Figure 3a, the current takes a single path. In a **parallel circuit,** shown in Figure 3b, the wire branches out into two or more paths and these paths subsequently join up again. In a parallel circuit, one should envision the current as being divided at the point where the wires branch out and then joining together again where the wires come together.

[3]We emphasize *basis*, since neither scientist took his observations to a practical conclusion.

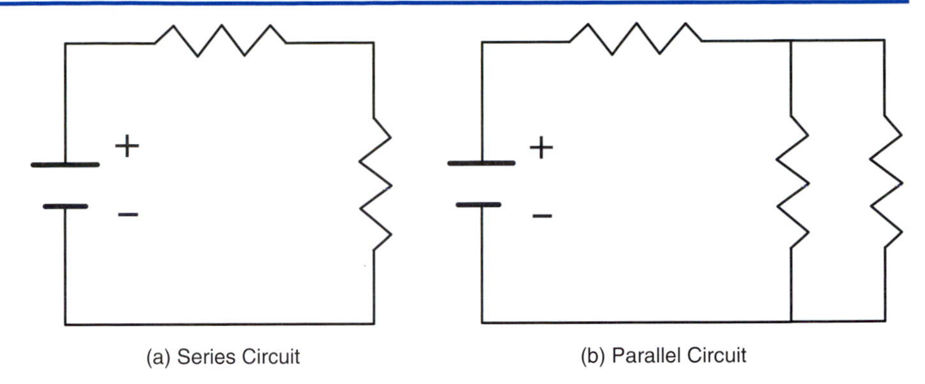

(a) Series Circuit (b) Parallel Circuit

Figure 3

Circuits

As previously stated, the current in each of the two branches has to add up to the amount that went in by the conservation of charge principle, and, conversely, the total that goes out of the two branches has to equal the total that went in.

The mathematics for determining the voltages and currents in series and parallel circuits, rests on two very simple principles: (1) two resistors in a series will have the same current passing through each of them, and (2) two resistors in a parallel will have the same voltage drop across each of them.

EXAMPLE 2

Consider the series circuit in Figure 3a, with the battery voltage taken as 12 volts and each resistor having a resistance of 100. ohms. What is the current in the circuit and the voltage drop across each resistor?

Need: Current in circuit and the voltage drop across each resistor.

Know: Battery voltage = 12 volts; resistance of each resistor = 100. ohms.

How: Call the unknown current I and the resistance in each resistor R. By the first principle above, the current I must be the same in each series resistor. So by Ohm's Law, the voltage drop across each resistor must be $I \times R$. Since the total voltage drop around the circuit must be zero, $V(\text{battery}) + V(\text{resistor}_1) + V(\text{resistor}_2) = 0$, or $V(\text{battery}) - I \times R - I \times R = 0$. Solve for the unknown I.

Solve: $12.0 \text{ volts} - I \times 100.0\ \Omega - I \times 100.0\ \Omega = 0$.
$\therefore 200.0 \times I = 12.0$ or $I =$ **0.06 Amps.**
The **voltage drop** across each resistor is $-100.0 \times 0.06 =$ **−6.0 V.** ■

Notice that we could have simply added the two 100.0 Ω resistances together and used their sum of 200.0 Ω to directly get the 0 current by Ohm's law in one step: $I = 12/200 = 0.06$ A. In a series circuit, the equivalent series resistance R is always the sum of the individual resistances: $R = R_1 + R_2 + R_3 + \ldots$.

EXAMPLE 3

Consider a parallel circuit in Figure 7.3b with the battery voltage = 12 volts and each resistor having a resistance of 100.0 ohms. What is the current drawn from the battery?

Need: Current drawn from battery.

Know: Battery voltage = 12 volts; resistance of each resistor = 100.0 ohms.

How: Call that current the unknown, I. By the second principle above, the voltage drop must be the same across each parallel resistor. Since the total voltage drop around the circuit must be zero, $V(\text{battery}) = V(\text{resistor}_1) = V(\text{resistor}_2)$. Now assume that the current divides between the two resistors with I_1 going through one of the resistors and I_2 going through the second. By Ohm's law, $V(\text{battery}) = V(\text{resistor}_1) = I_1 \times R = V(\text{resistor}_2) = I_2 \times R$. Solve for I_1 and I_2. Since the current coming out of the branches is the sum of the current in each branch, $I = I_1 + I_2$.

Solve: 12 volts $= I_1 \times 100.0\ \Omega = I_2 \times 100.0\ \Omega$.

$\therefore I_1 = 0.12\ \text{A} = I_2$ (by symmetry of the circuit resistors).

$\therefore \boldsymbol{I} = I_1 + I_2 =$ **0.24 Amps.** ■

Note that in this case the parallel circuit draws more current than the series circuit. This will generally be true for circuits with similar resistances. It's the same thing as adding a second hose to a garden faucet; provided the supply pressure remains constant, you will spray more water (analogous to an electric current) with two hoses than with one. Of course there is a well-known formula that sums the inverse of electrical resistances (and that you can now deduce yourself) that gives the equivalent inverse resistance of a parallel circuit. It is not necessary and tends to be used formalistically whether appropriate or not. The method used in Example 3 is more general and is recommended.

One particular form of series circuit enables us to add more realism to our simple model. Earlier, wires were modeled as conductors with zero resistance. However, a real wire has a small resistance, which is typically proportional to the length of the wire and inversely proportional to its cross-sectional area. In more realistic circuit models, this resistance of the wire is modeled by a resistance element (our familiar wiggly-line symbol) inserted at an arbitrary location into the wire. This resistance will henceforth be called the wire resistance, and the resistance of a component such as a lightbulb will be called the load resistance.

EXAMPLE 4

For the circuit shown, assume (a) the wire resistance (actually distributed along the whole length of the wire) totals 1.0 ohms, (b) the load resistance is 10.0 ohms, and (c) the battery voltage is 6.0 volts. Compute the efficiency of the circuit, where the efficiency is defined as the power dissipated in the in the load divided by the power produced by the battery.

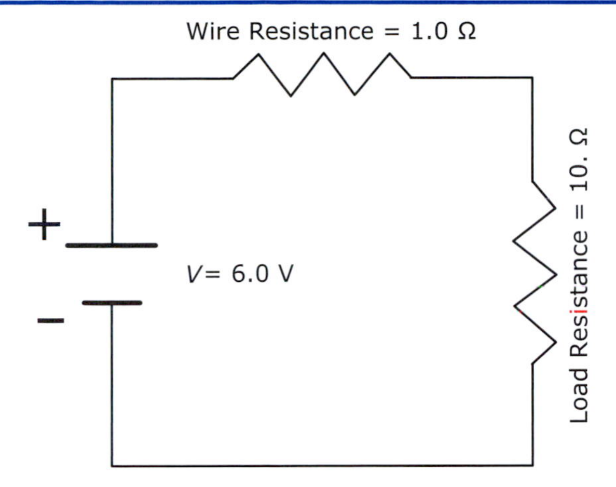

Need: Efficiency of circuit = (power dissipated in load)/(power produced by battery).

Know: Battery voltage = 6.0 volts; wire resistance = 1.0 Ω; load resistance = 10. Ω.

How: Computing efficiency requires two computations of power. Computation of power, using our "Power Law" $P = I \times V$ requires knowing the voltage across and the current through each circuit element. So first find the voltage across and current through each element using Ohm's Law. Then compute the power. Then compute the efficiency.

Solve: In a series circuit, the total resistance is the sum of the individual resistors.

So $R = R(\text{wire}) + R(\text{load}) = 10. + 1.0 = 11.\ \Omega$.
Battery voltage is 6.0 volts. So, by Ohm's Law, $I = V/R = 6.0/11. = 0.55\ \text{A}$.

Now, applying the "Power Law," $P(\text{battery}) = V(\text{battery}) \times I = 6.0 \times 0.55\ [V][A] = 3.3\ \text{W}$. This is the *total* power drained from the battery.

Also applying Ohm's Law to the wire's resistance its overall voltage drop is, $V(\text{wire}) = 1.0 \times 0.55\ [\Omega][\text{A}] = 0.55\ \text{V}$.

Applying the "Power Law" to wire losses, $P(\text{wire}) = 0.55 \times 0.55\ [V][A] = 0.30\ \text{W}$. $\therefore$ Load power, $= 3.3 - 0.30 = 3.0\ \text{W}$.

So efficiency = 3.0/3.3 = 0.91 (note that efficiency, being a ratio, has no units). ■

One task of an engineer might be to maximize the **efficiency** of application of power, since higher efficiency generally results in lower cost, and lower cost is generally desired by customers. The previous exercise gave one version of that challenge. The next example will give another.

It was Edison who in 1879 applied the "Power Law" and as a result made the first *useful*[4] lightbulb. He did so by recognizing that a lightbulb is simply a resistor that converts electricity first into heat and then renders part of that heat (typically only about 5%) into visible light. In the example that follows, treat the resistor (indicated by the zigzag line in the figure) inside the bulb (indicated by a circle) as an ordinary, if very high-temperature, resistor that obeys our two electrical laws. The lightbulb's resistor is called its *filament.*

EXAMPLE 5 In the diagram shown, assume the battery delivers a constant 10.0 A of electric current whatever the voltage across the filament may be. If one wishes to design the lightbulb to have a power of 100.0 W, what should the resistance of the lightbulb be? What is the voltage drop across the filament?

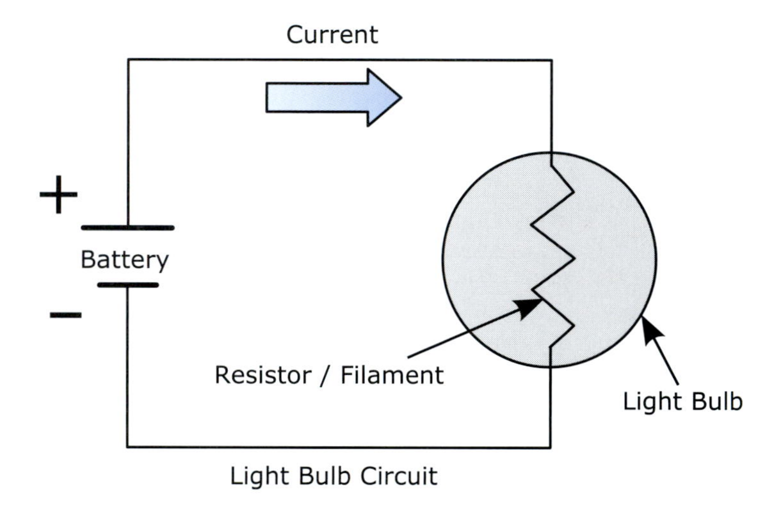

Light Bulb Circuit

Need: R and V.

Know: P and I.

How: $P = I \times V$ & $V = R\ I$.

Solve: $P = I \times V$ or $V = P/I = 100.0/10.0$ [W]/[A] = **10.0 volts** to three sig. figures

$\therefore R = V/I = 10.0/10.0$ [V/A] = $\mathbf{1.00\ \Omega}$ (to three sig. figures). ■

[4]Edison's major achievement came from considering the entire electrical production, distribution, and lighting problem as a single system design problem. The choice of a filament resistance was only a part of this greater problem. His rivals were unsuccessful because they were concerned only with producing lamps.

If the wire in this example also had a 1.0 Ω distributed resistance, as much power would be dissipated in it as in the bulb's filament. It would also have "consumed" another 10.0 volts of voltage, so the battery would have had to supply 20.0 volts (in other words, the efficiency would have been only 50% in this circuit!).

The fact of resistance in the wire comparable to that of the filament complicated the problem. The key to Edison's problem of maximizing the bulb's efficiency was in maximizing the efficiency of the *entire* circuit, rather than concentrating on the bulb alone. Higher efficiency meant lower costs, eventually low enough for electricity and the lightbulb to become mass consumer products.

In all the circuits we have discussed so far, the current always flows in the same direction (shown, according to convention, flowing from positive to negative around a circuit). These are called *direct current circuits.* However, there is another class of electrical circuits that allows the current to reverse its direction many times a second. These alternating current circuits are the basis for the vast number of electrical technologies beyond the lightbulb. These range from the 1000-megawatt generators that supply electricity, to the motors that put electricity to work, to the radio wave generators that enable us to send out radio signals or to thaw frozen food in minutes in our microwave ovens.

All of these alternating current applications rely on another property of electric circuits. A circuit can also act as a **magnet.** A wire carrying electricity is surrounded by a magnetic field. In this book, we will not define a magnetic field any more precisely than to say a magnetic field is something that can cause a piece of iron, such as a compass needle, to rotate or move when brought into proximity. We will also note that a way to strengthen the magnetic field in a circuit is to wind the wire in which current is flowing into a coil with multiple turns.

The fact that electricity is accompanied by magnetism is the basis of modern electric power generation, radio transmission, and much more. However, it is difficult to model this connection of electricity and magnetism using our water flow model. So this very important topic will be left for more advanced courses in electrical engineering or physics.

Switches

Instead, we will concentrate on a humble circuit element whose importance was appreciated by Edison: the switch. In our water flow model, a switch is equivalent to a faucet. It turns the flow on or off. Surprisingly, it is this simplest of components that is the basis of the most sophisticated of our electrical technologies, the computer. In its essence, a computer is nothing more than a box filled with billions of switches, all turning

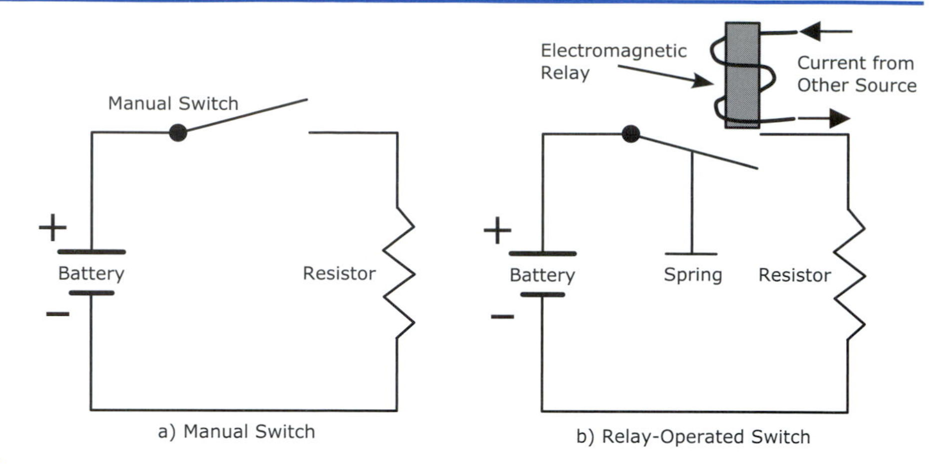

Figure 4

The Humble Switch

each other on and off. The faster the switches turn each other on and off, the faster the computer can do computation. The smaller the switches, the smaller the computer can be. The more reliable the switches, the more reliable the computer. The more efficient the switches, the less power the computer uses. So the criteria engineers have used in developing better switches are higher speed, smaller size, greater reliability, and increased efficiency.

In the late nineteenth century, a few farsighted individuals, such as the British economist William Stanley Jevons and the American mathematician Allen Marquand, proposed that the humble switch could be used to do something remarkable: carry out calculations in the field of philosophy and mathematics, called logic. (The mathematics of logic will be discussed in Chapter 8.) Here we will stick to the underlying electrical technology: the switch.

Unlike the elements we have discussed so far, the switch has two states. As indicated in Figure 4a, a switch can be open, in which case it presents an infinite resistance to the flow of electric current (that is, no current flows at all), or a switch can be closed, in which case ideally it presents zero resistance to an electric current.

Jevons and Marquand envisioned using switches to automate logic, but the technology of their time was not up to the task. One of the pioneers who turned their idea into twentieth-century reality was George R. Stibitz. Along with other inventors in the 1930s, he accomplished this automation of computation using a type of switch component called a **relay.**

A relay is a switch operated by an electromagnet, Figure 4b. The relay simply closes the switch against the spring when the external power source is activated. When the external power source is turned off, the spring returns the switch to its open position. It got its name from its original

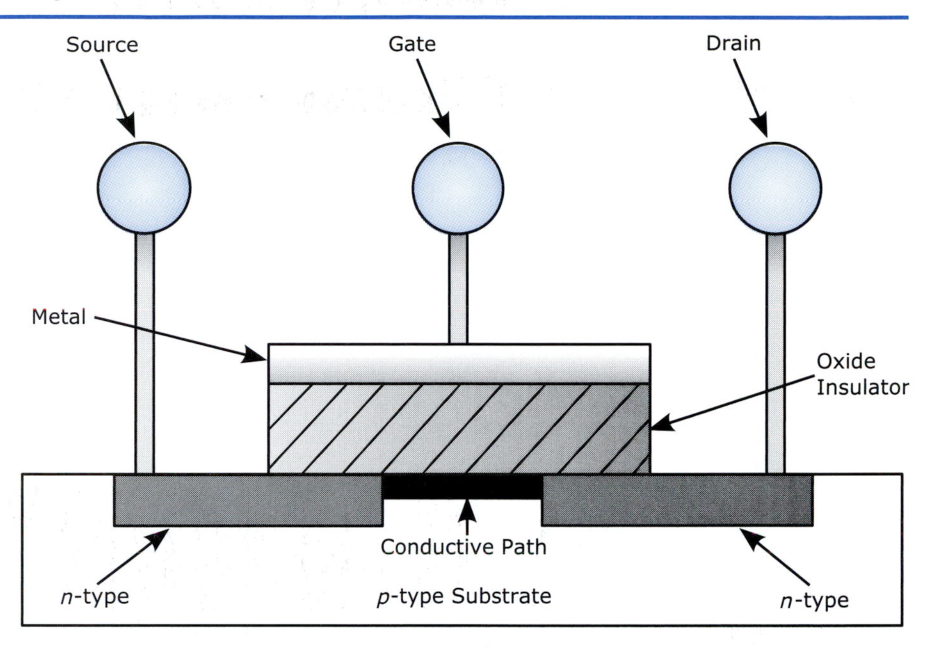

Figure 5

Principles of MOSFET Construction

use, relaying telegraph messages over longer distances than could be accomplished by a single circuit. The relay consists of two parts. The first is an electromagnet that is powered by one circuit, which we will call the driving circuit. The second part is another circuit with a metal switch in it, called the driven circuit. The driven circuit is placed so that when the driving circuit is on, its electromagnet attracts the moving metal part of the driven circuit's switch. This results in the closing of the gap in the driven circuit—turning it on.

Relays continue to be used in control systems because they are highly reliable, but they are high in power demands, slow, and bulky. In the twentieth century the basis of a faster switch emerged. It was based on an invention used in radios, a circuit element called the vacuum tube. Though drawing on knowledge of the physics of 1910 rather than the physics of 1882, its operation is conceptually much like that of a relay. Vacuum tubes were used as switches because they were fast. However, they were inefficient, bulky, and unreliable. By the 1940s, an even more revolutionary idea was already in the wings. It emerged in initial form in 1946, when three physicists at Bell Laboratories — John Bardeen, Walter Brittain, and William Shockley — invented the transistor, a solid state circuit element developed to replace the vacuum tube in communications. How the transistor works is a topic for an advanced course in electronic engineering or physics. For our purposes, we can carry on the analogy with the relay and the vacuum tube. Figure 5 shows one important type of transistor, the metal oxide semiconductor field effect transistor (MOSFET), a type that came along several years after the original one invented by Bardeen, Brittain, and Shockley. In the MOSFET, a driving circuit called a gate plays the role the magnet plays in the relay, and the source plays the role of a heated metal grid in the vacuum tube. Together they open and close a driven circuit that determines the state of the output.

A transistor is made of semiconductors—materials that have properties midway between those of a conductor and those of an insulator. Most of today's transistors use semiconductors made out of the element silicon. By "doping" silicon—that is, by adding trace amounts of other elements—its conductivity can be controlled. As a simple model of a particularly useful type of transistor called the "field effect transistor," one can imagine two such pieces of doped silicon, a "source" that serves as a source of electrical charge, and a "drain" that serves as a sink into which the charge flows. The source and drain are connected by a semiconductor channel, which is separated by an insulator from a metal electrode called a "gate." Some of the semiconductor is doped with material that sustains an excess of electrons and is called "*n*-type" (*n* for "negative"). Other semiconductors are doped with materials that sustain a deficiency of electrons called "*p*-type" (*p* for "positive"). When the gate is given a positive voltage in the *npn* MOSFET shown in Figure 5, this promotes the flow of charge from source to drain, effectively making a thin layer of continuous *n*-type conductive channel between the source and the drain. When the gate is given a negative voltage, current cannot flow from source to drain, and the transistor acts as an open switch.

Transistors were smaller, faster, more efficient, and more reliable than vacuum tubes. They made possible practical hearing aids, pocket-sized portable radios, and computers that were the size of refrigerators instead of filling entire rooms. But to get to the modern computer, one more step was needed. This was the **integrated circuit,** independently coinvented by Nobel Prize–winning engineer Jack Kilby and Intel cofounder Robert Noyce in the early 1960s.

The integrated circuit is just a whole lot of transistors (many millions or even billions) simultaneously manufactured and hooked up into a desired logic circuit on a chip of silicon the size of a fingernail. It is done by *photolithography,* which literally means "using light to write on a stone". To make an integrated circuit, a very large drawing of the circuit is made. Using lenses that shrink rather than enlarge this drawing, the image is projected by light onto that fingernail-sized chip of silicon (the "stone"), coated with special chemicals. The chemicals make possible the "writing" onto the chip of a working electrical circuit composed of millions of transistors. State-of-the-art integrated circuits might contain the level of detail that would be seen on a street map of the entire state of California! Packing all this complexity into a small space launched one of the most dramatic technological revolutions in history. Since 1960, the number of electronic computations possible per second and the number of switches that can be packed onto a single chip have doubled every one or two years (and since 1990, rather than slowing, the pace has actually speeded up). Integrated circuits are the hardware that made personal computers possible. But integrated circuits by themselves do not constitute a computer. A crucial distinction needed to understand computers is **hardware versus software.** Hardware is the collection of mechanical, electrical, or electronic devices that make up a computer. Among the hardware in a modern computer, two elements are crucial: (1) a system called a **central processing unit** or **CPU,** which carries out binary logic and arithmetic; and (2) memory, often called **RAM** (random access memory) or **ROM** (read-only memory), which store the ones and zeroes corresponding to inputs and outputs of the CPU. Both CPU and memory are integrated circuits composed of many millions of switches.

What enables that collection of switches to carry out computations? That is the job for a sequence of instructions, called **software.** Two important applications of software, control and computation, will be discussed in Chapters 8 and 9. In Chapter 9, we will model the fuel control system of an automobile using computer logic.

SUMMARY

The flow of electricity though wires can be modeled in the same way one might treat the flow of water through pipes. The model has as its basic variables **charge, current, voltage,** and **resistance.** The simplest model consists of two laws: **Ohm's Law** and the **"Power Law."** Using this model, one can analyze the operation of a particular class of electric circuits called (**direct) current** circuits and, in particular, two classes of direct current circuits called **series** circuits and **parallel** circuits. All circuits in which electric currents flow produce **magnetism.**

An important type of electric circuit component is the **switch.** On the hardware side, the development of the computer can be viewed as the search for ever-faster switches, based first on **relays,** then on **vacuum tubes,** then on **transistors,** and finally on the form of electronics called **solid state integrated circuits.** This form of electronics provides the **hardware,** which, in conjunction with a sequence of instructions called **software,** makes modern computation and control possible.

EXERCISES

If no circuit sketch is already given, it is highly recommended you *first draw each circuit* to answer these problems. It is generally useful to notate your

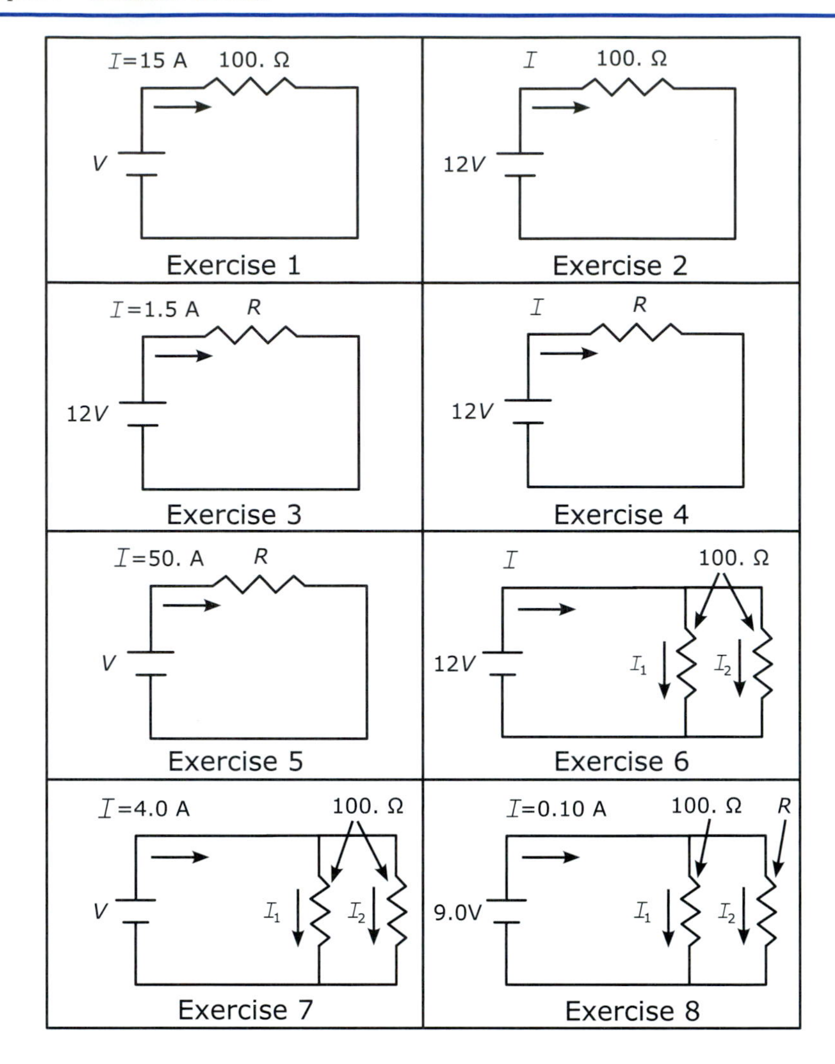

diagram with what you think you know, such as voltages and currents at each point.

Note the equivalence among these electrical units: $[\Omega]=[V]/[A]$, $[W]=[V][A]=[\Omega][A]^2=[V]^2/[\Omega]$ and all other combinations of these.

The circuits for Exercises 1–8 are given in the table.

1. For the circuit shown, find the voltage V. (**A**: **1500 V.**)
2. For the circuit shown, find the current I.
3. For the circuit shown, find the resistance R.
4. For the circuit shown, a power of 100. watts is dissipated in the resistor. Find the current I. (**A:** = **8.3 A.**)

5. For the circuit shown, a power of 100. watts is dissipated in the resistor. Find the resistance, R.
6. For the circuit shown, find the current I. (**A: 0.24 A**)
7. For the circuit shown, find the voltage V.
8. For the circuit shown, find the resistance R.

Use these circuits in conjunction with Exercises 9 and 20, respectively.

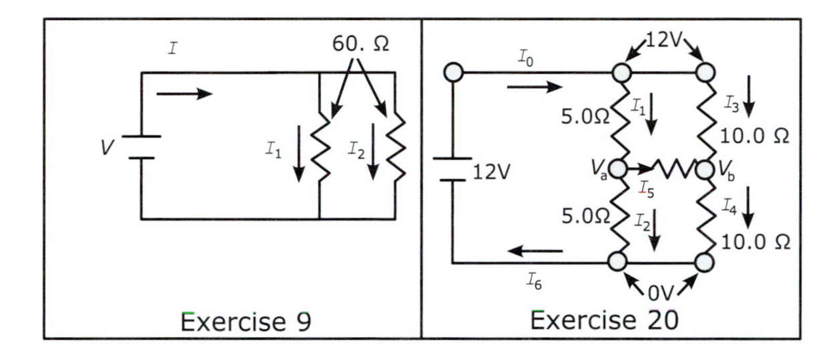

9. For the circuit shown, a power of 100. watts is dissipated in each resistor. Find the current I and the voltage V.

For Exercises 10–19 following, draw the circuit and solve for the unknown quantity requested. It is suggested that you include a small circle for each point where you can specify the local voltage.

10. A circuit consists of a 3.0 V battery and two resistors connected in series with it. The first resistor has a resistance of 10 Ω. The second has a resistance of 15 Ω. Find the current in the circuit. (**A: 0.12 A.**)
11. A circuit consists of a 12 V battery and a resistor connected in series with it. The current is 105 A. Find the resistance.
12. A circuit consists of a 9.0 volt battery and two parallel branches, one containing a 1500 Ω resistor and the other containing a $1.0 \times 10^3\ \Omega$ resistor. Find the current drawn from the battery. (**A: 0.015 A.**)
13. A circuit consists of a 12 volt battery attached to a $1.0 \times 10^2\ \Omega$ resistor, which is in turn connected to two parallel branches, each containing a $1.0 \times 10^3\ \Omega$ resistor. Find the current drawn from the battery.
14. An automobile's 12 V battery is used to drive a starter motor, which for several seconds draws a power of 3.0 kW from the battery. If the motor can be modeled by a single resistor, what is the current in the motor circuit while the motor is operating? (**A: 250 A.**)

15. An automobile's 12 V battery is used to light the automobile's two headlights. Each headlight can be modeled as a 1.00 Ω resistor. If the two headlights are hooked up to the battery in series to form a circuit, what is the power produced in each headlight? Why would you *not* wire car lights in series?

16. An automobile's 12 V battery is used to light the automobile's two headlights. Each headlight can be modeled as a 1.00 Ω resistor. If the two headlights are hooked up to the battery in parallel to form a circuit, what is the power produced in each headlight? Why is a parallel circuit preferred for this application?

17. An automobile's 12 V battery is used to light the automobile's two headlights. Each headlight can be modeled as a single resistor. If the two headlights are hooked up to the battery in parallel to form a circuit, and each headlight is to produce a power of 100 W, what should the resistance of each headlight be? (**A: 1.4 Ω per bulb.**)

18. Suppose one of the headlights in Exercise 17 suddenly burns out. Will be the power produced by the other headlight increase or decrease?

19. Suppose a car has headlights operating in parallel but with different resistances—one of 2.0 Ω and the other of 3.0 Ω. Suppose the headlight parallel circuit is connected in series with a circuit for a car stereo that can be modeled by a 1.0 resistor. What is the current being drawn from a 12 V battery when both the lights and the stereo are on? (**A:** $\boldsymbol{I} = \mathbf{5.5\ A}$.)

20. Consider the circuit shown before Exercise 9. What are the currents I_o through I_6? (**Hint:** What are the voltages V_a and V_b? Note the symmetry between $R_1/R_2 = 5.0/5.0$ and $R_3/R_4 = 10.0/10.0$.)

21. It is the last semester of your senior year and you are anxious to get an exciting electrical engineering position in a major company. You accept a position from company A early in the recruiting process, but you continue to interview, hoping for a better offer. Then your dream job offer comes along from company B. More salary, better company, more options for advancement. It is just what you have been looking for. What do you do?

    (a) Accept the offer from company B without telling company A (just don't show up for work).
    (b) Accept the offer from company B and advise company A that you have changed your mind.
    (c) Write company A and ask them to release you from your agreement.
    (d) Write company B thanking them for their offer and explain that you have already accepted an offer.

22. A female student in your class mentions to you that she is being sexually harassed by another student. What do you do?

    (a) Do nothing; it is none of your business.
    (b) Ask her to report the harassment to the course instructor.
    (c) Confront the student accused of harassment and get his or her side of the issue.
    (d) Talk to the course instructor or the college human resource director privately.

Chapter 8

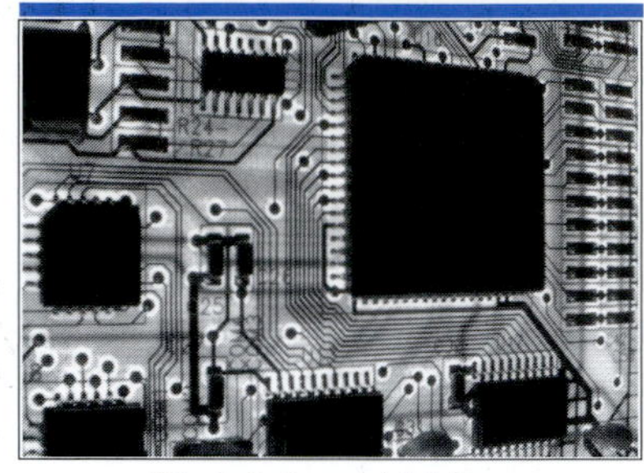

iStockphoto.com/ Calvinng

Logic and Computers

Introduction

Why do cars contain computers? For most of the first century of automobiles, from the 1880s through the 1960s, cars got along perfectly well without computers. As late as 1965, the question "Should a car contain a computer?" would have seemed ridiculous. In those days, a computer was not only much more expensive than a car, but also bigger.

In the late 1960s, the integrated circuit was invented, and computers began to shrink in both size and cost. In the 1990s also asserted that "if a Cadillac had shrunk in size and cost as fast as a computer did since 1960, you could now buy one with your lunch money and hold it in the palm of your hand." (Although no one has ever explained how you could drive around town in a Cadillac that you could hold in the palm of your hand.) The basic message, however was, and remains, valid. From the 1960s until the present, integrated circuits, the building blocks of computers, have doubled their computing power every year or two. This explosive progress is the result of a technological trend called *Moore's Law* that states that the number of transistors on an integrated circuit "chip" will double every year or two. The name honors electronics pioneer Gordon Moore, who proposed it in 1965. Moore's Law is not a law of nature, but an empirical rule of thumb, and it has held true for four decades.

Today's computers are so small that size is no longer an issue. For example, a computer for controlling the air/fuel mixture in an automobile is typically housed along with its power supply and communication circuitry in an assembly about the size of a small book. As a result, people reliably control not just the hundreds of horsepower of an automobile, but the delivery of energy routinely available to them in modern industrial societies while minimizing pollution emitted in the course of energy conversion.

The subjects of control and binary logic have taken on a life of their own in the form of computers and other information systems. Indeed, such systems have become so pervasive as to lead many people to assert that the industrialized world changed in the late twentieth century from an industrial society to an information society.

The computers that carry out these logical and arithmetical tasks are essentially collections of electrical/electronic circuits. In simpler terms, they are boxes of switches that are turning each other on and off. So the question for this chapter might be rephrased as follows: How can switches turning each other on and off respond to inputs and perform useful calculations in an automobile? We will give a modest answer to this question by the end of this chapter.

Answering that question introduces the topics of digital logic and computation, which are central parts of computer engineering. The term *digital* can apply to any number system, such as the base ten system used in ordinary arithmetic. However, today's computers are based on a simpler number system: the base two or binary system. Therefore, this chapter focuses on binary logic and computation.

Computation was implemented first by mechanical devices, then by **analog computers,** and today by **digital computers.** A particularly convenient way to use computation to accomplish control is through the use of **binary logic.** A means of summarizing the results of binary logic operations is through **truth tables.** This technique is useful not only in control, but also in other areas of computation, such as **binary arithmetic** and **binary codes.** Binary arithmetic and information are the basis of computer software. Binary logic also enables us to define the engineering variable *information.*

In this chapter you will learn how to (1) carry out **logic operations** on **binary variables,** (2) summarize the results of carrying out logic operations on binary variables in a truth **table,** (3) implement truth tables as **electrical logic circuits,** (4) perform **binary arithmetic,** and (5) communicate the inputs and outputs of binary processes efficiently through **binary codes.** These topics will be mostly illustrated by reference to an automotive application: deciding when a seat belt warning light should be turned on.

As an afterword, the process by which actual computers do these things, the **hardware-software connection,** will be qualitatively summarized.

The topics in this chapter extend far beyond automobiles. Millions of *embedded* computers (that is, computers placed within other systems and dedicated to serving those systems) now can be found in hundreds of applications. Electronic and computer engineers have the jobs of reproducing this computer population, which brings forth a new generation every two or three years. The jobs these engineers do range widely. Some develop the electronic circuits that are the basis of electronic computation. Others devise new computer architectures for using those circuits. Yet others develop the "software": the instructions that make computers perform as intended. Many other engineers apply these tools to everything from appliances to space vehicles.

Analog Computers

For hundreds of years, engineers proved ingenious and resourceful at using mechanical devices to control everything from the rotation of water wheels and windmills to the speed of steam engines and the aiming of guns on battleships. However, such mechanical systems had major disadvantages. They were limited in their speed and responsiveness by the mechanical properties of the components. And they had to be custom designed for every control challenge.

By the twentieth century, this had led engineers to seek more flexible, responsive, and general types of controls. As a first step, engineers put together standardized packages of springs, wheels, gears, and other mechanisms. These systems also proved capable of solving some important classes of mathematical problems defined by a differential equations: Because these collections of standard mechanisms solved the problems by creating a mechanical analogy for the equations, they were called analog computers. Bulky and inflexible, they often filled an entire room and were "programmed" by a slow and complex process of reconnecting components mechanically in order to do a new computation. Despite these drawbacks, in such applications as predicting the tides, determining the performance of electrical transmission lines, calculating the transient behavior of nuclear reactors, or designing automobile suspensions, analog computers marked a great advance over previous equation-solving methods.

From Analog to Digital Computing

Meanwhile, a second effort, underway since about 1800, sought to calculate solutions numerically, using arithmetic done by people. Until about 1950, the word *computer* referred to a person willing to calculate for wages. Typically these were selected members of the workforce whose economic status forced them to settle for relatively low pay. It was not until about 1900 that these human computers were given access to mechanical calculators and not until the 1920s that practical attempts at fully automated calculation were begun.

Meanwhile, as early as the 1820s, beginning with the ideas of the British scientist Charles Babbage, attempts had been made to do arithmetic accurately using machinery. Because these proposed machines operated on digits as a human would (rather than forming analogies), they were called *digital* computers. Digital techniques were also used to control machinery. For example, the French inventor Joseph Marie Jacquard (1752–1834) used cards with holes punched in them as a digital method to control the intricate manipulations needed to weave large silk embroideries.

Humans learn digital computing using their ten fingers.[1] Human arithmetic adapted this ten-digit method into the decimal system. Digital computers can be built on a decimal basis, and some of the pioneers, such as Babbage in the 1840s and the team at the University of Pennsylvania who, in the 1940s built a very early electronic digital computer, the ENIAC,[2] adopted this decimal system.

Binary Logic

However, computer engineers quickly found that a simpler system less intuitive to humans proved much easier to implement with electronics. This is the binary system, based on only two digits: one and zero. The Jacquard loom was such a binary system, with a hole in a card representing a 1 and the lack of a hole representing 0 (zero). In other applications, turning on a switch might represent a 1, and turning it off might represent a 0 (zero). The logic and mathematics of this system were developed mainly by the British mathematician George Boole (1815–1864), whose contributions were so important that the concept is often referred to as **Boolean algebra.**

Binary logic begins with statements that can be identified as variables by being symbolized using a letter such as X. The statement can assert anything whatsoever, whether it is "The switch is open" or "There is intelligent life on a planet circling the star Procyon." The variable representing the statement can be assigned either of two values, 1 or 0, depending on whether the statement is true or false. (We will adopt the convention of using the value 1 for true statements and the value 0 for false ones.)

Binary logic permits three, and only three, operations to be performed: **"AND," "OR,"** and **"NOT."**

AND (sometimes called "intersection" and indicated by the symbol • or *) means that, given two statements X and Y, if both are true, then $X \cdot Y = 1$. If either one is false, then $X \cdot Y = 0$. For example, the statement "It is raining and the sun is out" is true only if both it is raining and the sun is out.

[1]The word *digitus* is literally "finger" in Latin, indicating the discrete nature of such counting schemes.
[2]Whether ENIAC was the first electronic computer is open to debate. Tommy Flowers, a British post office engineer, designed and built "Colossus" to break German codes in WWII; many attribute this to be the first electronic computer.

OR (sometimes called "union" and indicated by the symbol $+$) means that, given two statements X and Y, if either one or both are true, then $X+Y=1$, while only if both are false is $X+Y=0$. For example, the statement "It is raining (X) *or* the sun is out (Y)" is true in all cases except when *both* are not true. (Technically, this operation is called an *inclusive or* to distinguish it from the operation *exclusive or*, which gives a value 0 when both X and Y are true, as well as when both are false. In our example, this is when it is raining and the sun is out as well as when it is not raining and the sun is not out.)

NOT (sometimes called "negation" and indicated by the postsymbol $'$ and sometimes indicated with an overbar, $\overline{X}$) is an operation performed on a single statement. If X is the variable representing that single statement, then $X'=0$ if $X=1$, and $X'=1$ if $X=0$. Therefore, if X is the variable representing the statement "It is raining," then X' is the variable representing the statement "It is not raining".

These three operations provide a remarkably compact and powerful tool kit for expressing any logical conditions imaginable. A particularly important type of such a logical condition is an "if–then" relationship, which tells us that **if** a certain set of statements has some particular set of values, **then** another related statement, often called the **target statement,** has some particular value.

Consider, for example, the statement "If a cold front comes in from the south or the air pressure in the north remains constant, but not if the temperature is above 50°F, then it will rain tomorrow." The target statement and each of the other statements can be represented by a variable. In our example, X can be the target statement "It will rain tomorrow," and the three other statements can be expressed as A ("a cold front comes in from the south"), B ("the air pressure in the north is constant"), and C ("the temperature is above 50°F"). The "if" can be represented by an equal sign. The connecting words can be expressed using their symbols. Thus, the long and complicated sentence can be expressed as a short and simple *assignment:*

$$X = A + B\cdot C'$$

This is not a direct *equivalence* in which information flows both ways across the equation. Specifically, in an assignment,[3] the information on the right-hand side is assigned to "X" but not vice-versa. This distinction is required because of the way that computers actually manipulate information.

However, as written above, the statement still presents a problem. In which order do you evaluate the operations? Does this make a difference? A simple example will show that it *does* make a difference! Consider the case

[3]In some programming languages such logic statements are written with an *assignment* command, ": =" and not just an "=" command so that our Boolean statement could be written as $X{:}= A + B\cdot C'$ to reinforce the fact that these are not reversible equalities.

$A=1, B=0, C=1$. Suppose the symbol $+$ is used first, the symbol $\cdot$ next, and the symbol $'$ last. Then in the preceding statement, $A+B=1+0=1$; $1 \cdot C=1 \cdot 1=1$; and $1'=0$—thus $X=0$. However, if the symbols are applied in the reverse order ($'$ first, $\cdot$ next, and $+$ last), then $1'=0$; $0 \cdot 1=0$; $1+0=1$, and thus $X=1$. So to get consistent results when evaluating logic statements, a proper order must be defined. This is similar to standard precedence rules used in arithmetic.

> That order is defined as follows: All NOT operators must be evaluated first, then all AND operators (starting from the right if there is more than one), and finally the OR operators (starting from the right if there is more than one).

If a different order of operation is desired, that order must be enforced with parentheses, with the operation within the innermost remaining parenthesis being evaluated first, after which that parenthesis is removed. In our example above, the proper answer with explicit parentheses would have been written $X=A+(B \cdot C')$ and evaluated as $1+(0 \cdot 1')=1$. However, the value of the expression $X=(A+B) \cdot C'$ is $X=(1+0) \cdot 1'=1 \cdot 1'=1 \cdot 0=0$.

EXAMPLE 1

Consider the following statement about a car: "The seat belt warning light is on." Define the logic variable needed to express that statement in binary logic.

Need: Logic variable (letter) = ". . . ." where the material within the quotes expresses the condition under which the logic variable has the value 1 = true and 0 = false.

Know–How: Choose a letter to go on the left side of an equation. Express the statement in the form it would take if the content it referred to is true. Put the statement on the right side in quotes.

Solve: $W=$ "the seat belt warning light is on." ■

This example is trivially simple, but more challenging examples can arise quite naturally. For example, if there are two or more connected constraints on a given action, then the methods of Boolean algebra are surefire ways of fully understanding the system in a compact way.

EXAMPLE 2

Consider the statement involving an automobile cruise control set at a certain speed (called the "set speed"): "Open the throttle if the speed is below the set speed and the set speed is not above the speed limit." Express this as a logic formula, and evaluate the logic formula to answer the question "If the speed is not below the set speed and the set speed is above the speed limit, then will the throttle be opened?" To illustrate your answer, use these variables: the car's speed is 50 miles per hour, the set speed is 60 miles per hour, and the speed limit is 45 miles per hour.

Need: A binary logic formula expressing the statement "Open the throttle if the speed is below the set speed and the set speed is not above the speed limit," and an evaluation of the formula for the situation when the speed is 50 miles per hour, the set speed is 60 miles per hour, and the speed limit is 45 miles per hour.

Know: That any statement capable of being true or false can be represented by a variable having values 1 = true and 0 = false, and these variables can be connected by AND (•), OR (+), and NOT (′).

How: Define variables corresponding to each of the statements, and use the three connectors to write a logic formula.

Solve: Let X = "throttle is open."
Let A = "speed is below set speed."
Let B = "set speed is above speed limit."

Then the general logic formula expressing the target statement is

$$X = A \cdot B'$$

If the speed is 50 miles per hour, and the set speed is 60 miles per hour, then $A = 1$. If the set speed is 60 miles per hour, and the speed limit is 45 miles per hour, then $B = 1$. Substituting these values, the general logic formula gives

$$X = 1 \cdot 1'$$

Evaluating this in the proper order gives

$$X = 1 \cdot 0 = 0$$

In English: If the speed is below the set speed, and the set speed is above the speed limit, the control will not open the throttle. ■

EXAMPLE 3

Suppose we want to find a Boolean expression for the truth of the statement "W = the seat belt warning light should be on in my car," using all of the following Boolean variables:

Case 1:

D is true if the driver seat belt is fastened.
Pb is true if the passenger seat belt is fastened.
Ps is true if there is a passenger in the passenger seat.

Case 2: For a late warning light, include the additional Boolean variable:
M is true if the motor is running.

Need: $W = ?$

Know–How: Put the W variable on the left side of an assignment sign W= and then array the variables on the other side of the assignment sign.

Case 1:

(a) Put D, Ps, and Pb on the right side of the assignment sign. Thus, the temporary (for now, incorrect) assignment statement is $W = D\ Ps\ Pb$.
(b) Connect the variables on the right side with the three logic symbols •, +, and ′ so that the relationship among the variables on the right side correctly represents the given statement.

A good way to do this is to simply put the symbols the way they should appear in the if statements. Thus, for example, since part of the if statement Pb contains the words "the passenger's seat belt is fastened," it should also contain a "not," the corresponding logic variable will appear as Pb'. Also, you only care about the passenger's seat belt if the passenger is sitting in the seat; that is the intersection between these two variables.

Solve: $W = D' + Ps \cdot Pb'$.
Once written, a logic equation can now be solved for any particular combination of variables. This is done by first plugging in the variable and then carrying out the indicated operation.

Case 2: For the late activation of the warning light, the light will go on if the motor is running, and if either the driver's seat belt is not fastened and if there is a passenger in the passenger seat and that seatbelt is not fastened.

$$W = M \cdot (D' + Ps \cdot Pb') \quad ■$$

Truth Tables

It is often convenient to summarize the results of a logic analysis for all possible combinations of the values of the input variables of an "if. . . then" statement. This can be done with a **truth table.** It is simply a table with columns representing variables and rows representing combinations of variable values. The variables for the "if" conditions start from the left, and their rows can be filled in systematically to include *all possible combinations* of inputs. The column at the far right represents "then." Its value can be computed for each possible input combination.

EXAMPLE 4

Consider the condition in Example 2: Open the throttle if the speed is below the set speed and the set speed is not above the speed limit.

The "If conditions" are A=speed is below the speed limit and B=set speed is above the speed limit. The "then" condition is "open the throttle." The truth table is set up as shown here.

A	B	B'	$X = A \cdot B'$

Need: All 16 entries to the truth table

Know: Negation operator, ′ and union operator •.

How: Fill in all possible binary combinations of statements A and of B.

Solve: One convenient way of making sure you insert all the possible input values is to "count" in binary from all zeroes at the top to all 1's at the bottom. (If you're not already able to count in binary, this is explained on the next page or so.) In this example, the top line on the input side represents the binary number 00 (equal to decimal 0), the second line is 01 (decimal 1), the third is 10 (decimal 2), and the fourth is 11 (decimal 3).[4]

A	B	B'	$X = A \cdot B'$
0	0		
0	1		
1	0		
1	1		

Next the value of the "then" ("open the throttle") is computed for each row of inputs (this is done here in two steps, first computing B′, then computing $X = A \cdot B'$).

A	B	B'	$X = A \cdot B'$
0	0	1	0
0	1	0	0
1	0	1	1
1	1	0	0

In English, this truth table is telling us that the only condition under which the control will open the throttle ($X=1$) for is when both the speed is below the set speed and the set speed is below the speed limit.[5] ■

[4]If the input had three columns, A, B, and C, you would count from 0 to 7 in binary (000 to 111). If four columns, the 16 entries would count from 0 to 15 in binary (000 to 1111), and so on.
[5]This is indeed the way we *should* want our cruise control to operate!

Truth tables can also be conveniently expressed as electric circuits. Indeed, this capability is the essence of computing. This capability is further explored in the exercises.

Binary Arithmetic

If we assign numeric values to "0" and "1" rather than the Boolean values of true and false, then we can use these symbols to form the digits in a binary, or base 2, number system. This binary number system can be used to carry out arithmetic by means of logic functions.

The binary number system is a positional number system similar to the decimal number system, except it contains only 2 digits, 0 and 1, rather than 10 digits, 0–9.

Table 1

Binary Numbers and Binary Points

Place	2^4	2^3	2^2	2^1	2^0	.	2^{-1}	2^{-2}	2^{-3}
Bit	-	-	-	-	-	.	-	-	-

The value of a **binary digit** (or **bit**) depends on its position relative to the "binary point." For example, the binary number 11010.101 has the decimal value of $16+8+0+2+0+.5+0+.125=26.625$ as indicated in the second and third rows of Table 2.

Table 2

Comparison of Binary and Decimal Numbers

Place	2^4	2^3	2^2	2^1	2^0	.	2^{-1}	2^{-2}	2^{-3}
Bit	1	1	0	1	0	.	1	0	1
Decimal Value	16	8	0	2	0		0.500	0	0.125

Counting in binary is quite similar to counting in decimal, except we count on two fingers instead of on 10 fingers. Every time we reach the second finger, we start over and shift the place value to the left.

To perform arithmetic in binary, we use the same methods of "carry" and "borrow" that we use in the decimal system. Can you complete the Table 4? You can self-check against their decimal equivalents. Binary subtraction is conceptually identical to decimal subtraction. However, instead of borrowing powers of ten, one borrows powers of two.

Don't worry if you find binary subtraction a little more difficult than the other binary operations. To reliably carry out this unfamiliar procedure, a systematic procedure is helpful. One such systematic procedure is shown here.

Table 3

Binary and Decimal Equivalents

Binary Count	Decimal Equivalent	Binary Count	Decimal Equivalent
0	0	110	6
1	1	111	7
10	2	1000	8
11	3	1001	9
100	4	1010	10
101	5	1011	11

Table 4

Simple Binary Arithmetic Examples

Example 1		Example 2		Practice 1		Practice 2	
Binary	**Decimal**	**Binary**	**Decimal**	**Binary**	**Decimal**	**Binary**	**Decimal**
1001	9	1001	9	1011		1011	
+101	+5	–101	–5	+110		–110	
1110	14	100	4				

EXAMPLE 5 Do the following binary subtraction, 1001 – 111.

Need: 1001 – 111 = ? (a binary number)

Know–How: The following table summarizes binary subtraction. The first and third rows are the given two numbers. The second row shows the results of "borrowing," as in digital arithmetic. The bottom row, the difference, is the answer and is the result of subtracting the smaller number from the "original plus borrow" number.

Original number	1	0	0	1
Original + borrow	0	1	10	0
Subtracted #		1	1	1
Difference	0	0	1	0

Solve: 1001 – 111 = 10.

To verify this is correct, convert each number to decimal and repeat the subtraction. Thus 9 – 7 = 2 (base 10). ■

The rest of the familiar arithmetic functions can also be carried out in binary. **Fractions** can be expressed in binary by means of digits to the right of a decimal point. Once again, powers of two take the role that powers of ten play in digital arithmetic. Thus, the binary fractional equivalent 0.1 is the binary expression for the decimal ½ and 0.01 is the binary fractional equivalent of the binary ¼, and so on. A fraction that is not an even power of ½ can be expressed as a sum of binary numbers. Thus, the decimal $\frac{3}{8}=$ binary $0.0011=\frac{1}{4}+\frac{1}{8}$.

Multiplication and **division** can be carried out using the same procedures as in decimal multiplication and "long division." The only complication is, again, systematically "carrying" and "borrowing" in powers of two, rather than powers of ten.

EXAMPLE 6

If a car has an $(A/F)_{mass}$ of 15 [kg air]/[kg fuel] and the air intake draws in 1.5 kg of air per second, how much fuel must be injected every second? Solve in binary to five significant binary digits.

Need: Fuel rate = ______? kg fuel/s.

Know–How: Fuel rate = $(F/A) \times 1.5$ kg air/s = (1/15) (kg fuel)/(kg air) × 1.5 kg air/s = 0.10 kg fuel/s.

To illustrate binary arithmetic, let's break down this problem into two separate ones. While unnecessary to do this, one will illustrate binary division and the other binary multiplication.
The division problem will be 1/15 (decimal) = 1/1111 (binary), and the multiplication problem will take the solution of that problem and then multiply it by 1.5 (decimal) = 1.1 (binary).

Solve: Start with 1/1111, then binary multiply that answer by 1.1.

Division	Multiplication
.00010001	0.00010001
1111 ⟌ 1.00000000	× 1.1
1111	0.00001000
10000	0.00010001
1111	0.00011001
1	

∴**Fuel rate = 0.00011001 kg/s(binary).**

Checking in decimal: (1/15) × 1.5 = **0.10** and 0.00011001 = 0/2 + 0/4 + 0/8 + 1/16 + 1/32 + 0/64 + 0/128 + 1/256 = **0.098** —to get it more exactly, we would need more significant binary values. ■

Many times each second, a computer under the hood of an automobile receives a signal from an air flow sensor, carries out a binary computation such as the one shown in this example, and sends a signal to an actuator that causes the right amount of fuel to be injected into the air stream in order to maintain the desired air-fuel ratio. The result is much more precise and reliable control of fuel injection than was possible before computers were applied to automobiles.

Binary Codes

We can now see how 0 and 1 can be used to represent "false" and "true" as logic values, while also of course 0 and 1 are numeric values. It is also possible to use *groups* of 0s and 1s as "codes." The now outdated Morse code is an example of a binary code, while the genetic code is based on just the pairings of four, rather than two, chemical entities known as bases.

Suppose we want to develop unique codes for the following nine basic colors: red, blue, yellow, green, black, brown, white, orange, and purple. Can we do this with a three-bit code (three 0s or 1s [bits])? No, since there are only eight combinations of three bits (note: $2^3 = 8$)—namely 000, 001, 010, 011, 100, 101, 110, 111.

There is a four-bit code that would work, but of course it is only one of several, since there are $2^4 = 16$ possible combinations of a four-bit code. If we have N bits, we can code 2^N different things.

Table 5

Assignment of Binary Numbers to Colors

Color	Binary Equivalent	Color	Binary Equivalent
red	0000	brown	0101
blue	0001	white	0110
yellow	0010	orange	0111
green	0011	purple	1000
black	0100		

How Does a Computer Work?

How can these abstract ideas of Boolean algebra, binary logic, and binary numbers be used to perform computations using electrical circuits, particularly switches (which we studied in the previous chapter)? The modern computer is a complex device, and any answer we give you here is necessarily oversimplified. But the principles are sufficient to give you some insight. For this discussion we will need to know what a **central processing unit,** or CPU, is and what a computer **memory** is (which can take such forms as read-only memory, or ROM, and random access memory, or RAM).

The "smart" part of the computer is the CPU. It is just a series of **registers,** which is nothing but a string of switches. These switches can change their voltage states from "off" (nominally no voltage above ground) or to "on" or +5 volts above ground. The early personal computers (or PC) used only 8-bit registers and modern ones use 64 or 128, but we can think in terms of the 8-bit registers. (The notation x, y in Table 6 means either state x or state y.)

This register can contain 2^8, or 256, discrete numbers or addresses. What the CPU addresses is the memory in the computer. You can think of memory as a pigeonhole bookcase with the addresses of each pigeonhole preassigned.[6] If we have just 256 of these pigeonholes in our memory,

Table 6

Eight-Bit Register

Bit #	7	6	5	4	3	2	1	0
Voltage	0, 5	0, 5	0, 5	0, 5	0, 5	0, 5	0, 5	0, 5
Bits	0, 1	0, 1	0, 1	0, 1	0, 1	0, 1	0, 1	0, 1

our CPU can address each of them. If we want to calculate something, we write a computer code in a suitable language (see following) that basically says something like: Add the number in pigeonhole 37 to that in pigeonhole 64, and then put the contents in pigeonhole 134. The binary bits can represent numbers, logic statements, and so on. All we then need to do is to be able to send out the results of our binary calculations in a form we can read. (We decide that the contents of pigeonhole 134 means—for example, an equivalent decimal number.)

This is an oversimplification, since the pigeonholes are also binary coded locations and to do even simple decimal arithmetic, there have to be enough registers connected up that we can properly represent these numbers.

If, in the above example, we identify the program input as what is in pigeonholes 37 and 64, and the program output is pigeonhole 134, the computer is constructed essentially as in Figure 1 to accomplish its mission.

The previous chapter introduced the concept that a computer is made up of **hardware** (electrical circuits containing such components as transistors). How the computer is instructed to execute its functions is managed by **software** (the instructions that tell the components what to do).

Part of the hardware of the computer that we have so far ignored is an internal clock. One might view the software as a list of instructions telling the computer what to do every time the clock ticks. This list of instructions includes the normal housekeeping functions that the computer carries out regularly (such as checking whether a user has entered in any keystrokes on the keyboard in the very short time interval since the last time

[6] M. Sargent III and R. L. Shoemaker, *The IBM PC from the Inside Out* (Reading, MA: Addison-Wesley Publishing Co. Inc., revised edition, 1986), p. 21.

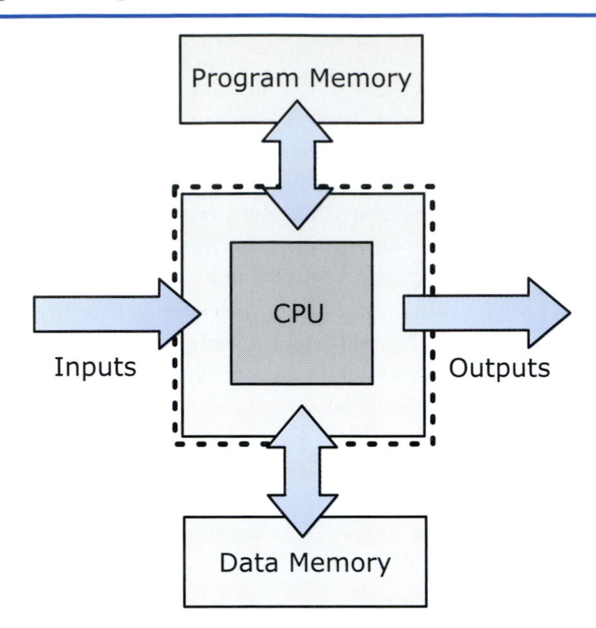

Figure 1

The CPU

this was checked). But it can also include downloading into the portion of the computer's memory called **program memory,** a special list of instructions called a **stored program,** and then executing that stored program. Examples of stored programs are a word processor and a spreadsheet program.

Once this stored program is downloaded into the memory, the CPU then carries out this stored program step by step. Some of those steps involve carrying out computations, which are executed in binary arithmetic by the CPU using the methods described in this chapter. In some cases, the result of the computation determines which of the stored program's instructions is the next one that should be carried out. This flexibility regarding the order in which instructions are carried out is the basis of the computer's versatility as an information processing system.

Software in this binary form is called **machine language.** It is the only language that the computer understands. It is, however, a difficult language for a human to write or read. So computer engineers have developed programs that can be stored in the computer that translate software from a language that humans understand into machine language.

The type of language that humans understand is called a **higher-level language.** This language consists of a list of statements somewhat (though not exactly) resembling ordinary English. For example, a higher-level language might contain a statement such as "if $x < 0$, then $y = 36$." Examples of higher-level languages are C++, BASIC, and Java. Most computer programs are originally written in a higher-level language.

The computer then translates this higher-level language into machine language. This translation is typically carried out in two steps. First, a computer program called a **compiler** translates the statements of the higher-level language into statements in a language called **assembly language** that is closer to the language that the computer understands. Then another computer program called an **assembler** translates the assembly language program into a **machine language program.** It is the machine language program that is actually executed by the computer.

The personal computers that we all use, typically by entering information through such devices as a keyboard or mouse, and receiving outputs on a screen or via a printer, are called **general-purpose computers.** As the name suggests, they can be used for a wide range of purposes, from game playing to accounting to word processing to monitoring scientific apparatus.

Not all computers need this wide range of versatility. So there is another important class of computers directed at narrower ranges of tasks. These computers are called **embedded computers** (Figure 2).

As their name suggests, these computers are embedded within a larger system. They are not accessible by keyboard or mouse, but rather receive their inputs from sensors within that larger system. In an automobile, there are about a dozen or more embedded computers. There might be, for example, an embedded computer for controlling a car's stereo system, another recording data for automated service diagnostics, and yet another

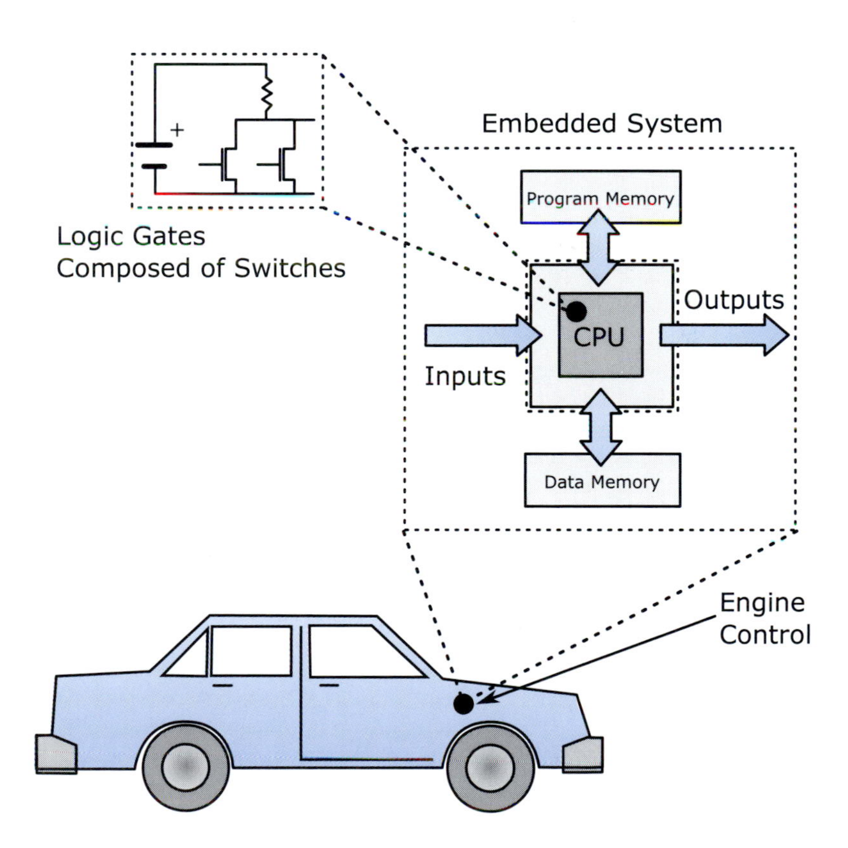

Figure 2

Schematic for an Embedded Automotive Computer

for fuel control. It is such a system for automobile fuel control based on an embedded computer that will be the subject of our next chapter.

SUMMARY

Although analog computers have been of some historical importance, digital computers do almost all important control jobs in today's advanced technologies. So an engineer must understand the principles of digital computation, which rest on the immensely powerful concepts of Boolean algebra, binary logic, truth tables, binary arithmetic, and binary codes. These concepts enable an engineer to make a first effort at defining the concept of information. Binary arithmetic and information are the basis of computer software. These concepts make it possible to move on to the challenge of implementing digital controls using computation.

EXERCISES

A number of these exercises use electrical circuits to effect logical statements. If you are uncertain about electrical circuits, you should review Chapter 7.

A popular ditty of the late nineteenth- and early twentieth-century railroad era was the following:

Passengers will please refrain
From flushing toilets
While the train
Is standing in the station
I love you.

(Sung to the tune of "Humoresque" by the nineteenth-century Czech composer Antonin Dvorak.)

1. For the preceding ditty, define a variable S expressing whether or not the train is in the station, a variable M expressing whether or not the train is moving ("standing" means "not moving"), and a variable F expressing the fact that the toilet may be flushed. (**A: S = "the train is the station," M = "the train is moving" (or you could use its negation, M' meaning the train is stationary), and F = "the toilet may be flushed"**)
2. For the preceding ditty, (a) express as a logic formula the conditions under which one may flush the toilet, (b) evaluate the formula you wrote in (a) for $M=1$ and $S=1$, and (c) express in words the meaning of your

answer to (b). (Assume that any behavior not explicitly forbidden is allowed.)

3. Consider the following logic variables for a car:

Db = "the driver's seat belt is fastened"
Pb = "the passenger seat belt is fastened"
W = "the seatbelt warning should be on in my car"

Write a sentence in English that expresses the logic equation $W = Db' + Pb'$.

(A: W = 'If either the driver's seat belt is not fastened or the passenger's seat belt is not fastened, the seatbelt warning light should be on.")

4. Consider the following logic variables for a car:

W = "the seatbelt warning light is on"
D = "a door of the car is open"
Ps = "there is a passenger in the passenger seat"
K = "the key is in the ignition"
M = "the motor is running"
Db = "the driver's seat belt is fastened"
Pb = "the passenger seat belt is fastened"

Write a logic equation for W that expresses the following sentence:

"If all the doors of the car are closed, and the key is in the ignition, and either the driver's seat belt is not fastened or there is a passenger in the passenger seat and the passenger's seat belt is not fastened, then the seatbelt warning light should be on."

5. Consider the following logic variables for a car:

M = "the motor is running"
Db = "the driver's seat belt is fastened"

In the early 1970s the government ordered all seat belt warnings to be tied to the motor in a manner expressed by the following sentence: "If the driver's seat belt is not fastened, then the motor cannot be running."

(a) Write a logic equation for M in terms of Db that expresses this sentence.
(b) Write a truth table for that logic equation.
(In practice, this seat belt light logic caused problems. Think of trying to open a nonautomatic garage door or pick up the mail from a driveway mailbox.[1] The government soon retreated from an aroused public.)

A: (a) $M = Db$

(b)

Db	**M**
0	**0**
1	**1**

6. Consider the following logic variables:

A = "a customer at a restaurant orders an alcoholic beverage"

I = "the customer shows proper identification"

M = "a customer at a restaurant is over 21"

Consider the sentence "If a customer at a restaurant is over 21 and shows proper identification, then she can order an alcoholic beverage." (a) Express this sentence as a logic equation, and (b) write a truth table for the logic equation.

7. Consider the following variables expressing a football team's strategy.

T = "It is third down"

L = "We must gain more than 8 yards to get a first down"

P = "We will throw a pass"

The team's strategy is expressed by this truth table.

T	**L**	**P**
0	0	0
0	1	1
1	0	1
1	1	1

Write a logic equation for P in terms of T and L.

(A: $P = T + L$)

8. Consider the following electric circuit:

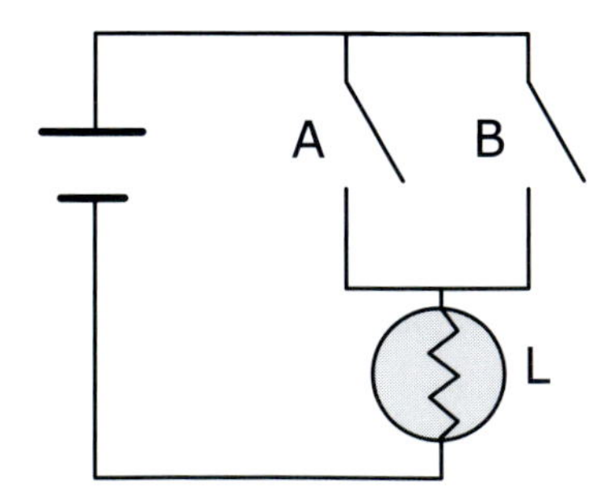

Consider the variables L = "the light is on," A = "switch A is closed," and B = "switch B is closed." Express the relationship depicted by the electric circuit as a logic equation for L in terms of A and B. (***A*: *L* = *A* + *B*.**)

9. Consider the following circuit diagram and the variables L = "the light is on," A = "switch A is closed," B = "switch B is closed," and C = "switch C is closed."

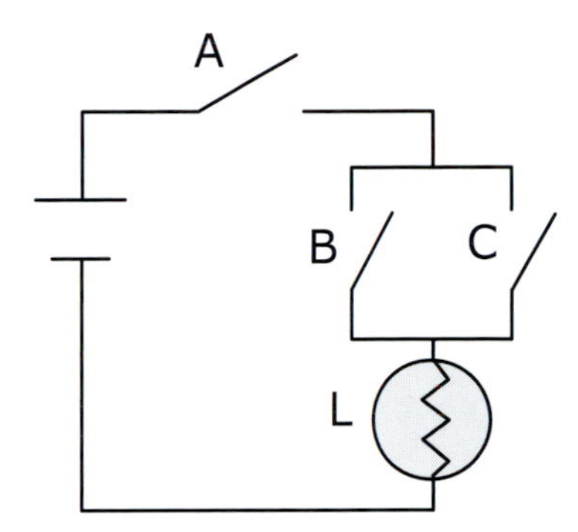

Write a logic equation for L in terms of A, B, and C.

10. Consider the following logic variables for a car:

 1. W = "the seatbelt warning light is on"
 2. Ps = "there is a passenger in the passenger seat"
 3. Db = "the driver's seat belt is fastened"
 4. Pb = "the passenger seat belt is fastened"

 Draw a circuit diagram for the logic equation $W = Db' + (Ps \cdot Pb')$.

11. Convert the following numbers from binary to decimal: (a) 110, (b) 1110, and (c) 101011.

Partial A:

—	2^5	2^4	2^3	2^2	2^1	20	Decimal equivalent
—	32	16	8	4	2	1	—
(a) 110	**0 × 32**	**0 × 16**	**0 × 8**	**1 × 4**	**1 × 2**	**0 × 1**	**6**
(b) 1110							
(c) 101011							

12. Convert the following numbers from decimal to binary: (a) 53, (b) 446, and (c) 1492.

Partial A:

Decimal	1024	512	256	128	64	32	16	8	4	2	1
Binary place	2^{10}	2^9	2^8	2^7	2^6	2^5	2^4	2^3	2^2	2^1	2^0
(a) 53	0	0	0	0	0	**1**	**1**	**0**	**1**	**0**	**1**
(b) 446											
(c) 1492											

13. Do the following binary additions. Check your answer by converting each binary number into decimal.

Partial A:

Binary	Decimal	Binary	Decimal	Binary	Decimal
1010	**10**	11101		10111	
+110	**6**	+10011		+ 10	
10000	**16**				

14. Do the following binary subtractions. Check your answer by converting each binary number into decimal.

Partial A:

Binary	Decimal	Binary	Decimal	Binary	Decimal
1010	**10**	11101		10000	
−110	**6**	−10011		−1	
0100	**4**				

15. If a car has an $(A/F)_{Mass}$ of 12.0 (kg air)/(kg fuel), and the air intake draws in 1.000 kg of air per second, how much fuel must be injected every second? **Solve in binary to three significant binary digits. (A: 0.000101 kg)**

16. Suppose we want to devise a binary code to represent the fuel levels in a car:

(a) If we need only to describe the possible levels (empty, $\frac{1}{4}$ full, $\frac{1}{2}$ full, $\frac{3}{4}$ full, full), how many bits are needed?
(b) Give one possible binary code that describes the levels in (a).
(c) If we need to describe the levels (empty, $\frac{1}{8}$ full, $\frac{1}{4}$ full, $\frac{3}{8}$ full, $\frac{1}{2}$ full, $\frac{5}{8}$ full, $\frac{3}{4}$ full, $\frac{7}{8}$ full, full), how many bits would be needed?
(d) If we used an 8-bit code, how many levels could we represent?

17. Construct a spreadsheet[7] that converts binary numbers from 0 to 111 to decimal numbers, print as formulae using the "control tilde" command. Check your spreadsheet against problem 11a. (**A: e.g., binary 11** $\equiv$ **1** $\times$ $\mathbf{2^1 + 1 \times 2^0}$ **= decimal 3.**)

18. Construct a spreadsheet that converts decimal number 53 to binary. Print as formulae using the "control tilde" command. Check your spreadsheet against problem 12a. (**Hint: 5 (decimal) can be divided by** $\mathbf{2^2}$ **to yield an integer "1" and remainder 1; 1 can't be divided by** $\mathbf{2^1}$ **(**$\therefore$ **integer "0") and "1" can be divided by** $\mathbf{2^0}$ **for last integer "1.")**

19. Construct a spreadsheet that does binary subtraction with borrowing. Test it on exercise 14. (**Partial A:** To complete your answer, you should also show the cell contents that produced this table!)

	B	C	D	E	F	G	H
8			**Binary ======>**				**Decimal**
9		2^4	2^3	2^2	2^1	2^0	
10	original	0	1	0	1	0	10
11	takeaway	0	0	1	1	0	6
12	borrow left	0	0	1	0	0	
13	borrow right	0	1	0	0	0	
14	**answer**	**0**	**0**	**1**	**0**	**0**	**4**

20. In the game "Rock, Scissors, Paper" we have the following rules:

Rock breaks Scissors
Scissors cut Paper
Paper covers Rock

If we were to create a code to represent the three entities, Rock, Scissors, and Paper, we would need two bits. Suppose we have the following code, where we call the first bit X and the second bit Y:

	X	Y
Rock:	0	0
Scissors:	0	1
Paper:	1	0

Now if we have two players who can each choose one of these codes, we can play the game.

[7]Spreadsheets have some built-in functions for decimal conversions to and from binary. It is recommended that you try first to use the actual mathematical functions described in this chapter and then *check* your answers using these functions to confirm your answers.

Examples:

Player 1	Rock (0 0)	Player 2	Scissors (01)	Player 1 wins
Player 1	Scissors (01)	Player 2	Scissors (01)	Tie
Player 1	Rock (10)	Player 2	Paper (10)	Player 2 wins

We see that there are three possible outcomes of the game: Player 1 wins, Tie, Player 2 wins.

Complete the table below that describes all the possible outcomes of the game.

Player 1 X, Y	Player 2 X, Y	Player 1 wins	Tie	Player 2 wins
0 0	00	**0**	**1**	**0**
0 0	0 1	**1**	**0**	**0**
0 0	1 0			
0 1	0 0			
0 1	0 1			
0 1	1 0			
1 0	0 0			
1 0	0 1			
1 0	1 0			
Totals				

21. A company purchased a computer program for your part-time job with them. The license agreement states that you can make a backup copy, but you can only use the program on one computer at a time. Since you have permission to make a backup copy, why not make copies for friends? What do you do?

(a) Go ahead, since your friends only use one computer at a time and these are backup copies.
(b) Make the backup copy, but sharing it with anyone clearly violates the license agreement.
(c) Ask your supervisor if you can use the backup copy at home, and then make as many copies as you wish.
(d) Use the program discretely, since software license agreements can't be enforced anyway.

Suggested answer: Fundamental Canons: Engineers, in the fulfillment of their professional duties, shall:

1. Hold paramount the safety, health, and welfare of the public. **Does not apply.**
2. Perform services only in areas of their competence. **Does not apply.**
3. Issue public statements only in an objective and truthful manner. **Does not apply.**
4. Act for each employer or client as faithful agents or trustees. **Options (a), (c), and (d) would put your employer in peril of being prosecuted for copyright violation. So your actions under either of these options would violate this canon.**
5. Avoid deceptive acts. **Excludes option (d).**
6. Conduct themselves honorably, responsibly, ethically, and lawfully so as to enhance the honor, reputation, and usefulness of the profession. **Copyright violation is theft, so options (a), (c), and (d) would be not only dishonorable, irresponsible, and unethical, but also unlawful.**

Solution: The canons rule out all but option (b).

22. You are a software engineer at a small company. You have written a software program that will be used by a major manufacturer in a popular product line. Your supervisor asks you to install a "back door" into the program that no one will know about so that he can monitor its use by the public. What do you do?

 (a) Install the back door, since it sounds like a fun experiment.
 (b) Tell your supervisor that you can't do it without authorization from the end user.
 (c) Install the back door but then deactivate it before the software is implemented.
 (d) Stall your supervisor while you look for another job.

Chapter 9

Control Systems Design and Mechatronics

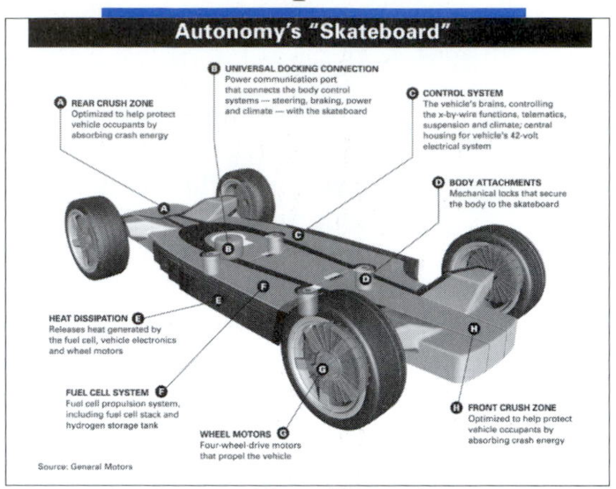

Courtesy of General Motors Corp.

Introduction

How does an automobile's cruise control system work? That question introduces the engineering topic of **control systems design,** a central part of the engineering discipline of mechatronics.

Mechatronics is the discipline of combining electronics, computer engineering, electrical engineering, and mechanical engineering into an *integrated* solution to a design challenge. Its uses extend far beyond automobiles. Mechatronics has made it possible for a person to zip around town on a Segway™ personal transporter (Figure 1), automatically keeping in perfect balance despite speeding up, slowing down, and maneuvering around obstacles.

Mechantronics enables a robot to tirelessly and accurately carry out an assembly task that would tax the skill, strength, and stamina of a human. Also, thanks to mechatronic controls, a solid-state camera can automatically sense light levels and adjust its settings to ensure perfectly exposed digital photographs.

As different as these applications are, they share a common topic, control systems design; a common discipline; mechatronics; and a common tool, the block diagram.

A block diagram, the analogue of an energy control surface or an electric circuit diagram, is a way of turning a conceptual sketch of a system into an engineering model that can be analyzed in terms of variables, numbers, and units. The block diagrams that are the focus of this chapter represent the key components of the conceptual auto that work together to make up a cruise control system.

Figure 1

Personal Transporter (Picture by kind permission of Segway Inc.)

Control system design consists of (1) modeling the system to be controlled as a block diagram; (2) translating that block diagram into a **mathematical model** and obtaining numerical results from the mathematical model, either in tabular, spreadsheet, or graphical form; (3) adding a mathematical model of the effects of external influences; (4) selecting a **control strategy** for the appropriate control of the whole system; and (5) **implementing** the chosen strategy in hardware.

The principal strategy used in control is **"feedback."** Feedback senses a current variable (such as speed = 70 mph) and brings it back to compare it to the desired or set variable (e.g., 65 mph). If the current variable is too low, a correction is made to increase it; if too high, then the correction is made to reduce it (e.g., by easing up on the accelerator if the current speed is 5 mph greater than the set speed).

Engineers often need to cause a variable of a system to stay automatically at a "set point"—the set point being the desired value. This is a principal concern in the wide-ranging subject of "control."

Control system design has wide-ranging applications, from controlling the darkness of toast in a toaster to controlling the flow rate of water to a hydroelectric plant. The thermostat in your home heating furnace is a familiar case in point. To be comfortable you want to keep the temperature of a room at 65–75°F, depending on your particular comfort level, even if no one is present in the room to turn the heat on or off.

Modeling the Control System as a Block Diagram

Developing a control system begins by converting a conceptual sketch of the system being controlled into a block diagram. However, it will be easiest to start with a physical sketch and develop the block diagram from what it tells you. Figure 2 is a conceptual sketch of a cruise-controlled car. In the sketch, the engine is shown as being controlled by a throttle.

The throttle in turn is connected to an accelerator and to a "microprocessor" or "microcontroller." The microprocessor is in turn connected to a speed sensor that measures how fast the car is going.

This conceptual sketch provides the basis for creating a block diagram. A block diagram is the next level of abstraction. It is both a visual and

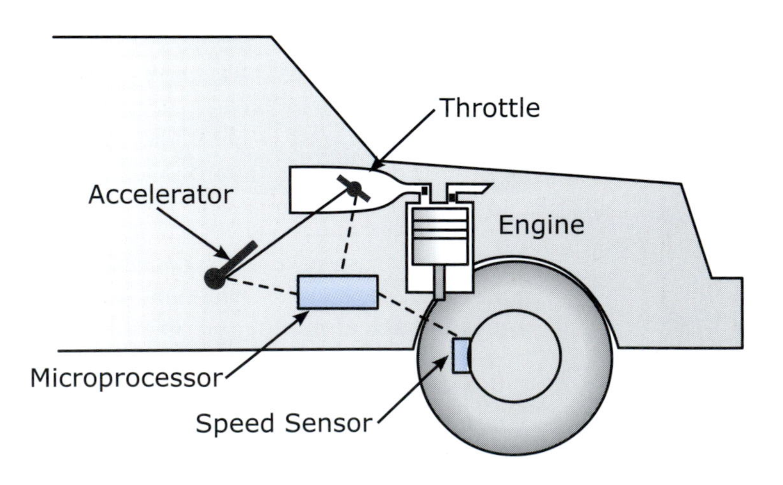

Figure 2

Speed Control in an Automobile

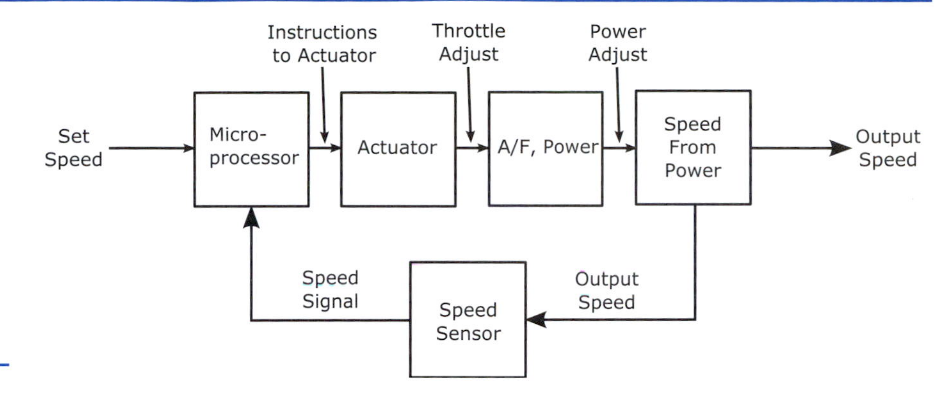

Figure 3

First Approximation to a Cruise Control

a mathematical tool for control analysis. It does for control analysis what control boundaries do for energy conservation analysis and what circuit diagrams do for the analysis of electric circuits—in other words, it focuses your attention on the essential core that you wish to analyze. A block diagram is connected together in the same way as the actual system it represents. To illustrate this, the portion of the conceptual car that is involved in cruise control is shown as a hardware block diagram (Figure 3) representing the cruise control.

This is designated "hardware" because at this point, it merely represents the physical rather than functional reality of the configuration. How engineers think about problems such as this starts with a block diagram. The next step is to turn that initial block diagram into a mathematical model of the system. Each block will now express, through a mathematical operation, what the corresponding component of the physical system actually does. For example, if a component did the job of converting one variable to another, it would be represented by the block shown here as Figure 4. That block consists of an **input variable**—say, "2" entering from the left, the block itself, —and an **output variable**—say, "6" exiting on the right. The interior of the block describes the mathematical operation performed on the input to produce the output. We will call the internal mathematical operation a **response function,**[1] and if it is a simple multiplication as here, it is called a **gain.** In Figure 4 the transfer function operates on the input "2" producing the output "6." The label below the block gives the block a descriptive name.

2 → 6 = 3 × 2 → 6
Gain

Figure 4

A Simple Block Operation

The current task is to translate each of the elements of the engine and its control system into blocks of this type. As we do, the blocks will be hooked

[1] In most texts on control theory, there is a more restrictive form of the response function called a **"transfer function."** It deals only with responses that are classified as "linear" that is concerned with variables only to their first power (i.e., no quadratic or square root terms, etc.).

up in a similar arrangement as the conceptual sketch of the physical system. For example, consider a model for the throttle.

EXAMPLE 1 Model the throttle of the conceptual car with a control block.

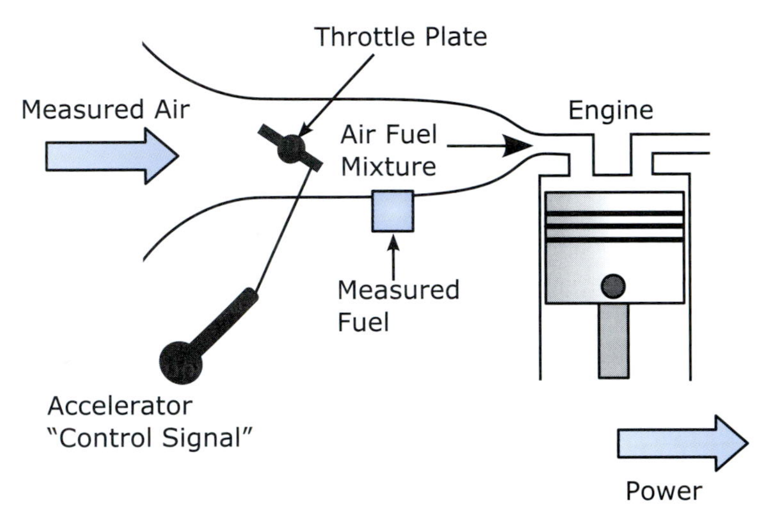

Schematic of Fuel/Engine System

Need: A control block corresponding to the throttle mechanism shown.

Know: The input to the throttle is a control signal from the accelerator (call it C). The output from the throttle is an air-fuel mixture that goes to the engine (call it AF[2], and it will be in units of flow—for example, in SI units AF would be expressed in kg/s). The relation between C and AF is determined by the amount a given control signal "opens" the throttle. The physical linkage could be a mechanical or electrical signal and a corresponding electrical relay-like actuator.

How: Assume that relation consists of the multiplication of C by G (G for gain). That relation can then be expressed by a block diagram.

[2]This is *not* the (A/F) mass or molar ratio that we previously used but the mass *rate* of air and fuel entering the engine.

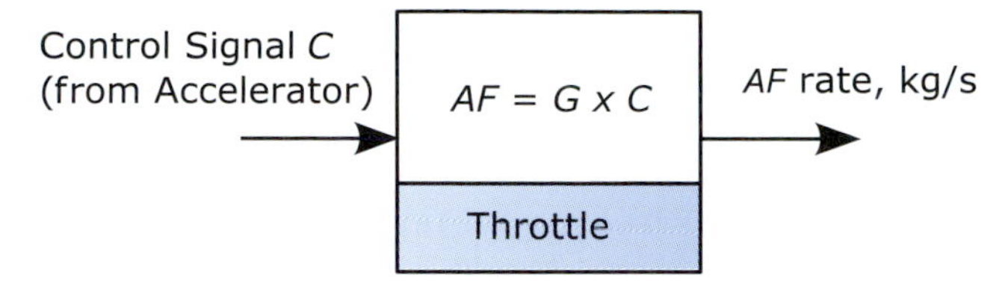

Controller Block for Fuel Injection Rate

For example, if the linkage were mechanical, perhaps you move the accelerator just 2.5 mm, but the throttle plate opens 10° of rotation. The gain is then 4°/mm. The units of gain are therefore quite specific to the particular operation we are mimicking.

Solve: The block diagram we are seeking is shown above. Our block for the air and fuel rate consists of the accelerator control signal variable C (basically the mechanical (or electrical) effect of the various mechanical (or electrical) linkages) entering from the left, the block itself, and an output air and fuel rate variable AF exiting on the right. The interior of the block describes the mathematical operation performed on the input to produce the output. When the control signal calls for more or less fuel, the amount is $G \times C$ and in SI units AF is in kg/s. ■

The output from this block is thus the relationship $AF = G \times C$. The transfer function described in this block is a simple gain. In addition, since the value of the output AF is proportional to C, it is called **proportional gain.** More complex transfer functions are also common. Since it is AF that we are interested in, you can theoretically achieve it by manipulating either G or C, or both, although normally the gain is fixed by the mechanical or electrical design.

For the special case of on/off control, the gain simply switches between $G = 0$ and $G = 1$. Many on-off control systems do exist (older home thermostats operated this way), but there are much better control systems for most applications (as you will soon learn). In this book, we will always assume that the gain is fixed at some value chosen to produce the required output for a given input signal.

The next required block is the engine's response to additional fuel and air. The power output of an automobile engine is approximately proportional to its intake rate of air and fuel (see Chapter 6). So engine power can be modeled by another simple block. In this model we have a measured or calculated, power output for our engine that says $P(engine) = K \times AF$, in which K is the gain (i.e., engine output for a given air and fuel input).

We will now depart from the convenient SI units of power (normally in W or kW). This is because "horsepower" has become an internationally recognized legacy term for automotive power. This is easy to preprogram into our design. Suppose, for example, that our engine produces 40.

HP from an input of $AF = 0.05$ kg/s of air and fuel. The engine's gain[3] is thus 800. [HP/kg/s].

EXAMPLE 2 Depict the portion of an automobile from air intake to power output as a block diagram.

Need: Overall block diagram for fuel system and for engine power.

Know: Block diagrams for each subsystem.

How: Add systems together at common point of intersection.

Solve:

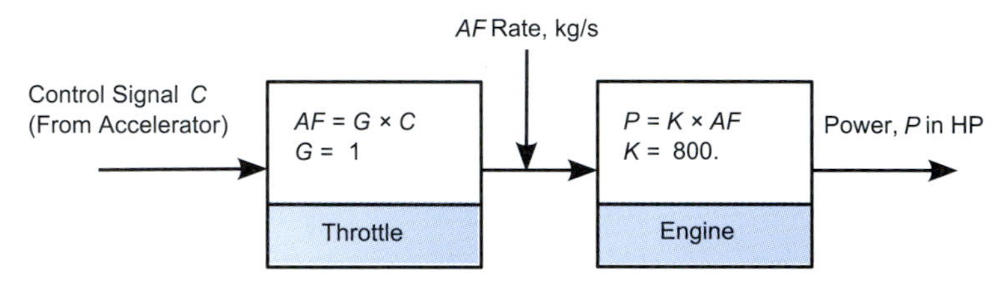

Block Diagram for Overall Engine Power

Note we have simplified the gain G to equal 1 (although any other numerical value could have been chosen.) The block diagram above can be simplified by combining its two blocks into one.

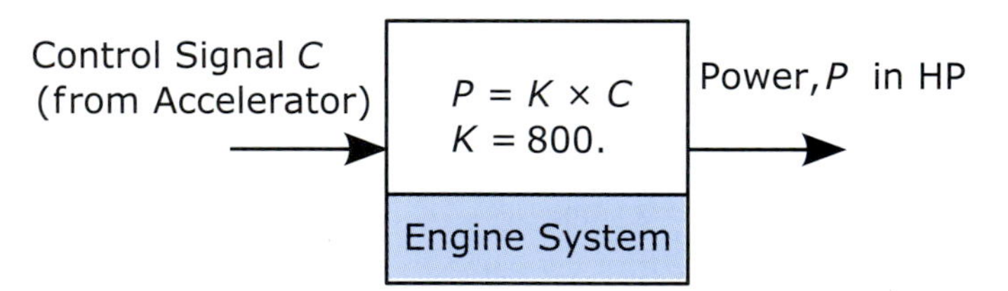

Simplified Block Diagram for Overall Engine System

■

Recognize that our operations in the blocks are "linear," meaning they can be simply multiplied together. For example, $AF = G \times C$ and $P = K_1 \times AF$ so that $P = K_1 \times G \times C \equiv K \times C$, defining K as the gain associated with the engine due to a change in the accelerator's position. If $C = 0.10$, then $P = 80.$ HP.[4]

[3]Since 0.05 [kg/s] × 800. [HP/(kg/s)] = 40. HP.
[4]Obviously assuming that the engine is capable of 80. HP.

Note one important feature of the model so far. Assuming we have the necessary physical setup to achieve these values, all we have controlled is the engine's *power*. Remember, however, that our objective is to control the *speed* of the automobile. That variable "speed" does not yet appear anywhere in the diagram! This is a typical situation in control system design. The variable the system directly modifies is often *not* the one that the system ultimately is intended to control. In this case, the system modifies power, but the goal is to control speed.

Translating the Block Diagram into a Mathematical Model

As presented so far, the model still does not answer the engineer's initial question: What car speed will result from the specified series of control actions? Answering that question requires modeling the way speed is determined by the driver and external influences. Such external influences might include, for example, the car's inertia (i.e., mass), the effects of gravity during hill climbs and descents (i.e., weight), the dynamics of the transmission, the internal and the road friction losses, and the air resistance to motion. Here we will restrict ourselves to just two of those influences: inertia and air resistance.

First, think through the physics of what these two phenomena will cause to happen: The addition of inertia means that the car will accelerate only at a rate given by Newton's Law of Motion (as we discussed in an earlier chapter), and, on a horizontal road, air resistance to motion will eventually balance the available engine power, thus limiting the vehicle's ultimate speed. So these are important variables in our analysis; together they will determine how long a called-for control change will take and what the value is when we have achieved the end state.

The response of the car's speed is described by Newton's Second Law of Motion. In an interval of time Δt (say, one second), the car's speed at time "t" given by S_t will accelerate to another speed—say, S_{t+1}. Equation 1 is an approximate relationship that expresses this in a mathematical function. (The proof of equation 1 is left as an advanced exercise.) Equation 1 will be used for a control block for an updated estimate S_{t+1} of the vehicle's speed:[5]

$$S_{t+1} \cong \sqrt{S_t^2 + \frac{2(\Delta t)(P(engine) - P(losses))}{m}} \tag{1}$$

in which we have assumed a flat road for a vehicle of mass m. *P*(*engine*) is the power the engine is producing and *P*(*losses*) is the power loss caused

[5]There is a more common way of deriving a response function for speed; it involves the use of "free body" diagrams and the concept of "torque." Even though such methods are common, we have not laid the groundwork for them in this text. Equation 1 is still reasonably valid.

by the wind resistance. We will eventually evaluate the term *P*(*losses*) from the wind resistance term, but for the moment we will ignore this term. (Please also note our equation is approximate, since it assumes that the wind losses are a constant independent of speed, but see below.)

Our next step is obtaining **numerical results.** These results can be expressed either in tabular, spreadsheet, or graphical form. For expressing the results, a spreadsheet is an ideal tool.

EXAMPLE 3

Assume all inertial effects can be represented by the mixed-units function $S_{t+1} = \sqrt{S_t^2 + 10. \times P \times \Delta t}$, with S in mph if P is in HP and Δt in seconds. Note that this model assumes *P*(*losses*) is zero. Provide a table of each variable at one-second intervals if $S_0 = 50$ mph.

Need: Transient response to control signal.

Know: Have all the necessary blocks for throttle, engine/transmission, and transient response. $G = 1$, $K = 800.$ HP/kg/s, $C = 0.05$ kg/s, $P = G \times K \times C$ in HP and $S_{t+1} = \sqrt{S_t^2 + 10. \times P \times \Delta t}$, with $\Delta t = 1.00$ s and S in mph.

How: Put all the blocks together to get the response function.

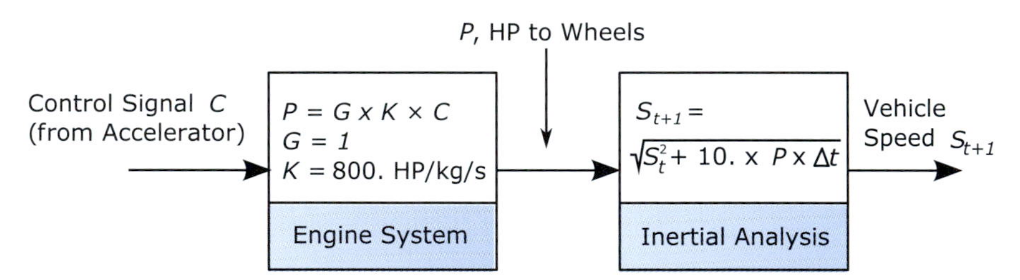

Block Diagram for Overall Vehicle Speed

Solve: Put blocks together and solve using a spreadsheet. A spreadsheet solution to this problem would look like this:

	B	C	D	E	F	G
17	*t*, sec	*C*	*G*	*K*	*P*, HP	*S*
18	0	0	1	800	0.0	50.0
19	1.00	0.05	1	800	40.0	53.9
20	2.00	0.05	1	800	40.0	57.4
21	3.00	0.05	1	800	40.0	60.8
22	4.00	0.05	1	800	40.0	64.0
23	5.00	0.05	1	800	40.0	67.1
24	6.00	0.05	1	800	40.0	70.0
25	7.00	0.05	1	800	40.0	72.8
26	8.00	0.05	1	800	40.0	75.5
27	9.00	0.05	1	800	40.0	78.1
28	10.00	0.05	1	800	40.0	80.6

Initially the car was coasting at 50.0 mph; then at $t = 1.00$ s the power was applied, and, in response, the vehicle begins to accelerate. Look at column G. Without a constraining counter balance to slow it, the car continues to accelerate (and it will forever in our model). We have not yet formulated a realistic control model!

	B	C	D	E	F	G
17	t, sec	C	G	K	P, HP	S
18	0	0	1	800	=D18*C18*E18	50
19	=B18+1	0.05	1	800	=D19*C19*E19	=SQRT(G18^2+10*F19*(B19-B18))
20	=B19+1	0.05	1	800	=D20*C20*E20	=SQRT(G19^2+10*F20*(B20-B19))
21	=B20+1	0.05	1	800	=D21*C21*E21	=SQRT(G20^2+10*F21*(B21-B20))

Here are the contents of a few cells[6] so you can see what is in this, our first nontrivial control model:
(Use the cell addresses to compare the output of the spreadsheet to the contents of each cell.) ■

How do we restrain the ever increasing vehicle speed? On a flat road it is the air resistance that slows down the car by providing the counter-balance. (You can feel the force of the wind on your hand if you put it outside the car's window at speed.) The force on an object moving in air is proportional to its (speed)^2; thus, the power[7] to maintain its speed is proportional to (speed).^{3} Suppose, for example, one determined by a series of measurements that at 60.0 mph a particular car required 40.0 HP to maintain it on a level road. Then we would model the power losses as $P = kS^3$, with $k = 40.0/60.0^3 = 1.85 \times 10^{-4}$ HP/mph^3 (admittedly a strange collection of units!).

EXAMPLE 4 Draw a block diagram for a wind resistance power loss model. What is its response function? How much HP will be lost to the wind at 30.0–100.0 mph in 10 mph increments?

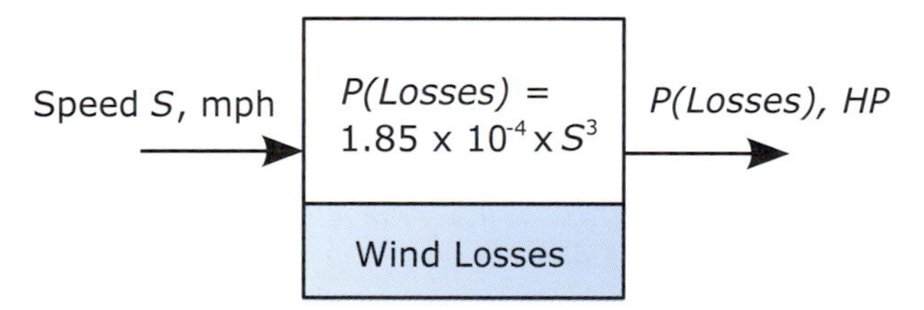

Block to Model Wind Losses

[6]Recall we can use the "^~ (control tilde)" to see the contents of any cell.
[7]Recall power is the rate of working or speed × force.

Need: Wind resistance block.

Know: $P(losses) = 1.85 \times 10^{-4} \times S^3$ (S in mph, $P(losses)$ in HP)

How: Wind speed in; power losses out.

Solve: The equation, $P = 1.85 \times 10^{-4} \times S^3$ (S in mph, $P(losses)$ in HP) is the required response function, since it converts the wind speed to power losses. It predicts the following losses to the wind: ■

S, mph=>	30.0	40.0	50.0	60.0
Losses HP	5.0	11.8	23.1	40.0
S, mph =>	70.0	80.0	90.0	100.0
Losses HP	63.5	94.7	134.9	185.0

Notice how quickly the wind losses mount as the speed increases. At 100. mph the losses are nearly 185 HP compared to just 40.0 HP at 60.0 mph and a mere 5.0 HP at 30.0 mph! If not safety, then the additional fuel (with its currency cost to you personally and environmental costs to all of us) should dissuade you from speeding!

The next piece of physics to include in our developing control model is the effect of the wind losses on the vehicle's speed. We will do this by writing $P(losses) = kS^3$ in equation 2:

$$S_{t+1} \cong \sqrt{S_t^2 + \frac{2(\Delta t)(P(engine) - kS_t^3)}{m}} \tag{2}$$

EXAMPLE 5

Assume a car is coasting at 50. mph. Assume that all inertial effects can be represented by the mixed-units function $S_{t+1} = \sqrt{S_t^2 + (10. \times P - kS_t^3) \times \Delta t}$ with S in mph, P in HP, Δt in seconds, and $k = 1.85 \times 10^{-4}$ HP/mph^3. Provide a table of each variable at one second intervals and plot the vehicle's speed and other parameters starting from an initial speed of $S_0 = 50.$ mph. The engine's gain is 800. HP/kg/s, and the accelerator is set at 0.05 after one second.

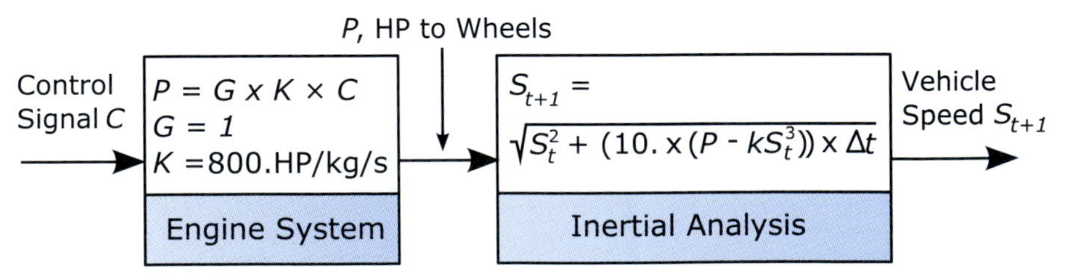

Block Diagram for Overall Vehicle Speed Including Wind Losses

Need: S_t as a function of time.

Know: From the previous examples, we have the complete series of blocks that describe this system.

How: Use the block diagram together and use a spreadsheet to answer the problem.

Solve: We start with the full block description of the control system from the throttle position to the car's speed.

	B	C	D	E	F	G	H	I
16		*k*	1.85E-04	HP/mph^3				
17		*K*	800	HP/kg/s				
18	t, sec	G	S_t, mph	C	P(*eng*), HP	P(*losses*), HP	ΔP, HP	S_{t+1}, mph
19	0	1	50.0	0.00	23.1	23.1	0	50.0
20	1.00	1	50.0	0.05	40.0	23.1	16.9	51.7
21	2.00	1	51.7	0.05	40.0	25.5	14.5	54.3
22	3.00	1	54.3	0.05	40.0	29.6	10.4	56.9
23	4.00	1	56.9	0.05	40.0	34.1	5.9	58.8
24	5.00	1	58.8	0.05	40.0	37.6	2.4	59.7
25	6.00	1	59.7	0.05	40.0	39.4	0.6	60.0
26	7.00	1	60.0	0.05	40.0	39.9	0.1	60.0
27	8.00	1	60.0	0.05	40.0	40.0	0.0	60.0
28	9.00	1	60.0	0.05	40.0	40.0	0.0	60.0
29	10.00	1	60.0	0.05	40.0	40.0	0.0	60.0

Spreadsheet: In addition, here are the contents of a few cells:

	B	C	D	E
16		*k*	=0.000185	HP/mph^3
17		*K*	800	HP/kg/s
18	t, sec	G	S_t, mph	C
19	0	1	50	0
20	=B19+1	1	=I19	0.05
21	=B20+1	1	=I20	0.05
22	=B21+1	1	=I21	0.05

	F	G	H	I
18	P(*eng*), HP	P(*losses*), HP	ΔP, HP	S_{t+1}, mph
19	=23.1	=$D$16*D19^3	=F19-G19	=SQRT(D19^2+10*(B19-B19)*H19
20	=$D$17*E20*C20	=$D$16*D20^3	=F20-G20	=SQRT(D20^2+(10*B20-B19)*H20
21	=$D$17*E21*C21	=$D$16*D21^3	=F21-G21	=SQRT(D21^2+(10*B21-B20)*H21
22	=$D$17*E22*C22	=$D$16*D22^3	=F22-G22	=SQRT(D22^2+(10*B22-B21)*H22

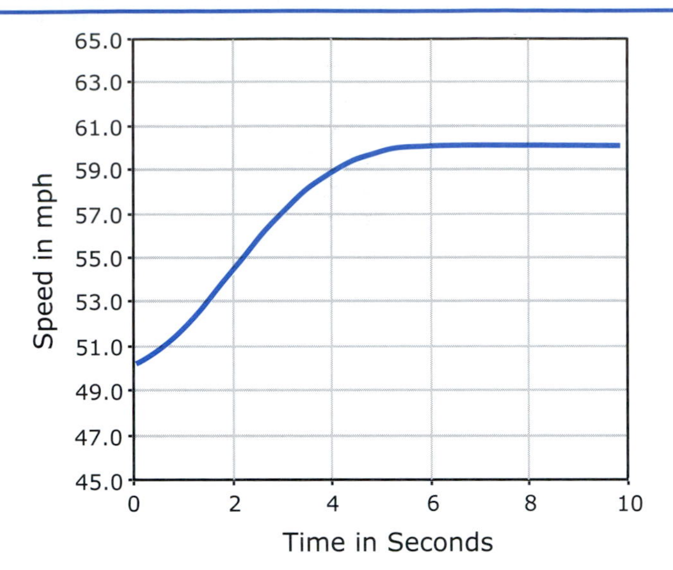

Let's analyze what we have seen here. Initially, the engine is putting out no power, so there is a deficit over what is needed and what is used (cell H19).

The control kicks in at one second, and from then we are applying a fixed 40. HP to the task (column F20:29). Eventually, as the car accelerates, the losses begin to balance the engine's output (cells H20:29) so that the car's speed asymptotes from its initial speed to a final road speed of ~60. mph—see the graph and cells I(19:29).

Unfortunately, we have let the twin variables of engine HP and of wind losses pick the final speed. What would happen if we *didn't* want ~60. mph as our final set speed? What happens if the external world changes on us? And why did we decide we just wanted a fixed throttle position when you probably know that at times you may need more or less throttle?

■

Selecting a Control Strategy

Now the entire system, car plus external influences, has been modeled. It is time to select a control strategy. A **control strategy** is a method of automatically enabling the system to control itself. One example of a control strategy is the method arrived at in the preceding section; it's the equivalent of putting a brick of a certain weight on the accelerator and *hoping* it gives us what we want. A successful outcome is obviously very unlikely.

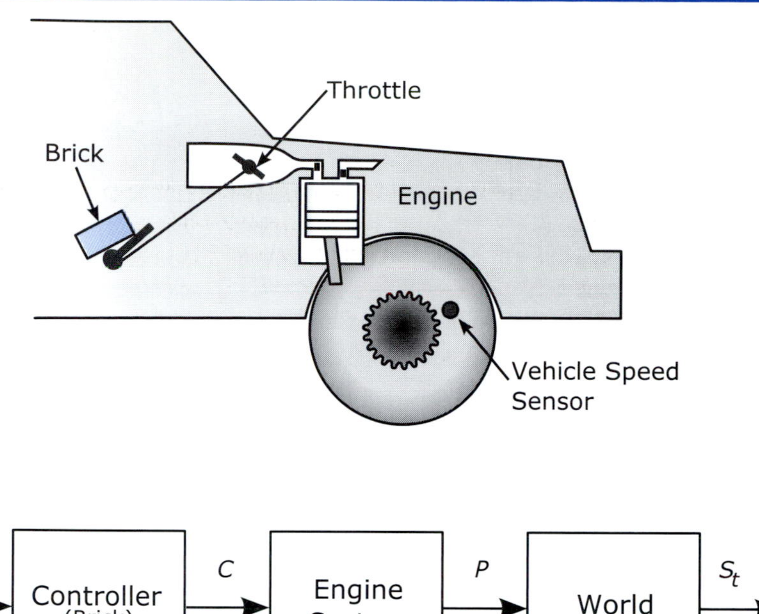

Figure 5

Open Loop Control

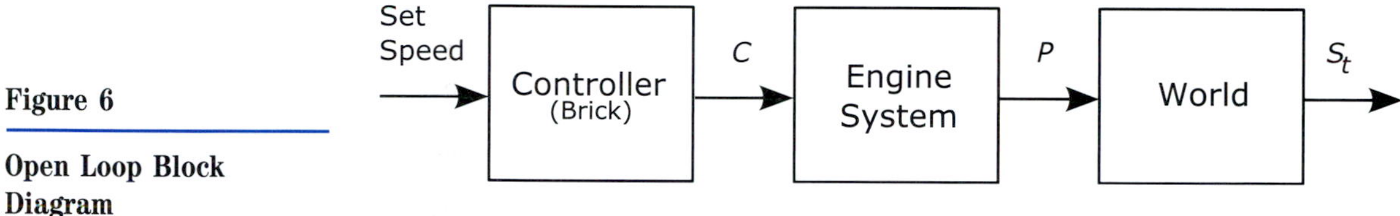

Figure 6

Open Loop Block Diagram

This is an example of a class of strategies called **open loop.** A conceptual sketch and a block diagram for an open loop controller implementing this strategy are shown as Figures 5 and 6, respectively.

We are sensing the vehicle's speed by a device attached to the wheels. Even though we are measuring the quantity we desire to control, in open loop strategies we are ignoring any new information relating to it after we have sought to control it. Another simple example of a system employing an open loop strategy is a light switch. Once it is turned on, the lights go on and stay on, whether it is day or night.

Often, however, the engineer wants the system to respond to changes in the world. For example, a lighting engineer might want to combine a light switch with a light sensor that automatically turns the lights off in the daytime and then turns them back on at night. Systems that sense and respond to the world in this way are called **closed loop** systems.

In the automobile example, closed loop strategies correspond to what a human driver does. A driver attempting to maintain a desired constant speed regularly compares that desired speed to the actual speed as displayed on the speedometer. If that actual speed is lower than desired, the driver steps on the accelerator. If the actual speed is higher than desired, the driver lets up on the accelerator.

How can this closed loop strategy be implemented automatically? This requires putting a new symbol into the control diagram to perform the observation and action done by a human driver.

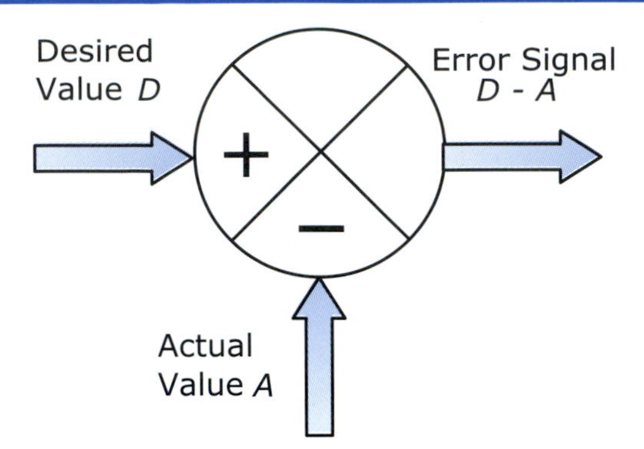

Figure 7

Summer

The new symbol is called a **summer.**[8]

The summer operation, Figure 7, is the essence of both how we use a speedometer on a car and how an automatic control system operates. The summer compensates for the error in the desired value by regularly comparing the measured speed and the desired speed.

Obviously the summer, unlike our previous blocks, has two (or more) inputs, the measured speed from the speed sensor and the desired speed called the "set point," and usually just one output. The output of the summer is the *difference* of the two input values and is called the **error signal** $(D - A)$.

Figure 8 shows the application of a summer to a cruise control. As can be seen, the loop literally has been closed. Here we will use the error signal, the difference between the desired and actual speeds, to directly set the AF amount and ignore the less visible throttle setting C altogether. For obvious reasons, any control strategy that uses a current signal compared to a prior set desired signal is called **feedback control;** it is the principal way that the great majority of controllers work.

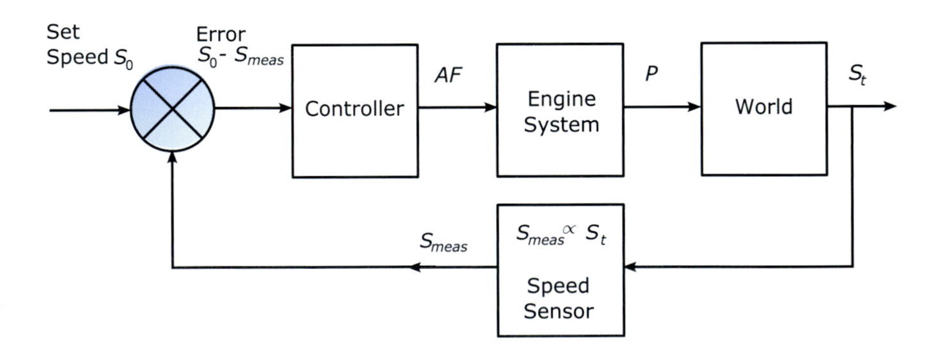

Figure 8

Closed Loop Control System

[8]This has nothing to do with vacations!

To use the summer, all we need is the speed signal converted into an appropriate format. For us, it's mph. For the car's control system it's usually an electrical signal proportional to speed. The speed sensor is modeled by a block that takes as its input actual speed as determined by the external response function (shorthand "world" in Figure 8) and produces as its output measured speed. We now need to define what the "controller" in Figure 8 will do.

We will deal here with just one type of controller, a proportional controller with the error signal as its input. It multiplies that error signal by a constant proportional gain (K_p) to produce as its output the *incremental* command signal. That signal, as before, becomes the input to the engine. Mathematically,

$$\Delta AF = K_p(D - M) \tag{3}$$

(i.e., desired, D, less measured, M).

Since the variable ΔAF is now the result of an *incremental* change, the engine block transfer function has to be modified accordingly, since it will receive just a small signal change. Let P_0 be the power needed by a car traveling at constant speed. Then the power is given by $P = P_0 + K \times \Delta AF$. We write it as: $\Delta P = P - P_0 = K \times \Delta AF$. The complete closed loop block diagram is shown as Figure 9.

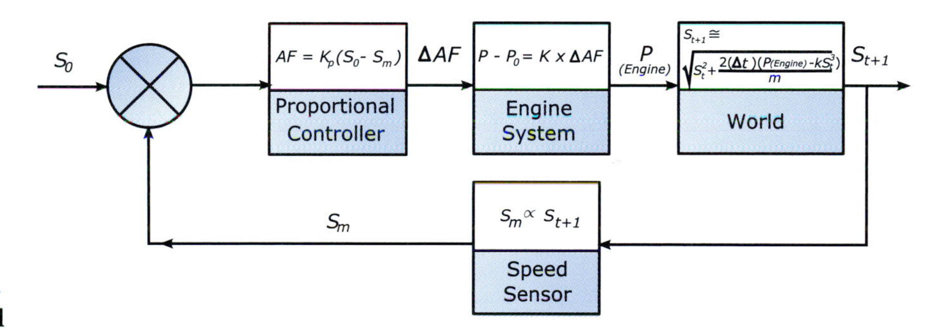

Figure 9

Closed Loop Proportional Control

EXAMPLE 6 Plot a vehicle's speed under proportional control starting from an initial 50. mph to a desired $D = 60.$ mph if the proportional gain is:

(a) $K_p = 1.5 \times 10^{-4}$ kg/s/mph
(b) $K_p = 5.0 \times 10^{-4}$ kg/s/mph
(c) $K_p = 1.5 \times 10^{-3}$ kg/s/mph

The engine's gain is 800. HP/kg/s. In addition, the inertial block is still $S_{t+1} = \sqrt{S_t^2 + (10. \times P - kS_t^3) \times \Delta t}$ in mph if $P(engine)$ is in HP and $k = 1.85 \times 10^{-4}$ Hp/mph^3 (S in mph).

Need: S_t as a function of time.

Know: The block diagram in Figure 9 has the complete series of blocks that describe this system.

How: Analyze Figure 9 by modeling the change in the air/fuel amount with K_p $(D - S)$ HP as the proportional control law.

Solve: We simply modify our spreadsheet with the new control law. Start with case (b) $K_p = 5.0 \times 10^{-4}$ kg/s/mph. Our spreadsheet can be set up with absolute constants, so we need only access one cell to change the effect of gain for the whole spreadsheet. Cell F22 contains the proportional gain. The attached graph is a composite of the curves for the three cases.

D, mph	60.0	K_p, kg/s/mph	5.00E-04	*K*, HP/kg/s	800		
t, sec	S_t, mph	(*D-S*) mph	Δ*AF*, kg/s	*P(eng)*, HP	*P(losses)*, HP	Δ*P*, HP	S_{t+1}, mph
0	50.0	0.0	0.0	23.1	23.1	0.0	50.0
1.00	50.0	10.0	0.0	27.1	23.1	4.0	50.4
2.00	50.4	9.6	0.0	31.0	23.7	7.3	51.1
3.00	51.1	8.9	0.0	34.5	24.7	9.8	52.1
4.00	52.1	7.9	0.0	37.7	26.1	11.6	53.2
5.00	53.2	6.8	0.0	40.4	27.8	12.6	54.3
6.00	54.3	5.7	0.0	42.7	29.7	13.0	55.5
7.00	55.5	4.5	0.0	44.5	31.7	12.8	56.7
8.00	56.7	3.3	0.0	45.8	33.7	12.1	57.7
9.00	57.7	2.3	0.0	46.7	35.6	11.1	58.7

	C	D	E	F	G	H
22	*D*, mph	60	K_p, kg/s/mph	0.0005	*K*, HP/kg/s	800
23	*t*, sec	S_t, mph	(*D-S*) mph	Δ*AF*, kg/s	*P(eng)*, HP	*P(losses)*, HP
24	0	50	=0	=E24*$F$22	=H24	=1.85*(0.0001)*D24^3
25	=C24+1	=J24	=$D$22-D25	=E25*$F$22	=$H$22*F25+G24	=1.85*(0.0001)*D25^3
26	=C25+1	=J25	=$D$22-D26	=E26*$F$22	=$H$22*F26+G25	=1.85*(0.0001)*D26^3
27	=C26+1	=J26	=$D$22-D27	=E27*$F$22	=$H$22*F27+G26	=1.85*(0.0001)*D27^3
28	=C27+1	=J27	=$D$22-D28	=E28*$F$22	=$H$22*F28+G27	=1.85*(0.0001)*D28^3
29	=C28+1	=J28	=$D$22-D29	=E29*$F$22	=$H$22*F29+G28	=1.85*(0.0001)*D29^3

In case (a) the graph shows how long it will take to achieve the speed we asked for—almost 40 seconds, possibly a dangerously long time. The reason is the incremental fuel addition is very small, since K_p is so small. The resulting increment in engine power is accordingly very small. We can call this case "overdamped" for reasons that will become clearer.

	I	J
23	ΔP, HP	S_{t+1}, mph
24	=G24-H24	=SQRT(D24^2+10*(C24-C24)*I24)
25	=G25-H25	=SQRT(D25^2+10*(C25-C24)*I25)
26	=G26-H26	=SQRT(D26^2+10*(C26-C25)*I26)
27	=G27-H27	=SQRT(D27^2+10*(C27-C26)*I27)
28	=G28-H28	=SQRT(D28^2+10*(C28-C27)*I28)
29	=G29-H29	=SQRT(D29^2+10*(C29-C28)*I29)

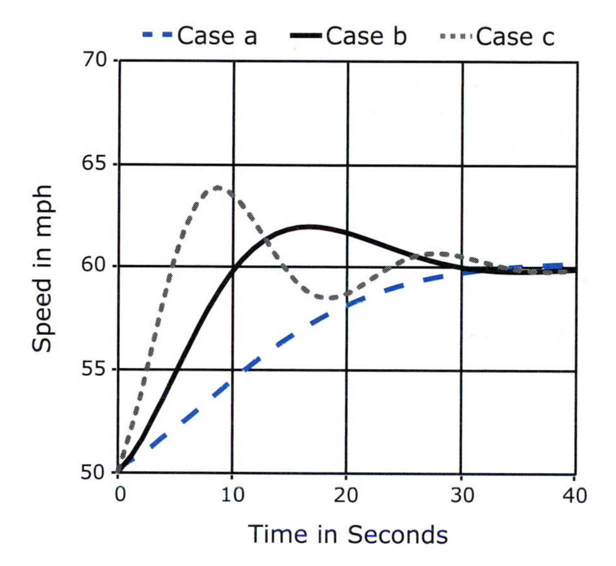

In case (b) we have tripled the proportional gain: The spreadsheet will only need to change cell F22 to the new gain, and we only need look at the resulting graph. In this case we achieved our 60 mph goal in about 10 s, but the controller visibly overshot the set point of 60. mph before settling. Still, this controller is more or less doing what we asked of it.

So what happens in case (c) at still higher gain? Case (c) is behaving like an oscillating spring; this is referred to as underdamped, since it "rings" for a long time before settling down. Such behavior would be very undesirable on the highway. It does this because the controller is calling for changes in *AF* large enough to cause wild swings in engine power and thus in speed. In general, if gains are too large, unsteady behavior will follow. ■

In this example, the tradeoff between rapidity of response and stability is accomplished by changing the proportional gain, K_p. In real-world examples, engineers apply control laws that go beyond the on-off and proportional controls so far introduced. With them, plus considerable mathematical analysis, most control systems will behave in acceptable ways.

The key to all of these control systems, however, remains the choice between open loop and closed loop systems. In practice, almost all control systems are closed loop. Most are as shown here—feedback in which a summer uses an error signal for subsequent control.

Implementing the Chosen Strategy in Hardware

The engineer's next step translates the blocks on the control diagram into hardware. Some of that hardware is already installed—for example, some actuators such as the throttle mechanism and the engine. This section will focus on two other hardware implementations that illustrate particularly well the mechatronic approach. Those implementations are the choice of a speed sensor and of a computer system to drive the controller.

For a long time automobiles have possessed a simple and reliable speed sensor. That sensor, the speedometer, traditionally used mechanical linkages to indicate speed. The turning of one of the car's axles was mechanically connected through a series of gears, cables, and shafts to an indicator on the dashboard. That indicator displayed the speed to the driver.

One could simply adapt this mechanical sensor to the cruise control. But the mechatronic approach suggests a different solution. The control system employs a digital controller, the microprocessor. So why not design a sensor *a priori* that is also digital? Such a digital speed sensor is shown in Figure 10. It is based on the fact that a magnet moving past a coil of wire induces a pulse of electric current in the coil.

In the speed sensor in Figure 10, a magnet is attached to, and turns with, one of the car's axles. A stationary coil is placed so that the magnet passes the coil each time the axle rotates. When the magnet passes the coil, it induces a pulse in the coil. So the number of pulses per second in the coil is proportional to the number of revolutions per second of the axle. This

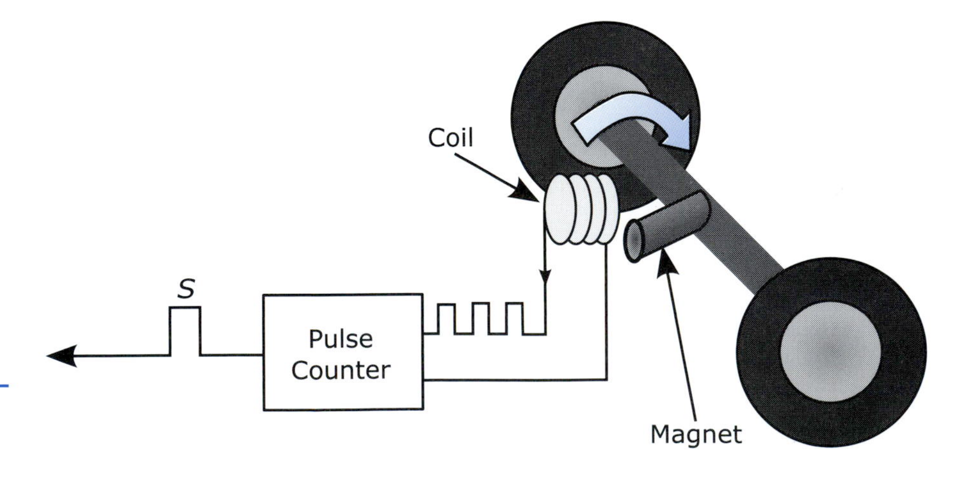

Figure 10

Electrical Digital Speed Sensor

provides a direct digital speed measurement for use by the cruise control system. A simple mathematical model of this speed sensor is the subject of one of the exercises.

In all modern cars the hardware implementation of the summer is just a small microcontroller, which also doubles as the element that carries out the proportional control block. This is a design choice, since we could have used a more traditional method of linkages and mechanical devices to achieve the control. But we know we will also need a microprocessor (or several) elsewhere in our cars, so why not design dual functionality into the microprocessor/microcontroller?

Drive-by-Wire

Automotive technology is being propelled by electronics, which are closely integrated into the mechanical operation of an automobile. Such close integration between mechanics and electronics is called mechatronics. "Drive-by-wire" is the direction of modern automobile technology that relies heavily on mechatronics. Eventually the present mechanical controls in a car (steering, gears, brakes, and accelerator) will all be replaced by integrated electronic ones. A present forerunner is how some cars connect the accelerator pedal to the fuel system.

In this text so far, we have used mechanical linkages to adjust the *A/F* system because that is conceptually easy to understand. However, some advanced vehicles now do this electronically.[9] One reason is that the rotation of a throttle plate by a mechanical linkage is inherently nonlinear (Figure 11).

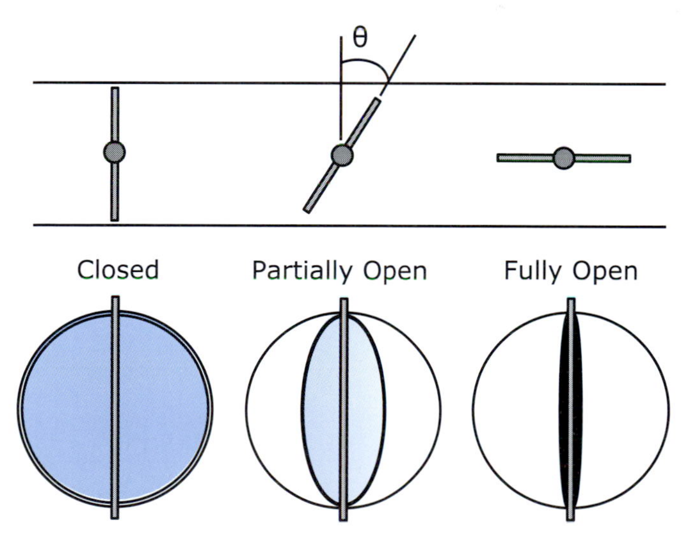

Figure 11

Throttle Position as a Function of Rotation

[9]http://www.findarticles.com/p/articles/mi_m3102/is_7_103/ai_n6137977#continue.

If you follow the geometry of opening the throttle plate, the % open area is equal to $(1 - \cos\theta)$ see Exercises. Table 1 shows the corresponding open area per rotation of the throttle plate.

Table 1

Incremental Opening Angle for 10% Change in Open Throttle Area

10%	20%	30%	40%	50%	60%	70%	80%	90%	100%
6°	6°	6°	6°	6°	7°	8°	9°	11°	26°

What this table means is that pushing your foot on the accelerator will produce different results depending on its current position. One solution to this lies in a specialized motor called a "stepper motor." Its operation is animated in some web references,[10] and one simplified configuration taken from there is shown as Figure 12.

The operation consists of pulses of electric power that are applied to the coils that actuate pole pairs 1, 2, and 3 (in that order), causing the rotor to increment counter clockwise. These "poles" are just pieces of iron that concentrate the local magnetic field. Typical commercial application stepper motors increment by much smaller amounts typically about 1°–2° per pulse

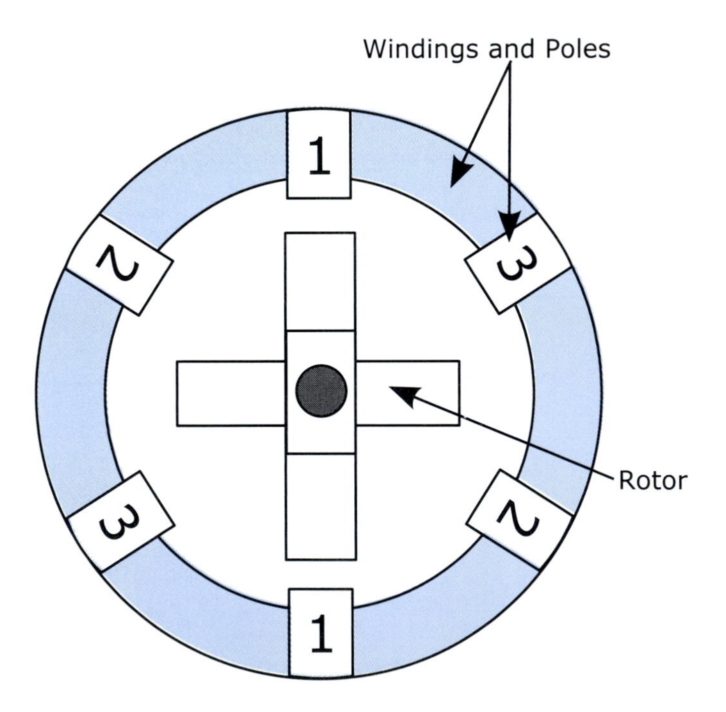

Figure 12

Principle of a Stepper Motor

[10]http://www.cs.uiowa.edu/~jones/step/types.html.

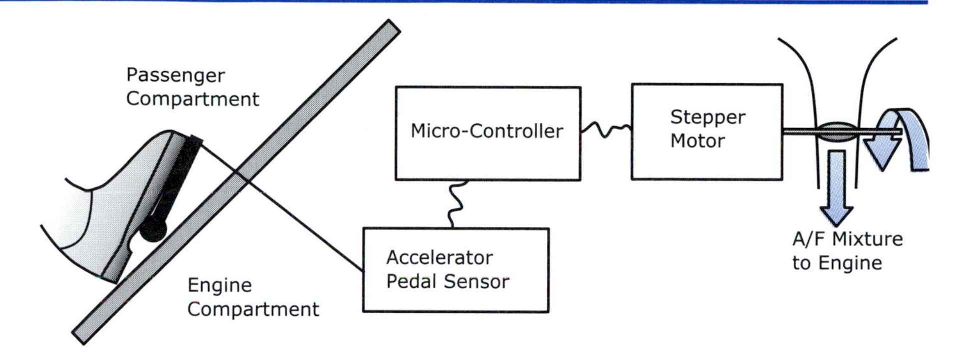

Figure 13

Drive-by-Wire

applied to the rotor. For this reason stepper motors are precise positioning devices and typically are used as actuators in robotic devices. In this automotive application,[11] they would replace the mechanical linkage as shown in Figure 13.

In essence, whenever the accelerator is displaced a fixed amount, the open area of the throttle can be made to respond by a corresponding amount.

EXAMPLE 7

A pulse motor increments 1.8° per pulse. You wish to use it to ensure that equal movement of the accelerator pedal results in equal rotation of the throttle plate. How many pulses must you supply to open a throttle an extra 10% from an initial position of (a) 10%, (b) 50%, and (c) 80%?

Need: Number of pulses to open throttle by 10% at initial positions of 10%, 50%, and 80% respectively.

Know–How: Table 1 has the opening response as a function of rotation.

Solve: (a) At 10% we need to open the throttle 6° or $6/1.8 = 3.3 = \mathbf{3}$ **pulses** (pulses being integral)

Case **(b)** Again we need $7/1.8 = \mathbf{4}$ **pulses.**

Case **(c)** This time we need $11/1.8 = \mathbf{6}$ **pulses.**

As long as the microcontroller will deliver the appropriate number of pulses in response to the signals from the pedal sensor, this will ensure the linear response that the driver expects. ■

[11]To achieve fast transient, a larger direct current motor may be used instead, but the basic reason to drive-by-wire still results.

The mechatronic approach would take additional advantage of the integration with electronics; presuming the vehicle also has cruise control, it will use the same stepper motor/throttle plate as shown in Figure 13; the difference will be the instructions (in the form of electrical pulses) transmitted to the stepper motor.

To conclude, the hardware implementation finishes the job of implementing cruise control, as initially introduced in Figure 3.

SUMMARY

The control design strategy outlined above is not just for cars. It can be applied to any situation where a desired value of a variable is to be maintained. The table[12] below gives a range of applications. Parts of these applications, as well as others, are discussed in the exercises.

Diverse as these applications are, they all implement similar principles of control systems design and can be represented through **block diagrams.**

Application	Controller	Process	Command Signal	Controlled Value
Home heat	Thermostat	Furnace	Heat	Temperature
Drill	Speed dial	Electric motor	Torque	Bit speed
Toilet	Float and lever	Water valve	Water flow	Water level
Economy	Government policy	Treasury and federal reserve	Interest rate, money supply	Gross domestic product (GDP)

The process of control design consists of the following steps: (1) modeling the control system as a **hardware block diagram;** (2) translating that block diagram into a **mathematical block diagram** model and obtaining numerical results from the mathematical model, either in tabular, spreadsheet, or graphical form; (3) adding a mathematical model of external effects or "world"; (4) selecting a **control strategy**—either **open loop,** but almost inevitably **closed loop**—for the appropriate control of system-plus-world; and (5) implementing the chosen strategy in hardware.

Most control strategies use **feedback.** Gain is an important variable in control strategies; too little and the desired result may be very slow in coming; too much and the system can become unstable. **Underdamping** and **overdamping** are inherent factors in many control strategies.

[12]Adapted from Raymond T. Stefani, et al., *Design of Feedback Control Systems* (New York: Oxford U, 2002), p. 4.

Mechatronics is the integration of mechanical and electronic systems into a design. If several objectives are considered simultaneously, the resulting system is usually less complicated than the sum of the individual systems. Control theory underlies the implementation of many mechatronic implementations.

EXERCISES

The answers to these exercises are not necessarily unique since there is some latitude in equivalent ways they can be formulated.

1. The circuit diagram below is an **open loop** control for tuning on or off a light. Draw a block diagram of the control. Why is it considered open loop?

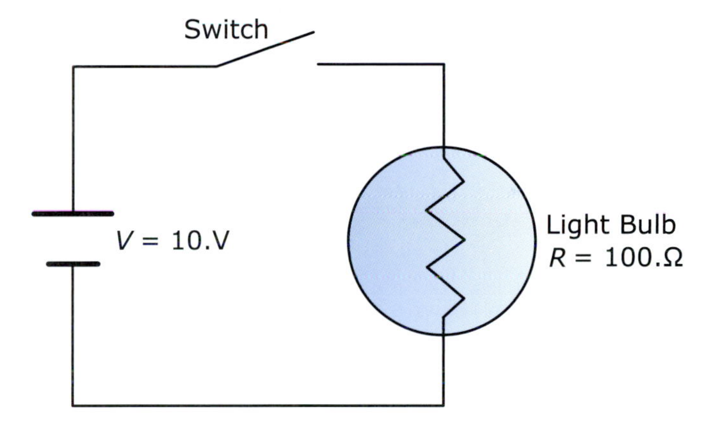

A:

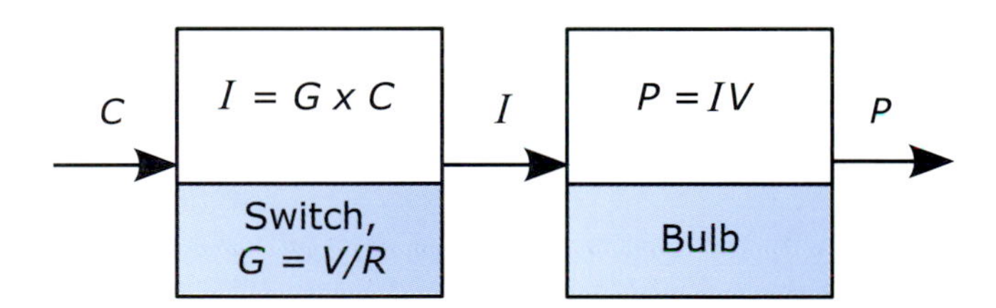

The bulb is on and stays on provided the switch is closed, day or night, needed or not, until the switch is manually turned off. Then the same reasoning applies in reverse. That is why this is considered an open loop control.

2. The circuit diagram represents a **closed loop** system for lighting control in an otherwise dark room. Assume you want a lightbulb to provide constant illumination of strength L_D even if the strength of the lightbulb

changes with age. Light control is achieved by a suitable light sensitive sensor that compares the bulb's measured output L_B to the desired light level. The sensor returns a measured light signal L_B. The command signal is the difference between L_D and L_B and is communicated to a controller that sets a variable circuit voltage from 0 to 100 V. The bulb's intensity is set by this voltage. Draw a block diagram of the control, indicating on the diagram which parts of the conceptual sketch correspond to the blocks.

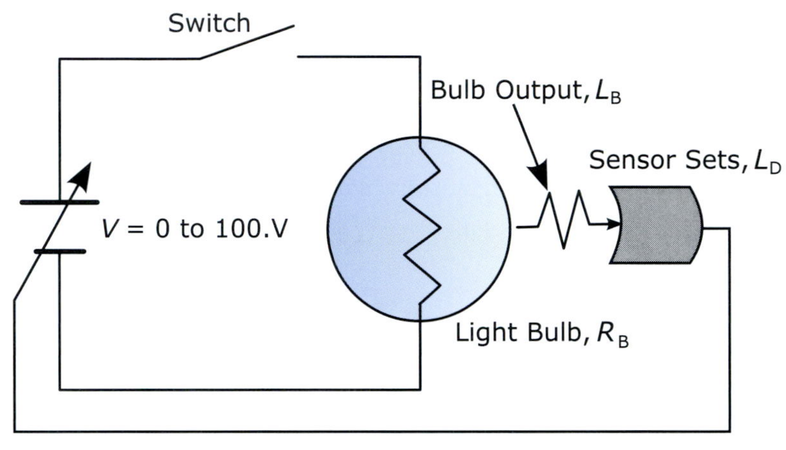

3. Part of the mechanism of a toilet is the system that refills the toilet tank after a flush. A conceptual sketch is shown below. Water enters the tank

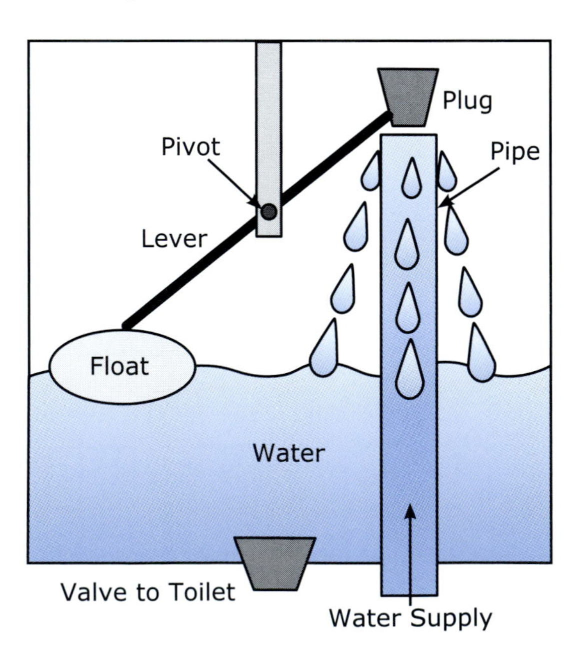

through an annular opening around the plug; eventually the water level, *WL*, in the tank rises and the float also rises, pushing down the plug and turning off the supply of fresh water. Draw a block diagram of the system, indicating on the diagram which parts of the conceptual sketch correspond to the blocks.

A:

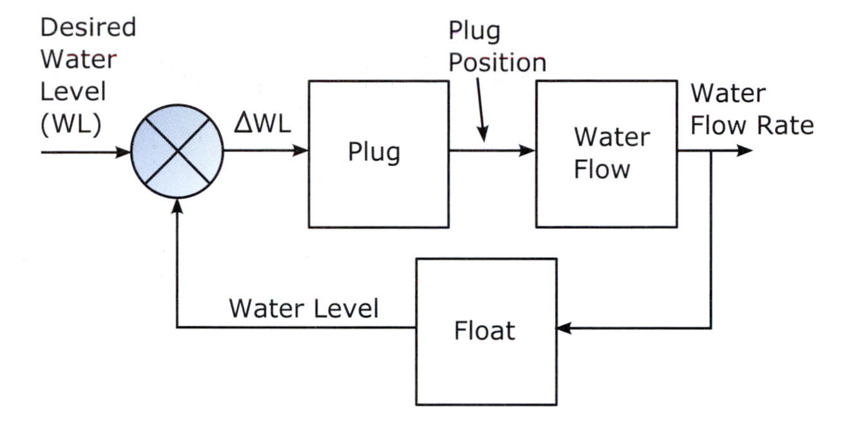

4. The conceptual sketch for an open loop control for a toaster is shown. Draw a block diagram of the system, indicating on the diagram which parts of the conceptual sketch correspond to the blocks.

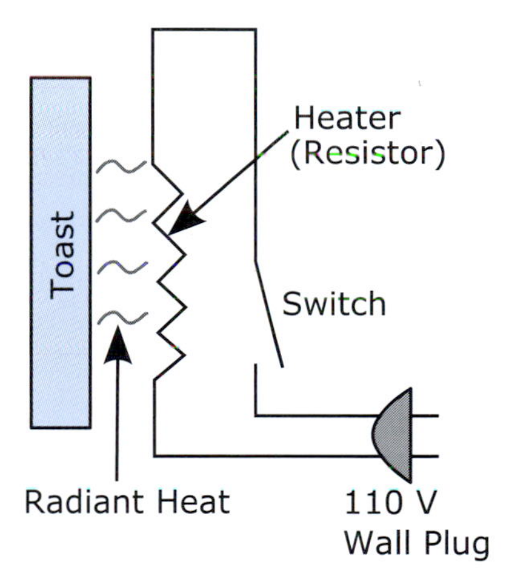

5. The conceptual sketch for a closed loop control for a toaster is shown. The controller is a bimetallic switch. It works by combining two metals of different thermal expansion so that, when heated, it bends away from the more expansive metal and can then open a pair of contacts. The heat is interrupted when the contacts are opened and is reestablished when the bimetallic strip cools.

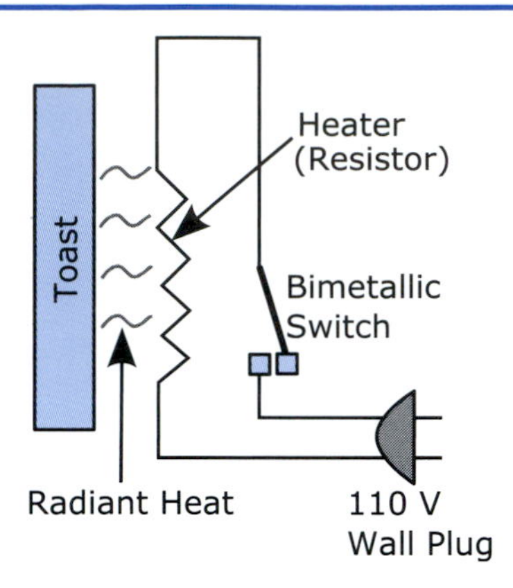

The heat source is "waste" heat generated in the resistor that also heats the toast. The stationary contact is adjustable so that the clearance between the two contacts determines the set point as to when the heater is on or off.

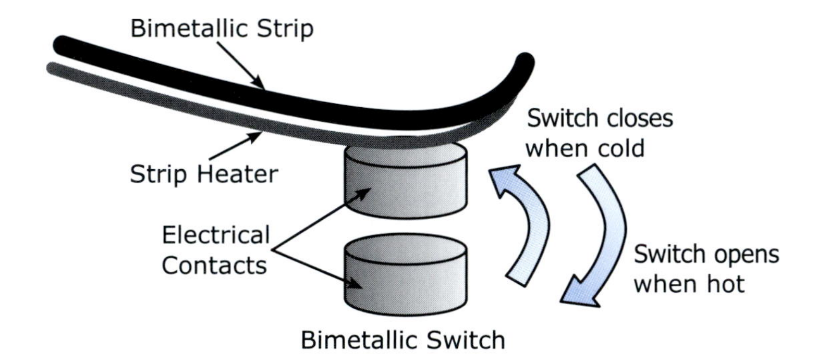

Draw a block diagram of the system, indicating on the diagram which parts of the conceptual sketch correspond to the blocks.

6. The conceptual diagram of an open loop control for the heating of a room is shown. The room achieves a given temperature as a function

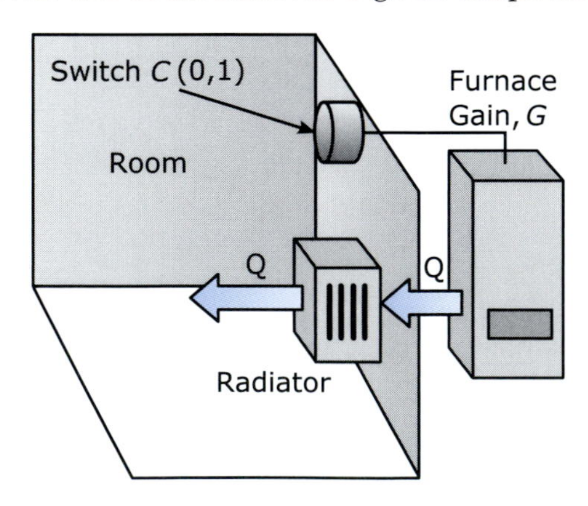

of the added heat from the radiator and the heat loss, Q_L, from the room but for simplicity ignore Q_L as being too slow to influence the immediate response.

Draw a block diagram of the system that determines T_{Room}, indicating on the diagram which parts of the conceptual sketch correspond to the blocks. Assume the controller switch is a simple off/on (0,1) and the furnace responds with a gain G to produce heat at a rate Q.

A:

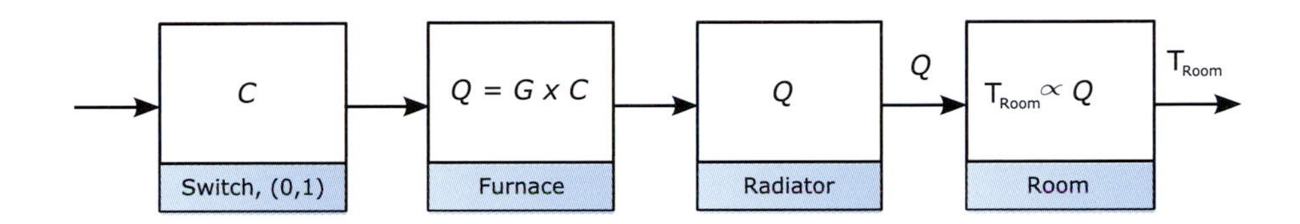

7. The conceptual diagram and block diagram of an open loop control for a lawnmower engine are shown below. Assume the control is turned on at time $t = 0$ and left on thereafter. Using the values of the proportional gains shown in the block diagram prepare a spreadsheet and graph showing the power of the engine each second for a total of 20 seconds.

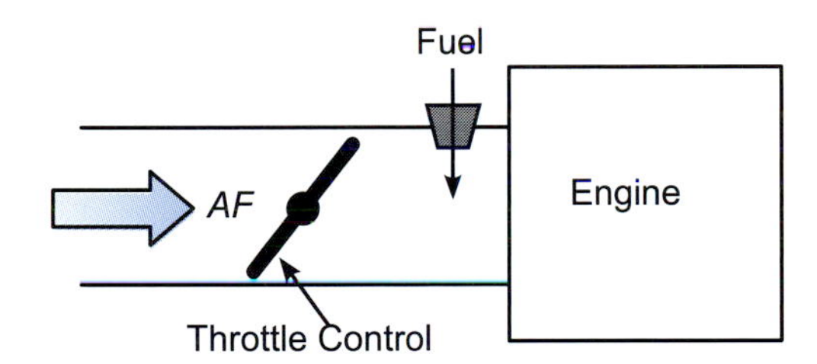

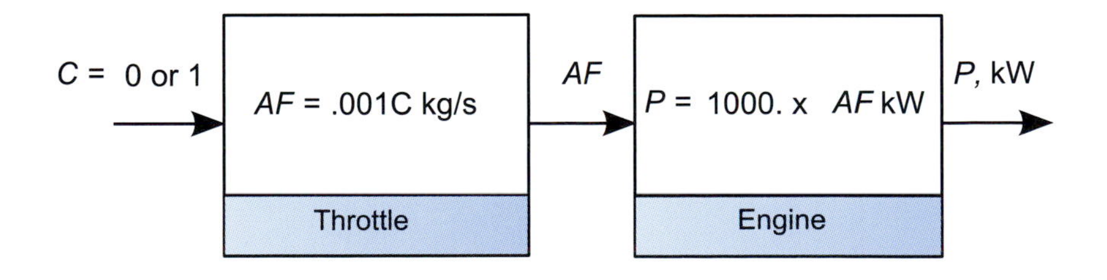

A:

	B	C	D	E
17		Throttle gain	=0.001	
18		Engine gain	=1000	
19				
20	*t* s	*C* (0 or 1)	*AF* kg/s	*P* kW
21	0	0	=$D$17*C21	=$D$18*D21
22	0	1	=$D$17*C22	=$D$18*D22
23	=1+B22	1	=$D$17*C23	=$D$18*D23
24	=1+B23	1	=$D$17*C24	=$D$18*D24
25	=1+B24	1	=$D$17*C25	=$D$18*D25
26	=1+B25	1	=$D$17*C26	=$D$18*D26

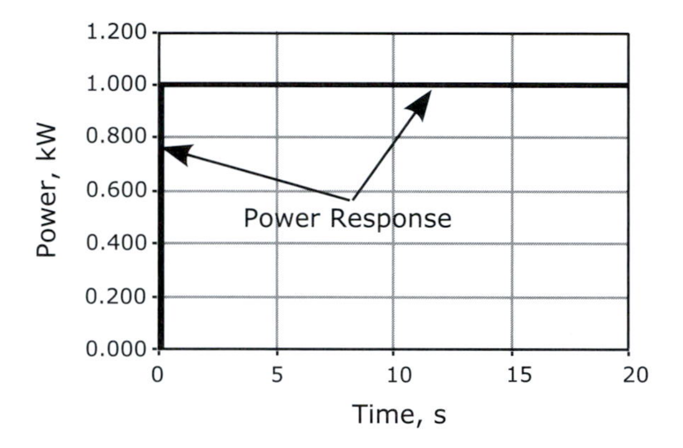

8. The conceptual diagram and block diagram for open loop control of a 1000. W microwave oven are shown below. (The "magnetron" is the device that delivers the microwaves into the oven's cavity.) Assume the control is turned on to value $C = 1$ at time $t = 0$ and left on thereafter. Prepare a spreadsheet and graph showing the power delivered to the food each second for a total of 6 seconds.

A:

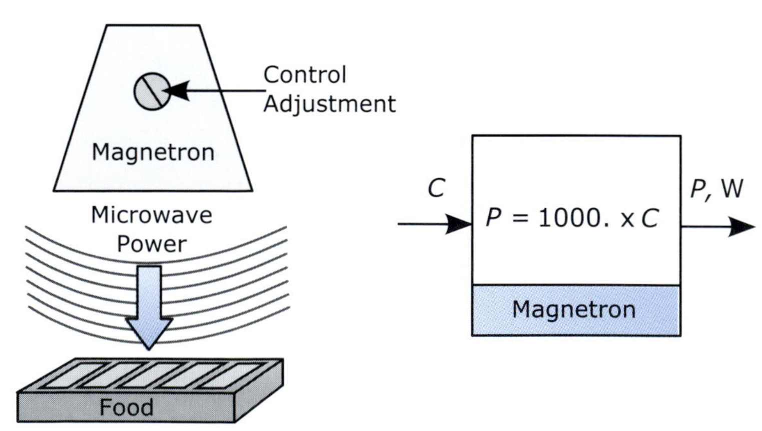

9. Suppose a simplified thermostatic room controller operates a furnace that can deliver heat to a room at the rate of 1000. watts. The initial temperature of the room is $T_R = 20.0°C$ and the outside temperature is $T_0 = 10.0°C$.

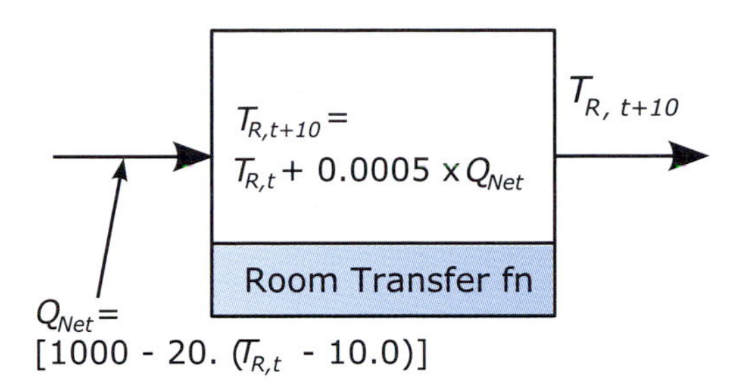

For this example assume the room will lose some heat over a relatively long period when you are tracking the temperature. To simplify, assume the furnace delivers the net heat being the difference between what the furnace produces and the heat lost from the room, $Q_{Net} = Q - Q_L$.
Draw a single block of a block diagram modeling the "world" of the heating system under the assumptions that the room gains heat at the rate of $Q_{Net} = Q - Q_L = (1000. - Q_L)$ where $Q_L = 20. \times (T_R - T_o)$ watts, and that the temperature of the room at time $t + 10$ minutes is the temperature at time t minutes plus $0.0005 \times Q_{Net}$. Give the room temperature at intervals of 10 minutes up to 100 minutes. (**Partial A:** at 100. minutes, $T_R = \mathbf{23.8°C}$.)

10. The conceptual sketch below shows a model for a closed loop control of a lawnmower engine. Unlike the lawn mower motor of Exercise 7, there is an additional feedback device. It takes advantage of a cooling fan that is directly connected to the output shaft of the lawn mower. This fan blows on a flap indicated in the diagram, which is preset to a desired position and that is kept in that position against a spring tension if the fan is rotating at a desired speed. The desired speed is set by initializing the spring with a tension S_{SP} corresponding to a desired operating condition such as 3,000 RPM.

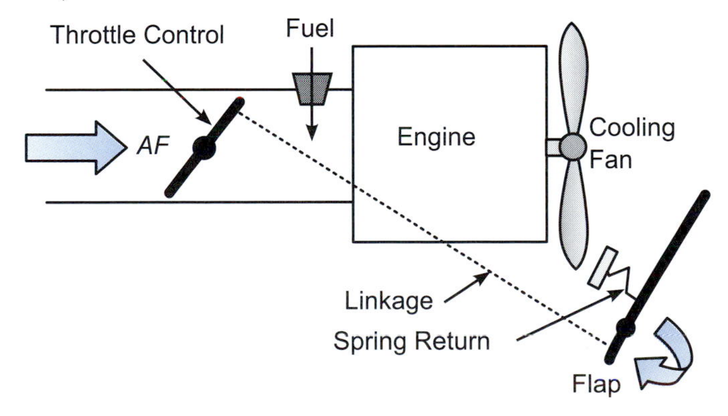

If the fan rotates too fast (i.e., the engine speed is too high), the flap is blown clockwise causing a mechanical linkage to close the throttle and reduce engine speed. If the speed is too low, the flap is pulled counterclockwise by the spring.
Draw a block diagram for the this control system, assuming that (1) the control unit sends an error signal in terms of the spring tension $\{S_{SP}$ (set point)$-S\}$ to meter a proportional incremental amount of air/fuel $K_p(S_{SP}-S)$ kg/s, (2) the incremental power is $\Delta P = K_2 \times \Delta AF$, and (3) the response function for the speed of the motor is:

$$N_{t+1} = N_t + \frac{a(P_{SP} + \Delta P)}{N_t} - bN_t^2$$

in which P_{SP} is the nominal engine power corresponding to the nominal set point. Finally the controller converts engine speed to spring tension by the relationship $S = K_3N$.

11. Suppose that for the open loop toaster control you sketched in exercise 4, the variable that you really want to control is d (for darkness), where d is the energy absorbed by the toast, in units of joules. Assume the initial energy of the toast (in units of joules) is $d = 0$, and $d_{t+1} = d_t + K \times P \times t$, where P is the power in watts delivered by the heating coil at time t and K is the proportional gain relating d to P.

   (a) Fill in the world block to correspond to this model.
   (b) If the line voltage is 115 V and the heater resistor is 13.2 Ω, and the darkness gain is 2.5, plot d in joules for 10. s.

   (Partial A: At 10. seconds, darkness function for toast = $\mathbf{2.5 \times 10^4}$ **J.)**

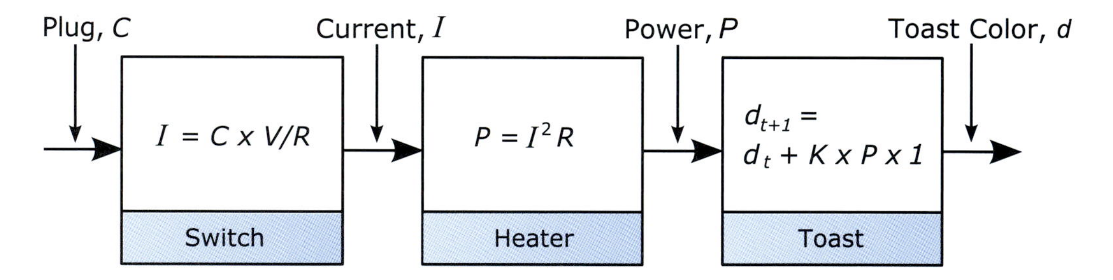

12. The conceptual sketch below shows a model of a *closed* loop control for heat in a room. Prepare a block diagram for this closed loop control. Label each block on the block diagram with the name of the corresponding component or components on the sketch. To simplify, assume the furnace delivers the net heat being the difference between what the furnace produces and the heat lost from the room, $Q_{Net} = Q - Q_L$.

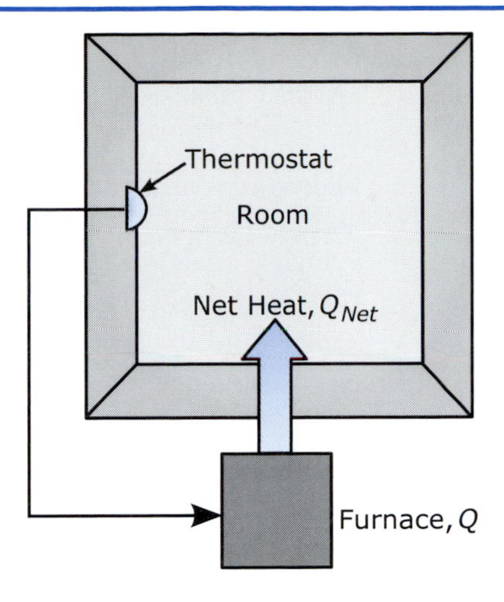

The thermostat operates by comparing T_{SP} (the thermostat's set point to T_M (the measured room temperature) and sending that signal to the furnace. The furnace is "on" if that difference is positive, and vice-versa.

A:

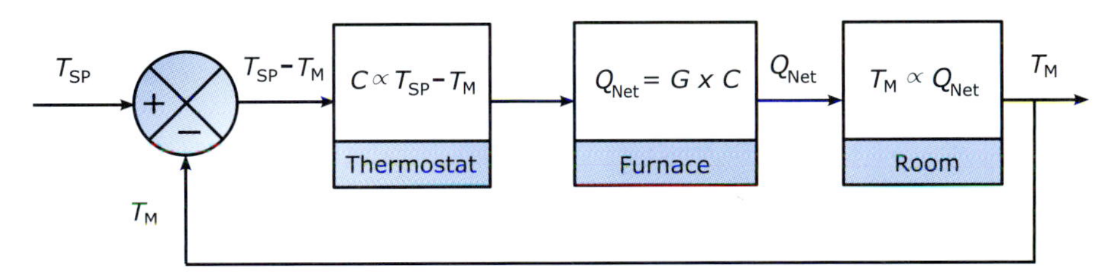

13. Use the block diagram developed in exercise 12. Assume the room is initially at $T_0 = 15°C$ and the set point for the furnace is 22°C. The furnace heat rate is $Q = 1000.$ W and the room losses are $Q_L = 20. \times (T_M - T_0)$.

    The response function for the room is $0.003 \times Q_{Net}$°C per minute when the furnace is on and $0.003 \times Q_L$°C per minute when it is off.

    Plot the temperature response of the room for 10 minutes. (**Hint**: Use "IF" statement, IF (test, value if true, value if false) to indicate when the furnace is on and when it is off.)

14. Suppose a car employs a digital speed sensor as shown in Figure 10, and the car's wheels are each 0.60 meters in diameter. If the speed sensor consists of the magnet and coil and the magnet induces one pulse in the coil for every revolution of the axle, how many pulses per second will the

coil send to the pulse counter if the car is traveling at 65 miles per hour? (**A:** 15.4 pulses/s if there is one pulse per revolution.)

15. Do you really need a cruise control? Suppose you are driving in the Midwest on a flat road. You are bored with the monotony of the terrain and you set a brick against the accelerator pedal that maintains the vehicle's speed to 60. mph. Unexpectedly, you drive into a construction road that is 10. m deep. Ignoring friction and wind losses, will your vehicle still be at 60. mph at the bottom of the construction road, and, if not, what will its speed S be? (**A:** 68 mph.)

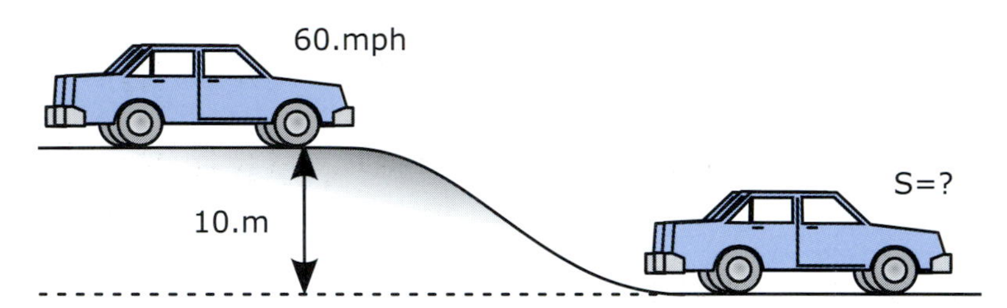

16. In this chapter, we have used equation 1 for the increase in the vehicle's speed when the engine power is *P(engine)* and the *constant* wind losses are *P(losses)*. Starting with Newton's Law of Motion, derive $S_{t+1} \cong \sqrt{S_t^2 + \frac{2(\Delta t)(P(engine)-P(losses))}{m}}$. *(You will need elementary calculus to solve this problem.)*

17. Referring to Figure 11, when a throttle plate rotates in a circular plenum, its effective blocking area is πab where a, b are the principal axes of an ellipse. Show that $\frac{\Delta A}{A} = 1 - \cos\theta$, where A is the fully open area of the pipe and ΔA is the open area when the plate has been turned through an angle θ.

18. You work for the control and guidance systems division of a major aircraft manufacturer with a major government contract. You observe employees who regularly leave work early while being paid for time not worked. What do you do?

    (a) Ignore it, since there is nothing you can do anyway.
    (b) Talk to your supervisor about it.
    (c) Report it to the government representative on the contract.
    (d) Blow the whistle and talk to a newspaper reporter.

19. You are a new engineer working for a motorcycle manufacturer that produces a bike with a known control system instability. This instability can cause the rider to lose control at high speed and crash. Your supervisor says that you should ignore it because everyone knows about it and it would be too expensive to fix. Besides, a new control algorithm

would cause more harm, since the drivers expect the bikes to behave in a certain way. What do you do?

(a) Attempt to convince your supervisor that it will be cheaper to fix the flaw than pay the subsequent law suits.
(b) Suggest that a warning label be put on the bikes about riding at high speeds.
(c) Talk to your corporation's legal office about your professional obligations.
(d) Blow the whistle and talk to a newspaper reporter.

Chapter 10

Kinematics and Traffic Flow

Introduction

How long should the on-ramp to a superhighway be? How many cars can that superhighway deliver to a city at rush hour? These are some of the sorts of problems faced by civil engineers. This chapter introduces some ways civil engineers solve traffic problems. Much of this chapter is also about how all engineers understand the relationships among the important variables of distance: speed, acceleration, and time. These relationships are collectively called *kinematics* when they do not also involve the forces on objects and the inertia of the objects.

Civil engineers are concerned with **distance, time, speed,** and **acceleration.** These are **kinematic** variables and are basic variables that are used to understand traffic flow. In the simplest cases they are related to each other by geometric methods—in more complicated situations by the methods of the calculus. These methods can be directly applied by civil engineers to design on- and off-ramps for highways and for other highway applications.

The **Highway Capacity Diagram** is a visual way to assess the capacity of a highway. It relates three major factors in traffic flow analysis: **capacity** (cars per hour), **car speed** (miles per hour, mph), and traffic **density** (cars per mile).

Of course, most civil engineers must also solve problems other than road building. They design tunnels that can be drilled into mountains or under the sea from points miles apart and have the tunnels meet exactly in the middle. They also build beautiful and useful bridges, skyscrapers, and dams.

Civil engineering is one of the oldest engineering disciplines.[1] The number of millennia you date it back depends on how you define *civil engineering*. You could say that the first civil engineers built pyramids in Egypt nearly 3,000 years ago. Or you could date civil engineering to the Ziggurat builders of ancient Iraq, or to the Chinese designers of the Great Wall, or to the Romans, who 2,000 years ago built superb roads, aqueducts, and stadiums using concrete.

The problems addressed in this chapter deal with only a small part of the civil engineer's challenges. But that small part, road design, is important to all of us, whether we are on our way to work, on a family outing, or rushing to the hospital emergency room.

Distance, Speed, and Acceleration

Determining how long an on-ramp should be on a highway requires understanding what an on-ramp is required to do. That requirement might be expressed as follows: An on-ramp should be long enough to enable a driver to speed up gradually and safely from the off-highway speed (probably about 15–30 mph) to the highway speed (probably above 60 mph). As always, the engineers translate this qualitative statement into variables, numbers, and units.

"Long enough" becomes the variable distance. It is measured in units of miles or meters, or whatever scales are appropriate to the problem at hand. As a mental exercise, at this point you might want to estimate in meters, based on your memory, how long you think a typical highway on-ramp is. (Do it before you read the next sentence.)

In this chapter, only motion in a single direction (often called in technoese **"one-dimensional"**) will be considered. For convenience, think of positive distance as from left to right, and negative distance as from right to left, just as in Cartesian geometry.

Speed is a variable commonly measured in miles per hour (mph) in the United States and in kilometers per hour "kph" in most of the rest of the world. But for engineering design, the SI units of m/s are often more useful (65. mph ~ 105. kph ~ 29. m/s). Speed is calculated by measuring the distance traveled and the time spent and dividing distance by time. In mathematical terms, speed $= \Delta x/\Delta t$, with the uppercase Greek delta Δ symbol meaning "difference." In Cartesian geometry, this is the average

[1]The term *civil* preceding *engineer* reminds you how civil engineers were originally distinguished from the other engineers of their day—*military* engineers.

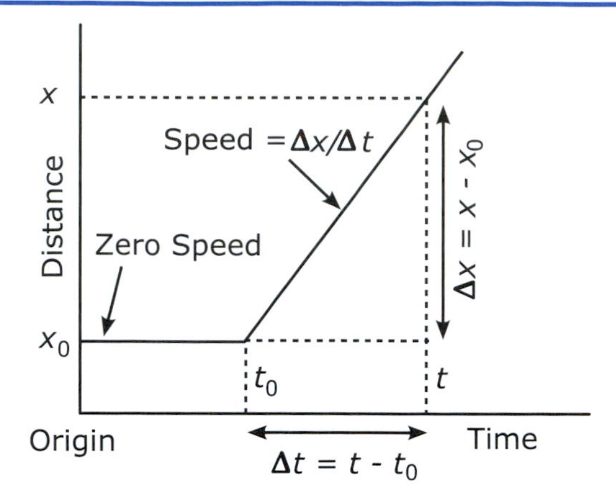

Figure 1

Constant Speed

slope of the line $x(t)$. In this case speed means the final position less the initial position divided by the final time less the initial time (Figure 1).

Again, only one direction will be considered. We will use the symbol v (as in velocity) for speed. (If we used the symbol s, it might be confused for the symbol s for "seconds".) There is an important differentiation between the concepts of *speed* and of *velocity*. Speed is the *magnitude* of velocity. Velocity may have more than one component, which is so much in the "y" direction (as in Cartesian geometry) and so much in the "x" direction, and so on. Of course, in our one-dimensional analyses, speed and velocity are functionally synonymous. For acceleration we use a, except for that due to gravity, when we use g.

Adding a word to "speed" gets us to the phrase "speed **up.**" We have already met the speed up variable, *acceleration*. It is defined as change of speed per unit time or, more mathematically precisely, as the rate of change of speed, and is measured in m/s^2 or ft/s^2. Please resist the temptation to express acceleration in such ugly and ultimately less useful hybrid forms as "miles per hour per second."

Finally, there is a fourth variable. It is mentioned only implicitly in the preceding description, but it is so important that, without formally introducing it, we have already used it. This is the variable *time*, measured in seconds in computation but, in the context of highway analysis, often in hours as in miles per hour, and familiarly contracted as "mph."

Strictly speaking, two of the preceding variables should be defined in two ways: in terms of *instantaneous* speed and acceleration and in terms of *average* speed and acceleration. These are important distinctions, as shown in Figure 2. An instantaneous variable, here the acceleration, is changing with the change in slope. Your physics courses will enable you to understand how to deal with nonconstant accelerations, but in this introductory treatment we can, and will, get along without nonconstant accelerations.

In an earlier chapter you learned the simplest principles of **kinetics,** the study of motion caused by impressed forces. This chapter is concerned

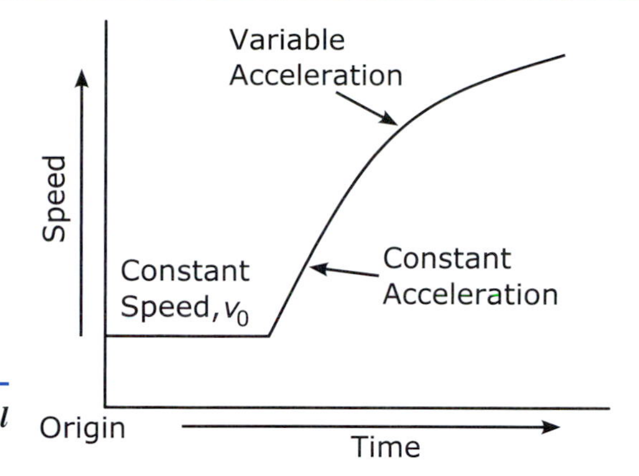

Figure 2

Acceleration is the *Local* Slope

with the closely related subject of **kinematics,** the relationship between distance, speed, acceleration, and time, without asking the questions of how can we achieve particular values of these variables. (Although, of course, one can calculate the associated momentum changes, forces, and energy given information on the mass or inertia of the bodies being accelerated and so forth.)

The only one of the four concepts of distance, speed, acceleration, and time that gives the typical student any difficulty is acceleration. So before proceeding further, please follow and solve this example.

EXAMPLE 1

A car enters an on-ramp traveling 15 miles per hour. It accelerates for 15 seconds. At the end of that time interval it is traveling at 60. miles per hour. What is its average acceleration? How does that compare to g, the acceleration due to gravity?

Need: Acceleration = ______ m/s^2.

Know–How: Acceleration = (change in speed)/time.
Initial speed = 15 miles per hour = 6.7 m/s.
Final speed = 60. miles per hour = 26.8 m/s.

Solve: Average acceleration $= \frac{(26.8 - 6.7)}{(15 - 0)} = \text{W} = 1.34 = 1.3\ m/s^2$ (in terms of a fraction of g, the acceleration due to gravity, this is $1.3/9.8 = 0.13$, so you know this is a mild acceleration). ■

The Speed Versus Time Diagram

One could deal with problems of distance, speed, acceleration, and time using words and equations, but there is a much more intuitive tool. It makes possible insightful description of motion problems. It frequently enables an engineer to solve the problem merely by inspecting the diagram.

Formal calculation may be needed only to achieve the correct number of significant figures.

This versatile and essential tool is the speed-versus-time diagram (Figure 2). For brevity, we will call our diagram the $v-t$ diagram. The $v-t$ diagram is just what it says: The horizontal axis represents time, with zero being the instant the situation being considered began. For each instant of time, the speed (or one-dimensional velocity) is plotted in the vertical direction.

EXAMPLE 2

Plot the $v-t$ diagram for Example 1.

Need: $v-t$ diagram describing the situation "a car enters an on-ramp traveling 15 miles per hour. It accelerates constantly for 15 seconds. At the end of that time interval it is traveling at 60. miles per hour."

Know–How: First, convert all speeds to SI. Then plot speed for $t=0$ (when $v=15$ mph $=6.7$ m/s) and $t=15$ s (when $v=60.$ mph $=26.8$ m/s). Join the two points with a *straight line*. (What tells us the line is straight is the phrase "constant acceleration.")

Solve:

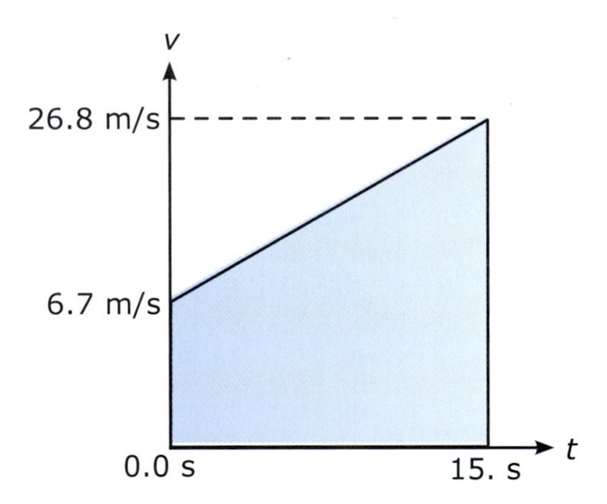

Why is this particular graph so important? There are two reasons.

1. The slope of the $v-t$ graph measures acceleration.
2. The shaded area under the $v-t$ graph measures distance traveled. ■

No other graph involving the four variables of interest summarizes so much information in so intuitive a manner. (If the concepts "slope" and "area" aren't intuitive to you, make them so. Review your high school geometry, if necessary!)

You should recall *slope* being defined as the quotient of a vertical "rise" divided by a horizontal "run." For the standard Cartesian x, y diagram, the slope is $\Delta y/\Delta x$. That the slope of the $v-t$ graph gives you acceleration is evident from considering dimensions. The "rise" of the graph has units of speed, m/s. The "run" of the graph has units of time, s. So slope = rise/run = $\Delta v/\Delta t$ in units of [m/s]/[s] or m/s^2, valid dimensions of acceleration.

The second statement relating area under the curve to distance traveled is less obvious. In Example 2, the *average* speed over the 15-second period of acceleration is $0.5 \times (26.8 + 6.7) = 16.8$ m/s, and so the vehicle will have covered $16.8 \times 15. = 250.$ m. Notice the shaded area is a trapezoid whose area is 0.5 × (sum of parallel sides) × perpendicular distance between them. It is identical to the calculation of distance covered. It also can be approached dimensionally. The "height" or ordinate has units of [m/s]. The horizontal axis has units of seconds. So length × height has units of [s × (m/s)] = [m], a unit of distance.

Those of you familiar with elementary calculus will recognize that saying "the area under the $v - t$ graph measures distance" is the same as saying "the integral of speed over time is distance." Saying that "the slope of the $v - t$ graph measures acceleration" is the same as saying "the derivative of speed is acceleration." These statements are true even when the lines are curved—that is, for the case of nonconstant acceleration. In the general case, calculus is required to evaluate the slopes or areas under curves.

We will, however, get along without calculus. Solving Examples 1 and 2 using the $v - t$ diagram has done more than enable us to visualize that problem. It has provided a bonus. It has already answered a second question.

EXAMPLE 3

What is the distance traveled in steadily accelerating from 15 mph to 60. mph in 15 sec.?

Need: Distance = ________ m.

Know–How: The distance is the area (shaded in the diagram for Example 2) beneath the $v - t$ graph. Geometry tells us we can break the area of a trapozoid into a rectangle with length 15 s and height 6.7 m/s and a triangle with base 15 s and altitude (26.8 – 6.7) s.

Solve: Distance = area = area of rectangle + area of triangle

$$= 6.7\,[\text{m/s}] \times 15.\,[\text{s}] + (1/2) \times 15.\,[\text{s}] \times (26.8 - 6.7)\,[\text{m/s}]$$

$$= 100.5\,\text{m} + 150.75\,\text{m} = 251.2\,\text{m}.$$

∴ **Distance = 250 m.** ■

You may already know some formulae you could have used to solve this problem ($x = v_0 t + \frac{1}{2}at^2$, for example), but it is recommended that you use the $v - t$ diagram on the problems and exercises of this chapter.

Use of this tool develops a visual and intuitive appreciation of motion problems that cannot be obtained merely by manipulating equations. Formulae are useful once you fully understand the process, but the $v-t$ diagram is the best way to achieve that understanding.

Applying the Tool to the On-Ramp Problem

Let's now apply the tool to our original problem. Recall its verbal statement: "An on-ramp should be long enough to enable the driver to speed up gradually and safely from the off-highway speed (probably about 15–30 mph) to the highway speed (probably above 60 mph)."

A few words remain untranslated into numbers. In particular, consider the words "gradually" and "safely." What constitutes a gradual and safe acceleration? This is a matter for judgment, for experience, for experiment, but not for definition. Judgment and experience can begin from a well-known acceleration that we have earlier considered. This is the acceleration due to gravity. This acceleration is often called "$1\,g$" or colloquially "1 gee" of acceleration (where $g = 9.8\,\text{m/s}^2$ and is not to be confused with the symbol g for "gram.")

Is $1\,g$ too high, too low, or just right for an automobile accelerating along an on-ramp? Experience reveals that a $1\,g$ acceleration is appropriate for amusement park rides, but not for on-ramps. A driver accelerating onto an on-ramp should accelerate with only a fraction of a g.

What fraction? This presents another problem of engineering optimization. If consideration of the value of the driver's time, the cost of real estate, and the cost of concrete did not matter, the fraction could be very small. The driver could accelerate at a leisurely and safe rate along an on-ramp that stretched for many kilometers!

Such considerations do matter, however. These considerations suggest a short on-ramp and high acceleration.[2] Safety suggests a long on-ramp and correspondingly low acceleration. The engineer's job is to compromise between the two. In order to do so, the engineer needs a tool that relates length of the on-ramp to acceleration, and this is exactly what the $v-t$ diagram provides.

EXAMPLE 4

Provide a table and graph of the length of an on-ramp as a function of acceleration for accelerations ranging from $1.0\,\text{m/s}^2$ (about 0.1 g) to $10.\,\text{m/s}^2$ (about $1\,g$) in increments of $1.0\,\text{m/s}^2$. Assume that the vehicle enters the on-ramp with a speed of 15. mph (6.7 m/s) and leaves the on-ramp at a speed of 60. mph (26.8 m/s).

[2]Of course, high acceleration wastes additional fuel, and, anyway, not all vehicles are capable of high acceleration.

Need: A table and a graph with entries length of on-ramp d in m vs. a m/s^2.

Know–How: The $v-t$ diagram gives the relationships among a, d, and t. Let τ be the (unknown) time you spend on the on-ramp accelerating to highway speed.

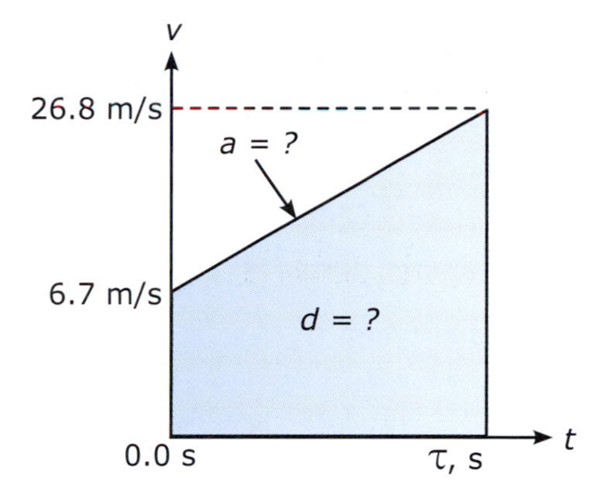

From the definition of acceleration as the slope of the $v-t$ line,

$$a = \frac{(26.8 - 6.7)}{\tau} \quad \text{or} \quad \tau = \frac{(26.8 - 6.7)}{a}[m/s][s^2/m] = \frac{20.1}{a} \; [s]$$

d is the area under the $v-t$ curve

$$d = 6.7\tau + 1/2 \times (26.8 - 6.7)\tau \;\; [m/s][s] = 6.7\tau + 10.05\,\tau \; [m]$$

Solve: Now substitute for $\tau = 20.1/a$.
$\therefore\; d = (135/a) + 202/a = 337/a$ in meters.

Take these relationships to a spreadsheet and prepare the table, and from the table, prepare graphs of a vs. τ and d vs. τ:

a, m/s^2	d, m	τ, s
1.00	337	20.10
2.00	169	10.05
3.00	112	6.70
4.00	84	5.03
5.00	67	4.02
6.00	56	3.35
7.00	48	2.87
8.00	42	2.51
9.00	37	2.23
10.00	34	2.01

■

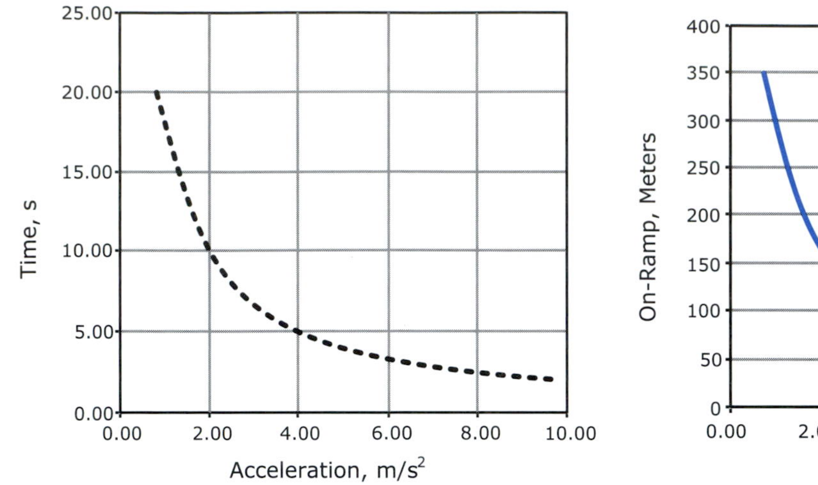

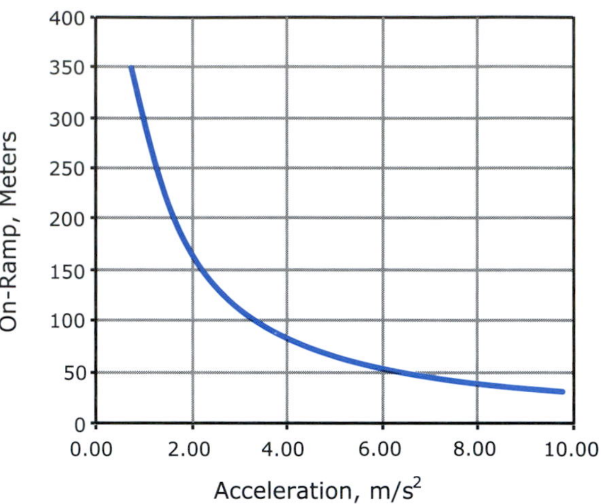

This table quantifies the trade-off between acceleration and on-ramp length. That trade-off is then combined with the considerations mentioned earlier. Those considerations, to repeat, include the value of the driver's time, the cost of real estate, and the cost of concrete. Such considerations give which point on the d vs. a curve the engineer will actually choose. Experience shows that the trade-off results in a point somewhere around $a = 3.\ \text{m/s}^2$ (about one-third of a g) for cars, and the on-ramp is then $d = 110\ \text{m}$. In somewhat more automotive terms, with $a = 3\ \text{m/s}^2$, the time for a standing quarter (of a mile) is a relatively leisurely 16 s and time to reach 60 mph is 9 s. Obviously, many cars can do much faster than that—which is another trade-off, this one between gasoline consumption and fuel economy.

If the appropriate acceleration for a heavy truck is one-half that for most cars at $1.5\ \text{m/s}^2$, the on-ramp has to be 250 m. This value of about 250 m is probably pretty near the one you guessed from personal experience when you began this chapter. But now you know the right answer for a better reason.

The right answers to a number of other civil engineering problems are also contained in the $v - t$ diagram. These range from the design of intersections, stoplights, and traffic lights, and even to the design of a cannon that might be used to shoot astronauts to the moon. These subjects are explored in the exercises.

General Equations of Kinematics

The key variables are:

t, the time, typically measured in seconds
x, the distance traveled typically measured in meters

Table 1

Basic Kinematic Equations

Constant	**General Kinematic Equations**	**Simplified Forms**
Velocity, $v = v_0$	$x = x_0 + v_0 (t - t_0)$	$x = v_0 t$ if $x_0 = 0$ at $t_0 = 0$
Acceleration, a	$x = x_0 + v_0(t - t_0) + 0.5\, a(t - t_0)^2$	If at $t_0 = 0$, $x_0 = v_0 = 0$, $x = 0.5\, at^2$
Acceleration, a	$v^2 = v_0^2 + 2a(x - x_0)$	If x_0, $v_0 = 0$, $v^2 = 2ax$

v, the velocity (or speed in one dimension), typically measured in m/s
a, the acceleration, typically measured in m/s^2

If you have taken high school physics, you probably will recognize the principal kinematic equations for simplified cases. They are summarized in Table 1.

More general relationships for nonconstant a can be derived using calculus. In any case, we will not be deriving them here—all cases, constant a or not, will be covered in your first-year physics classes.

The Highway Capacity Diagram

How many cars can a superhighway deliver to a city at rush hour? Consider a single lane of such a highway. Imagine, at rush hour, you stood beside the highway and counted the number of cars in any one lane that passed you in the space of one hour (assuming you have sufficient patience). What sort of answer do you think you might get? (Again, before reading on, try it.)

It's unlikely that your answer was less than 100 cars per hour, or more than 10,000 cars per hour. But how can one narrow down further and justify an answer? As in the previous parts of this chapter, the goal is not a single answer but the basis for a trade-off between the key variables. The key (and closely interrelated) variables in the case of highway capacity are:

Capacity (cars per hour). The number of cars that pass a certain point during an hour.

Car speed, (miles per hour, mph). In our simple model we will assume all cars are traveling at the same speed.

Density (cars per mile). Suppose you took a snapshot of the highway from a helicopter and had previously marked two lines on the highway a mile apart. The number of cars you would count between the lines is the number of cars per mile.

You can easily write down the interrelationship among these variables by using dimensionally consistent units:

$$\text{Capacity} = \text{Speed} \times \text{Density} \tag{1}$$

$$[\text{cars/hour}] = [\text{miles/hour}] \times [\text{cars/mile}]$$

For simplicity, we will consider these variables for the case of only a single lane of highway. Once again, our initial goal is a tool that provides insight into the problem, and helps us solve it intuitively and visually. As a first step toward that tool, solve Example 5.

EXAMPLE 5

Three of you decide to measure the capacity of a certain highway. Suppose you are flying in a helicopter, take the snapshot mentioned above, and count that there are 160 cars per mile. Suppose further that your first partner determines that the cars are crawling along at only 2.5 miles per hour. Suppose your second partner is standing by the road with a watch counting cars as described earlier. For one lane of traffic, how many cars per hour will your second partner count?

Need: Capacity = ______ cars/hour.

Know–How: You already know a relationship between mph and cars per mile (equation 1).

Capacity = Speed × Density
[cars/hour] = [miles/hr] × [cars/mile]

Solve: Capacity = [2.5 miles/hr] × [160 cars/mile] = **400.cars/hour.** ■

Let's now use this result to plot a new graph. It's one of two related graphs, each of which can be called a **highway capacity diagram** (Figure 3). It is made by plotting car density on the vertical axis and speed, in mph, on the horizontal axis. In this case, it gives us the highway capacity.

How might Figure 3 be made even more useful? If we knew a *rule*[3] relating car density and mph, we could plot a whole curve instead of just one point. This would make the highway capacity diagram a lot more useful.

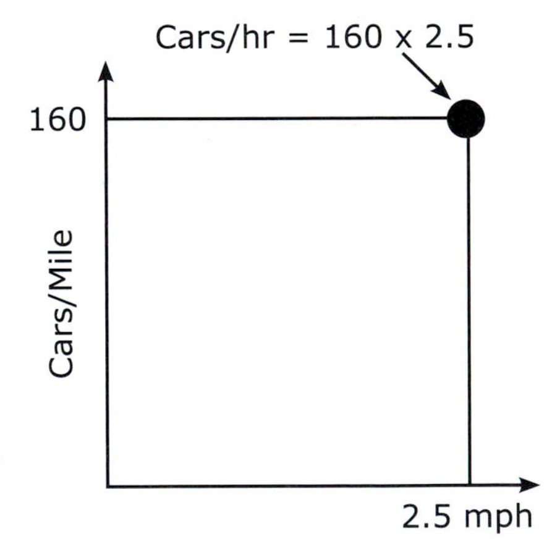

Figure 3

Highway Capacity Diagram

[3]A *rule* is something guided by experience as being useful for a given objective but not based on one of the hard-and-fast laws of nature.

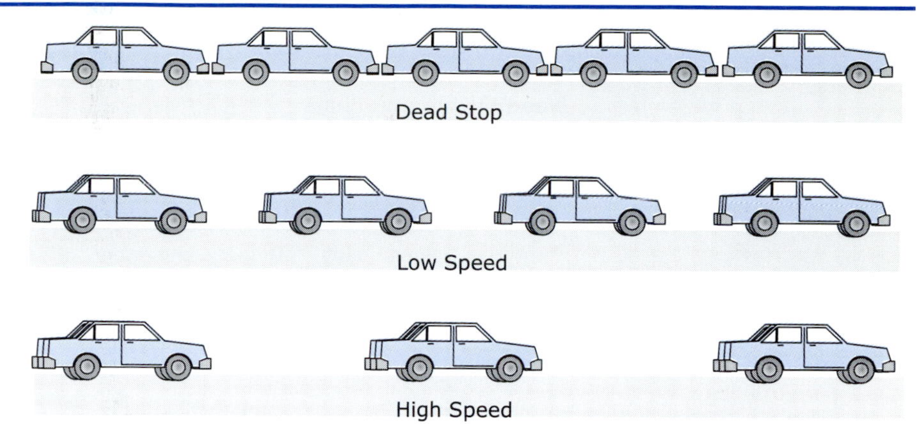

Figure 4

Follow Rule in Terms of Car Length Separation

You possibly already know such a rule either explicitly or implicitly. Let's consider two such rules. One is traditional but no longer recommended. Its implications are, however, worth exploring as a first pass at highway design. That rule is called the *follow rule*. You can see it at work visually in Figure 4.

In words, one version of this rule states that when following another car, leave one car length between vehicles for every ten miles per hour. Write the follow rule as equation 2:

$$\text{Number of car lengths between cars} = [\text{speed in mph}]/[10\,\text{mph}] \qquad (2)$$

The effect of this rule is graphically illustrated in Figure 4. The separation distance per car is more as the car speed increases. At a dead stop the cars are bumper to bumper. Thus, the car density falls with increasing speed.

In order to use this rule to calculate a highway capacity in cars/hour, it needs to be put into terms compatible with equation 1. The independent variable (subject to the constraints of common sense and the law) is the vehicle's speed. Thus, we need to use the follow rule to relate to the unknown car density. We know the length of a car—say, about 4.0 m on the average. If we do the arithmetic, this means we could pack about 400. vehicles/mile bumper-to-bumper. The follow rule relates the distance between these vehicles to their speed. Each car effectively occupies its own footprint plus the separation distance between them. The car density is thus:

$$\text{Density} = \frac{1}{\left(\dfrac{\#\,\text{car lengths} + 1}{400}\right)} = \frac{400}{\left(1 + \dfrac{\text{speed, [mph]}}{10\,[\text{mph}]}\right)}[\text{cars/mile}]$$

Hence, applying equation 1,

$$\text{Capacity} = \frac{400 \times \text{speed, mph}}{\left(1 + \dfrac{\text{speed, [mph]}}{10\ [\text{mph}]}\right)}[\text{cars/hr}]$$

EXAMPLE 6 Using the follow rule, draw the full highway capacity diagram.

Need: Table and chart.

Know–How: Put the follow rule into a spreadsheet and graph.

Solve:

Speed, mph	Density, cars/mile	Capacity, cars/hr
0	400	0
10	200	2,000
20	133	2,667
30	100	3,000
40	80	3,200
50	67	3,333
60	57	3,429
70	50	3,490
80	44	3,556
90	40	3,600
100	36	3,636

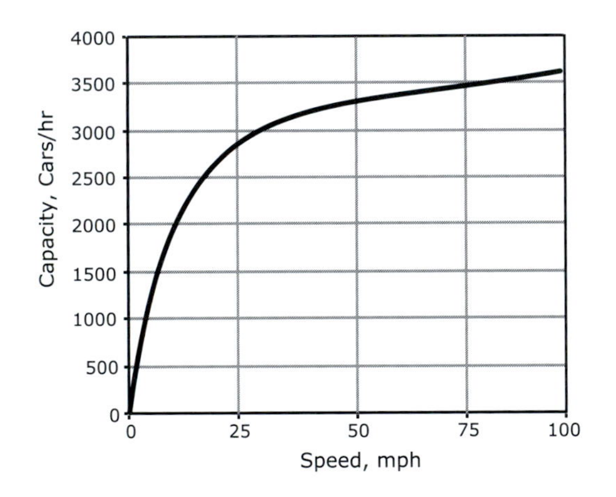

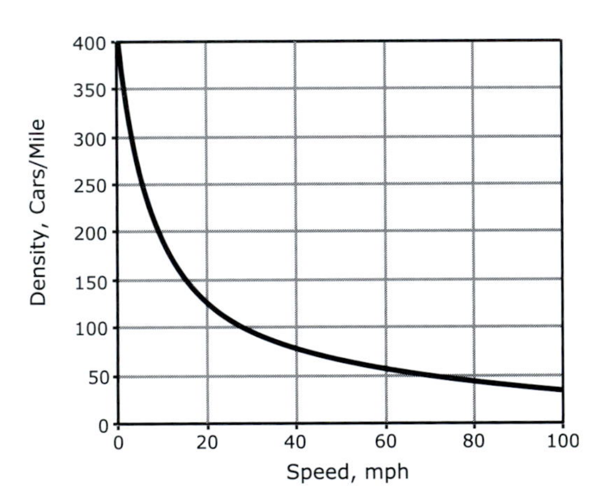

Either of these two curves could be appropriately called a "highway capacity diagram." Notice what happens to the capacity as the car gains speed: It begins to level off and approach a constant value of 4,000 cars/hr. There is a gain of only 300 cars/hour as we increase the speed from 50 to 100 mph.

The equation for capacity developed above shows why this is occurring: The denominator of the equation is dominated by the spacing between vehicles and not by the footprint of the car. The former was mathematically expressed as speed in mph/10 mph » 1 so that at high speeds,

$$\text{Capacity} \rightarrow \frac{400 \times 10}{\text{speed, [mph]}} \times \text{speed [mph]} = 4{,}000 \text{ [cars/hr]}$$

Notice too that you can always deduce the third member of the triumvirate of speed, density, and capacity by manipulating the other two from equation 1. For example, multiplying the density by the speed gives the capacity at every point (as per Figure 3), or you can get the density by dividing the capacity by the speed.

An up-to-date version of the follow rule is the *two second rule.*[4] It is practiced as follows. First, mark the point on the side of the road that the car in front of you is passing right now. Next, count the number of seconds[5] that it takes you to reach that point. If you count less than two seconds, you are following too closely.

The two-second rule can be expressed mathematically by making the distance between cars a function of speed.

Miles/car $= v \times t$ [miles/hr][s][hr/s] $= v$ [mph] $\times$ 2.0/3600.
or, inverting this, the car density, [cars/mile] $= 1800/v$.

Equally, speed [mph] = 1800/density [cars/mile].

Hence, using equation 1, capacity = speed $\times$ density, or dimensionally as [cars/hour] = [miles/hr] $\times$ [cars/mile] = 1,800—a fixed capacity independent of the car speed!

EXAMPLE 7 Using the two-second rule, draw the full highway capacity diagram.

Need: Table and chart.

Know–How: Put the two-second rule into a spreadsheet and graph.

Solve:

Speed, mph	Density, cars/mile	Capacity, cars/hr
10	180	1,800
20	90	1,800
30	60	1,800
40	45	1,800
50	36	1,800
60	30	1,800
70	26	1,800
80	23	1,800
90	20	1,800
100	18	1,800

[4]This is less arbitrary than it sounds; the average human reaction time is about 0.7 s, leaving 1.3 seconds to decelerate from 65 mph (29 m/s) to zero. The average deceleration rate is thus $(0. - 29.)/1.3 \approx -2\frac{1}{4}\ g$. This means you will feel as though you weigh 2¼ times your normal weight for 1.3 s—an unpleasant prospect!

[5]Count as "one, one thousand, two one thousand"—this will take about 2 seconds total to say in a normal speech pattern.

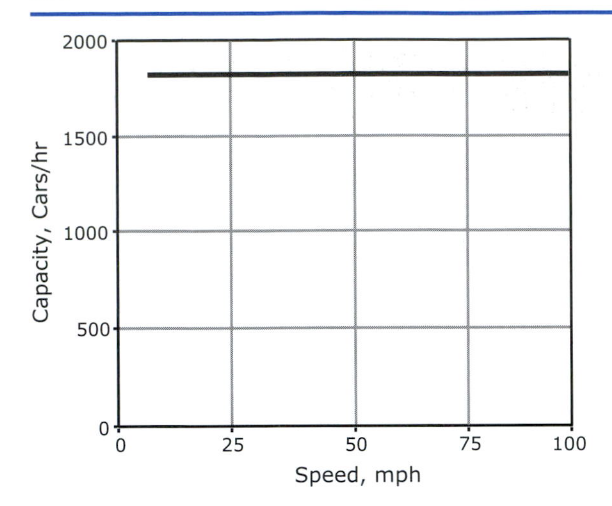

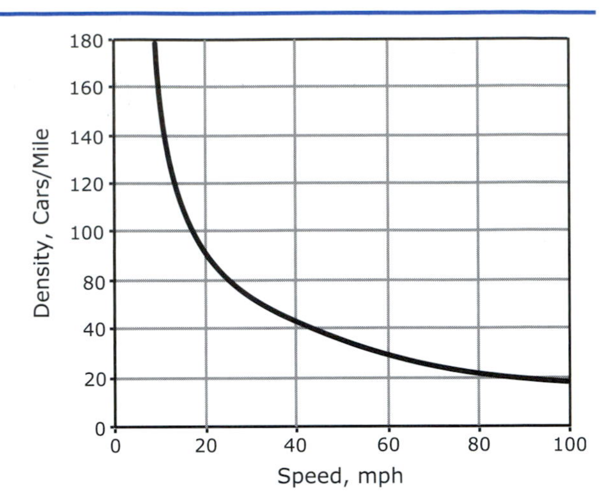

Notice that the two-second rule requires a greater distance between cars (as expressed by the calculated car density) than does the follow rule and, as just calculated, cuts the high speed highway capacity roughly in half! By the two-second rule, no matter how fast the cars are going, the number of cars that arrive at the destination each hour is the same! All that drivers accomplish by speeding up is to lengthen the distance between them and the surrounding cars (assuming they are all obeying either the follow rule or the two-second rule) and to lower the density of cars.

The implications are that for any superhighway, if the drivers observe the follow rule, the capacity of each lane for delivering cars is a little less than 4,000 cars per hour, or 1,800 cars per hour with the two-second rule, *no matter what the speed limit might be.* This insight enables civil engineers to solve practical problems such as the following.

EXAMPLE 8

You are a highway engineer hired by the New York State Thruway Authority. About 5,000 citizens of Saratoga, New York, work in the nearby state capital, Albany, New York. They all have to arrive in Albany during the hour between 8 A.M. and 9 A.M. These workers all travel one to a car, and all drivers obey the two-second rule. How many highway lanes are needed between Saratoga and Albany?

Need: Number of highway lanes = _____ lanes.

Know–How: We have (apparently) just discovered the universal law of highways: A highway lane can deliver 1,800 safe drivers per hour to their destination.
So, number of highway lanes = number of drivers/hr/(1,800 drivers/hr/lane).

Solve: Assume the number of drivers/hr is uniform at 5,000. Then the number of highway lanes = 5,000/1,800 = 3 lanes (to the proper number of significant figures!). ■

For actual civil engineering, life is much more complicated. Not all drivers obey the two-second rule (surprise!). Highways (such as Interstate 87 that links Saratoga and Albany) do not merely link two cities but many cities. (I-87 runs from the Canadian border to New York City.) Even within a small city, a highway typically has more than one on-ramp and off-ramp. While some highways run between cities, others are beltways that ring a city.

To meet these more realistic conditions, more complicated versions of the traffic rules, as well as many additional considerations, are needed. Civil engineers have devised criteria for rules called "level-of-service" from A (major interstate highways) to F (rutted, damaged, under repair, etc., where one cannot exceed 10 mph).

SUMMARY

Civil engineers have designed everything from pyramids to proposed space colonies. One important thing they have designed is roads. Roads designed by civil engineers made the Roman Empire possible.[6] Railroads designed by civil engineers made it possible for the United States to remain one nation, from the Atlantic to the Pacific oceans. Superhighways designed by civil engineers help make our modern automobile-based civilization possible. Perhaps, in the future, a radically different transportation system will make even safer and more convenient travel possible.

This chapter introduced one side of civil engineering. In fact, kinematics is necessary for several kinds of engineers as part of their kitbag of knowledge. It has presented kinematic relationships among the distance, speed, and acceleration of moving vehicles. It has presented a versatile tool, the speed versus time ($v - t$) diagram. It has shown how to apply that tool to a typical civil engineering problem, the design of on-ramps. It has also introduced a second new tool, the highway capacity diagram. Using that highway capacity diagram and knowledge about the characteristics of cars and drivers, such as embodied in the follow rule and the two-second rule, civil engineers can determine the carrying capacity of highways.

[6]Indeed, the remains of some can still be seen in Europe where some modern roads follow the same course originally laid by Roman engineers.

EXERCISES

It is strongly suggested you use the $v - t$ diagram for most of these exercises.

1. A car is alone at a red light. When the light turns green, it starts off at constant acceleration. After traveling 100. m, it is traveling at 15 m/s. What is its acceleration?

   Need: $a =$ ____ m/s^2.

   Know: $v_0 = 0$ and $v = 15\,\text{m/s}$ after $x = 100.$ m.

   How: a is slope of $v - t$ diagram; d is area.

   Solve: We need the time, τ, from the area under the acceleration curve.
   $d = ½ \times (15 - 0) \times \tau$ [m/s][s] = 100.
   $\therefore \tau = 2 \times 100./15 = 13.3\,\text{s}$

   $\therefore$ ***a*** $= \Delta v/\Delta t = (15 - 0)/(13.3 - 0)$ [m/s][1/s] = 1.13 = **1.1 m/s²**

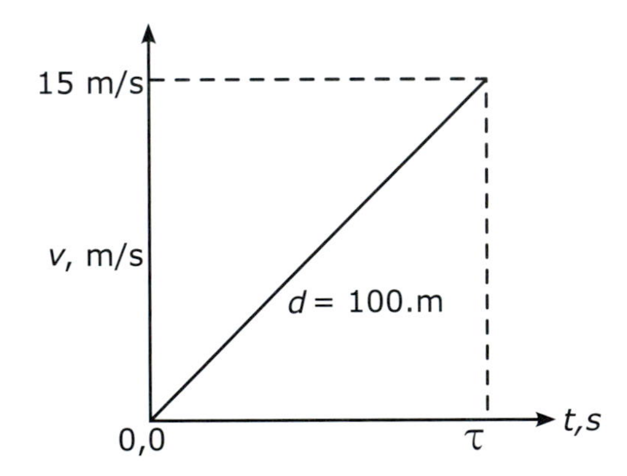

2. A car is alone at a red light. When the light turns green, it starts off at constant acceleration. After traveling exactly one-eighth of a mile, it is traveling at 30. mph. What is its acceleration in m/s^2?
3. An electric cart can accelerate from 0 to 60. mph in 15 sec. The world champion sprinter Tim Montgomery can run 100.0 m in 9.78 s. Who would win a 100.0 m race between this cart and the world champion sprinter? (**A: The sprinter wins by 15.9 m!**)
4. A car starts from a stop at a traffic light and accelerates at a rate of $4.0\,\text{m/s}^2$. Immediately on reaching a speed of 32 m/s, the driver sees that the next light ahead is red and instantly applies the brakes (reaction time = 0.00 s). The car decelerates at a constant rate and comes safely to a stop at the next light. The whole episode takes 15 seconds. How far does the car travel? (**Partial A:** See inset figure.)

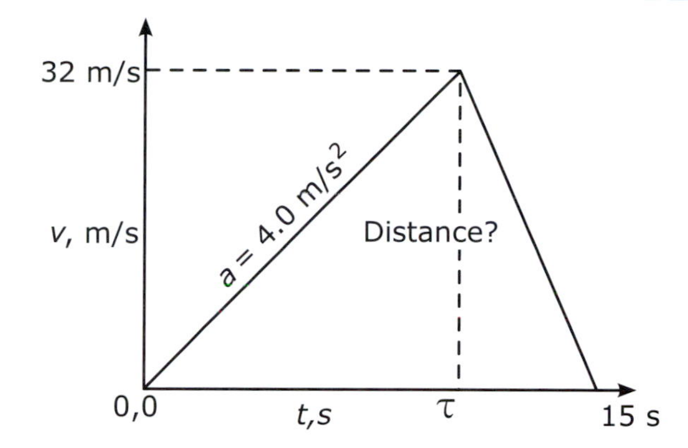

5. For the previous problem, supply a table and draw a graph showing the distance, velocity, and acceleration of the car versus time.
6. Based on your experience as a driver or a passenger in a car, estimate the maximum deceleration achieved by putting maximum pressure on the brakes when traveling at 30 mph.
7. A car leaves a parking space from a standing stop to travel to a fast-food restaurant 950 meters away. Along the journey it has to stop after 325 m at a stop sign. It has a maximum acceleration of 3.0 m/s ^{2} and a maximum deceleration of $-10.$ m/s^2. It never exceeds the legal speed limit of 15 m/s. What is the least possible time it can take until the car comes to a full stop in front of the fast-food restaurant? (**A: 69.8 s.**)
8. You are an engineer designing a traffic light. Assume a person can see a traffic light change color from red to yellow, and it takes one second to respond to a change in color. Suppose the speed limit is 15 m/s. Your goal is to enable drivers always to stop after seeing and responding to the yellow light with a maximum deceleration of -5.0 m/s^2. How long should the yellow light last?
9. You are a driver responding to the traffic light in the previous exercise. If it was correctly designed according to that problem, at what distance from the light should you be prepared to make your "to stop or not to stop" decision? Assume you are a safe driver who neither speeds up to get through the yellow light nor stops more suddenly than the deceleration rate of -5.0 m/s^2.
10. Why are red lights on the top of traffic lights and green lights on the bottom?
11. The distance d in meters from which a person can see a stop sign is given by the formula $d = D/30$ (where D is the diameter of the stop sign in meters). Assume it takes a person one second to step on the brakes after seeing the stop sign. Assume that the brakes decelerate the car at a rate of -5.0 m/s ^{2}. If the speed limit is 15 m/s, what should the diameter of the stop sign be? (**A: 1.3 m.**)
12. Assume that the purpose of an on-ramp is to allow cars to accelerate from 15 mph to 60 mph. An off-ramp will allow cars to decelerate from

60. mph to 15 mph. Should an off-ramp be longer than, shorter than, or the same length as an on-ramp? Give a reason for your answer. (**Hint:** Do cars bunch more when accelerating or decelerating?)

13. Suppose the deceleration of a car on a level off-ramp is $-3.0\,\text{m/s}^2$. How long would the off-ramp have to be to allow a car to decelerate from 60. mph to 15 mph? (**A: 110 m.**)
14. An early proposal for space travel involved putting astronauts into a large artillery shell and shooting the shell from a large cannon.[7] Assume that the length of the cannon is 30. m and that the velocity needed by the shell to achieve orbit is 15,000 m/s. If the acceleration of the shell is constant and takes place only within the cannon, what is the acceleration of the shell in gees?
15. Suppose that a human body can withstand an acceleration of 5.0 gee, where $1g$ is $9.8\,\text{m/s}^2$. How long would the cannon have to be in the previous exercise to keep the acceleration of the humans within safe limits? (**A: 4.6×10^6 m.**)
16. You wish to cover a two-mile trip at an average of 30.0 mph. Unfortunately, because of traffic, you cover the first mile at just 15.0 mph. How fast must you cover the second mile to achieve your initial schedule?
17. You are a TV reporter who rides in a helicopter to advise commuters about their travel time from their bedroom community to a city 20. miles away. You notice from your helicopter there are 160 cars per mile. Assume that all the drivers are observing the rule-of-thumb "follow rule" (see Figure 6). What should you tell your audience that the travel time between the cities will be? (**A: 1 hr, 11 min.**)
18. This chapter has shown that for the two-second rule the number of cars per hour completing a trip is independent of speed. Consider now a slightly more realistic follow rule: $\text{mph} = 1800/(\text{cars/mile})^{0.9} - 4$. Find the speed for which the number of cars per hour is a maximum. (Ignore the speed limit up to 100 mph!) (**A: ~ 40 mph.**)
19. If you want to achieve a flow of 2,600 cars per hour in each direction on a highway, and the two-second rule applies, how many lanes must the highway have in each direction.
20. Imagine a smart car of the future with radar and computer control capabilities that make it safe to operate on today's highways with twice the speed *and* half the spacing of today's human-driven cars. Using the follow rule, compare the capacity in terms of cars/hr for smart cars with cars/hr for human-driven cars. (**A: four times as much as for human driven cars.**)
21. You are a civil engineer with the responsibility for choosing the route for a new highway. You have narrowed the choice to two sites that meet

[7] A number of military satellites have been delivered to space using specially adapted naval cannons from the Vandenberg Air Force base in California.

all safety standards and economic criteria. (Call them Route A and Route B.) Route B is arguably slightly superior in terms of both safety and economics. However, Route B would pass by the site of an expensive new house recently built by your favorite niece, and the proximity of the highway would severely depress the value of the house. Since your niece's last name is different from yours, and you have never mentioned her at the office, you would be unlikely to be "caught" if you chose Route A in order to save the value of her house. What do you do?

(a) Choose Route A.
(b) Choose Route B.
(c) Ask to be relieved from the responsibility because of conflict of interest.

22. You are a civil engineer on a team designing a bridge for a state government. Your team submits what you believe to be the best design by all criteria, at a cost that is within the limits originally set. However, some months later the state undergoes a budget crisis. Your supervisor, also a qualified civil engineer, makes design changes to achieve cost reductions that he believes will not compromise the safety of the bridge. You are not so sure, though you cannot conclusively demonstrate a safety hazard. You request that a new safety analysis be done. Your supervisor denies your request on the grounds of time and limited budget. What do you do?

(a) Go along with the decision. You have expressed your concerns, and they have been considered.
(b) Appeal the decision to a higher management level.
(c) Quit your job.
(d) Write your state representative.
(e) Call a newspaper reporter and express your safety concerns.

Chapter 11

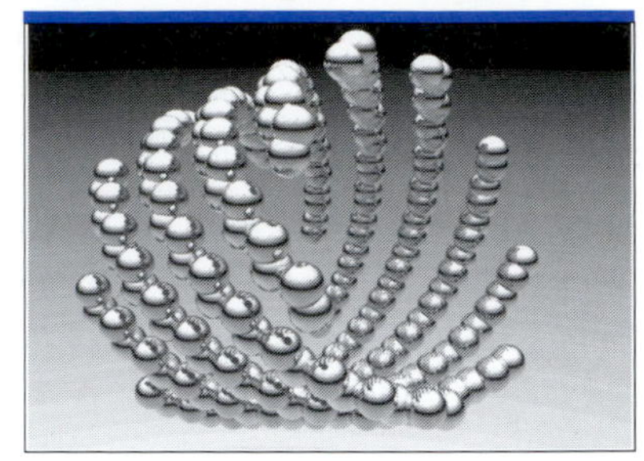

© iStockphoto.com/ Alwyn Cooper

Introduction to Materials Engineering

Introduction

Engineers select materials. Should a refrigerator case be made of steel or plastic? Should armor plate be a single sheet of steel or a lighter layered composite alloy? Should a transistor be made out of germanium or silicon? Should a telephone cable be made of copper or fiber optic glass? Should an artificial hip joint be made out of metal and, if metal, should it be titanium or stainless steel, or would a polymer composite be better? These materials selection problems are further examples of constrained optimization. The engineer finds the solution that best meets given criteria while also satisfying a set of constraints.

In materials selection, the criterion to be optimized when selecting the material might be cost, weight, or performance or some composite index such as weight × cost to be minimized with a requirement of a particular performance. The constraints (also called design requirements) typically involve such words as **elastic modulus** (also called **stiffness**), **elastic limit, yield strength** (sometimes abbreviated as plain "strength") and **toughness.** While we all have loose ideas of what is meant by these terms, it is necessary to precisely express them as engineering variables. Those variables must, in turn, contain the appropriate numbers and units.

To help develop appropriate variables, numbers, and units, this chapter will introduce a new tool: the **stress-strain diagram.** It is a tool for defining elasticity, strength, and toughness as engineering variables, as well as a tool to extract the numerical values of those variables in materials selection.

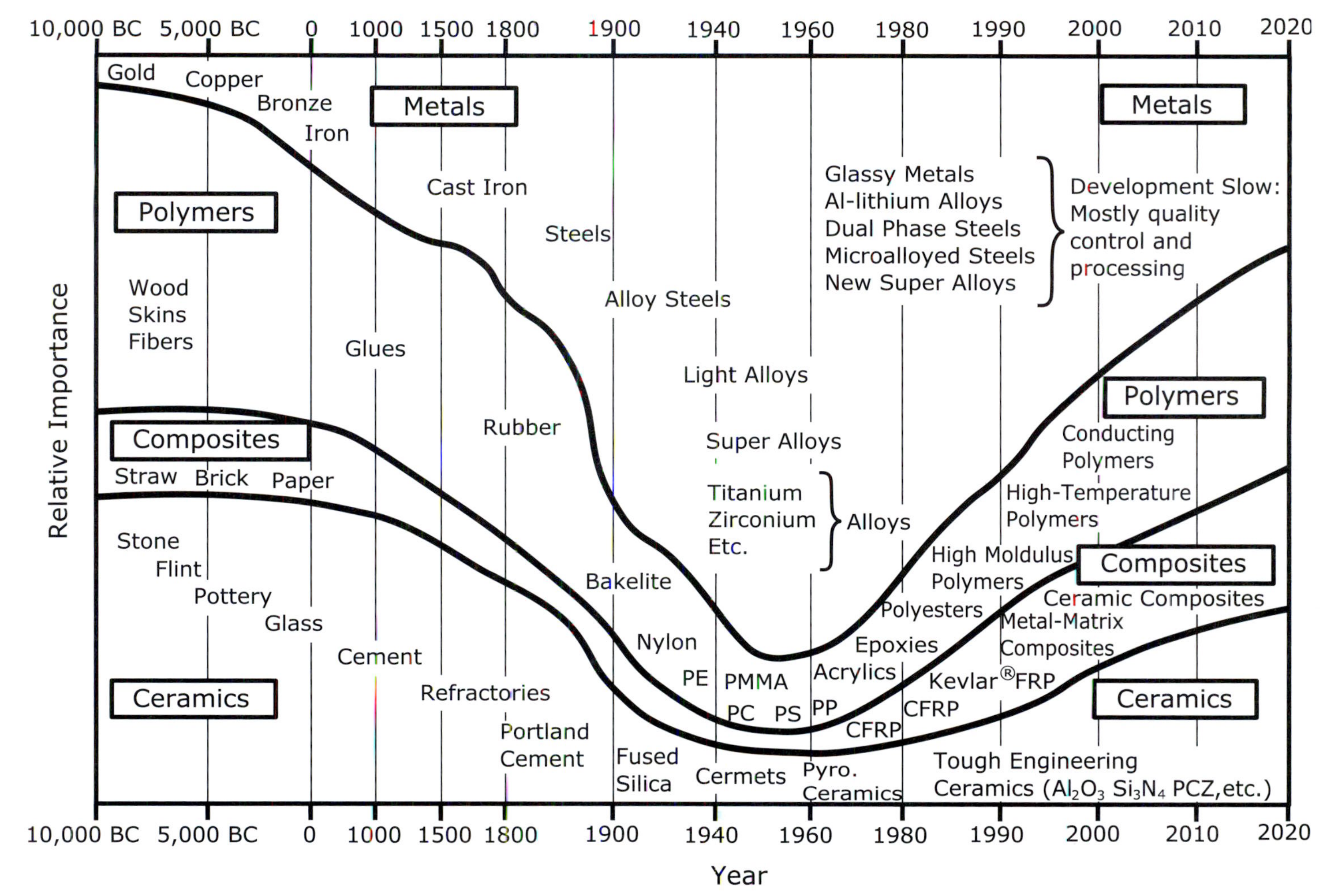

Figure 1 Materials Since the Ascension of Homo Sapiens. (Reprinted by kind permission of the Royal Society (London) and of Professor Ashby.)

Working through the examples and problems in this chapter will enable you to (1) define **material requirements;** (2) relate two important classes of materials, — **metals and polymers;** (3) understand to a first level, the **internal microstructure of materials** that can be **crystalline and/or amorphous;** (4) use a stress-strain diagram to express materials properties in terms of the **five engineering variables:** stress, strain, elastic limit, yield strength, and toughness; and (5) use the results determined for those properties to carry out materials selection.

We will use a specific example of materials selection. Consider the choice of material for a car's bumper. Over the last decade or two, the materials used in car bumpers have changed from metals to a composite made of a series of "plastics."

Mankind's reliance on the properties of materials for various applications, from weapons to shelter, has been around since the dawn of our species.[1] Notice in Figure 1 how the various materials have evolved over time. The earliest were naturally occurring elements[2] and compounds, the next were found by what today are called "Edisonian"[3] methods (i.e., the method of trial and error), and our modern materials are designed by systematic investigations. Our interest is to discover the principles behind new materials. Notice how the modern materials are classified by type: metals, polymers, composites, and ceramics. For the sake of brevity, we will confine ourselves to just the first two of these classes: metals and polymers.

Presumably you already have a picture of what is meant by a *metal.* Generally, metals are strong and dense. Metals also reflect light, and conduct heat and electricity. On the other hand, polymers, popularly known as "plastics," are generally weak, sometimes opaque and sometimes transparent, and generally do not conduct heat or electricity. We will eschew the term *plastics* in favor of *polymer* in describing these materials, since, as we shall see, the word *plastic* is reserved to describe a particular phenomenon mostly studied in metals.

Strength

What are the reasons for a particular material's strength? Material properties are directly related to their molecular properties and hence to the properties of their constituent atoms. The only truly fundamental example we will calculate is the inherent breaking strength of a pure material, such as a piece of pure iron as set by its molecular structure. All other failures must be

[1]Ashby, M. F., *Technology of the 1990s: Advanced Materials and Predictive Design.* Phil. Trans. R. Soc. Lond., A322, 393, (1987).
[2]Most metals were discovered in ascending order of their melting points.
[3]Another homage to that inventive giant of the nineteenth century, Thomas Edison.

due to some defect of their aggregated structures. Our proposed calculation thus represents the upper limit on its strength.

We will make a quick tour of material structures such as whether the material is **amorphous** (meaning no structure observable at the micro level—10^{-6} meters and smaller) or whether it is **crystalline** (which means it consists of definite microstructures that can be seen under a low power microscope or even by the naked eye).

Suppose you clamped the ends of a piece of pure iron and pulled it apart as shown in Figure 2 until it eventually breaks.

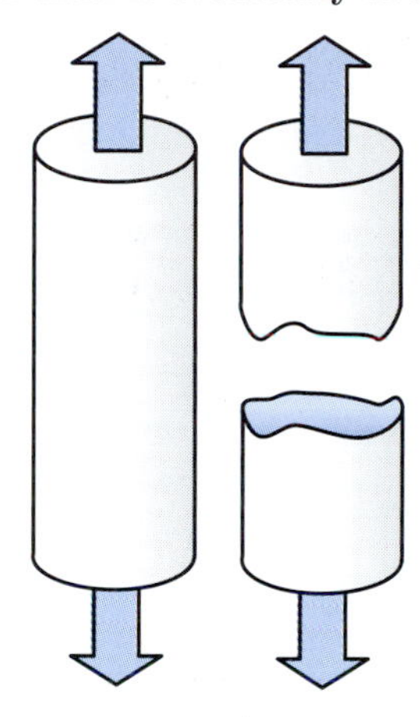

Figure 2

Failure Under Tension

The straight pull is called **tension.** If the nominal cross-sectional area of the break is known, and the number of atoms in that plane is also known, one can, in principle, calculate the force to separate the atoms at the break zone and also the work expended (i.e., force × distance) to break it.

In a crystal of iron the geometry is particularly simple. A **crystal** is a representative repeat pattern of the atoms that make up the structure. Figure 3 shows a plane view of the arrangement of the atoms in a perfect crystal of iron. If the crystals are large enough (i.e., there are trillions

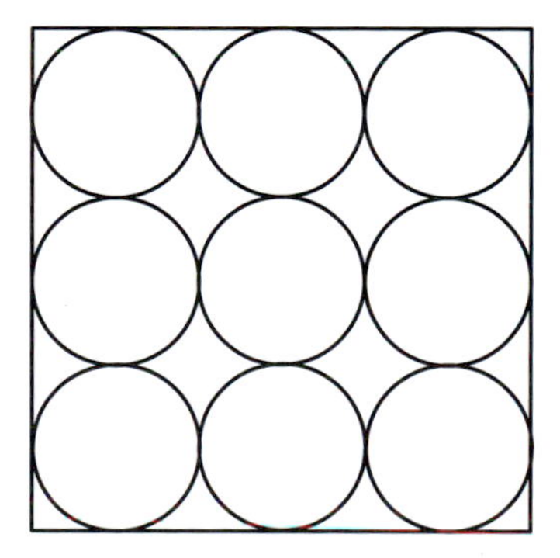

Figure 3

Structure of an Iron Crystal (Centers of the atoms are 0.234 nm apart.)

and trillions of atoms all arranged in the same pattern), these crystals can often be seen with low-power optical microscopes, and their atomic structure can be deduced using X-rays.

The crystal structure will arbitrarily extend in every direction in this idealized model (as in a single crystal). At the plane shown in Figure 3, this works out to contain 1.8×10^{19} atoms per m^2. All we need to know now is the strength of each bond and we will have a model of the theoretical strength of this material. We can estimate this quite easily by doing a simple thought experiment: Imagine you started with a lump of iron and then you heated it until it evaporated (just like boiling water). The total energy absorbed represents the energy to break all the bonds in the solid iron and make individual atoms from the solid, which we will use as a crude measure of the bond strength of all of those atoms in the original piece of iron. This works out to be about 6.6×10^{-19} J/atom of iron. The *work* to rip apart a piece of iron is therefore just about 12 J/m^2. This sounds quite modest.

Calculating the force required to break the material is another matter. Recall work done is force $\times$ distance. What we are asking is how far the atomic planes in iron should be pulled apart to consider them fully separated. Atomic forces fall off very rapidly with distance and a reasonable estimate is 0.1 – 0.2 nm *beyond* the equilibrium center to center separation of the centers of the individual iron atoms (0.234 nm), as shown in Figure 4. At that additional separation, the individual iron atoms will no longer interact.

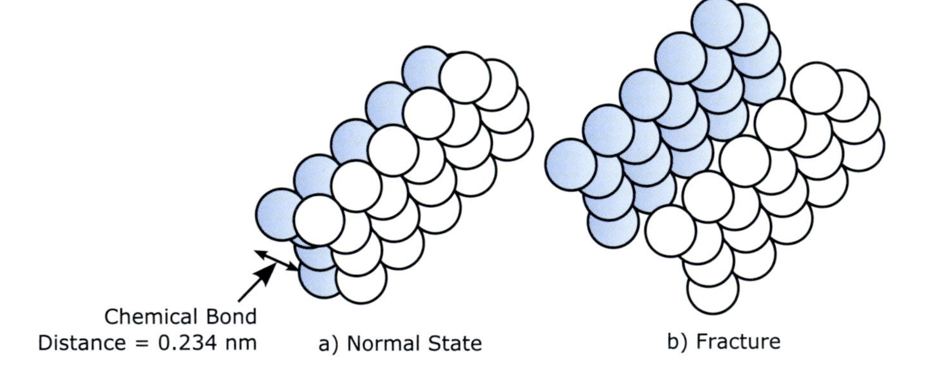

Figure 4

Before and After View of Fracture Along an Atomic Plane in Iron

The origin of the attractive force in metals is found in their atomic structure; this structure promotes their ability to conduct heat and electricity. The picture for a metal (Figure 5) is that there are free (negatively charged) electrons inhabiting the spaces between (positively charged) metal ions. This electronic structure provides for strong forces between the iron atoms as well as a picture to explain why it is easy to get electrons to flow in metal wires.

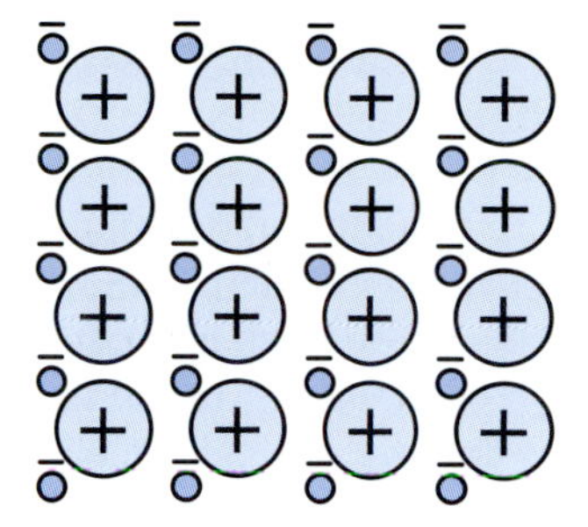

Figure 5

Metallic Bonds

We will equate the work done in separating the atoms beyond their equilibrium positions to the force × an *assumed* distance of 0.15 nm of separation: Per unit area,

$$\text{Energy } \left(12\ \text{J/m}^2\right) = \text{Force } (\text{N/m}^2) \times \text{Distance } \left(0.15 \times 10^{-9}\ \text{m}\right)$$

$$\therefore\ \text{Force} = 8 \times 10^{10} \approx 10^{11}\ \text{N/m}^2$$

The units N/m^2 also have the name **Pascal** (abbreviated as **"Pa"**), but we are generally interested in large numbers and therefore use the *giga* prefix of 10^9. (See Chapter 2, Table 3, for convenient units.) Our answer is thus ~100 GPa (G being the prefix for giga). This force per unit area is equivalent to piling 100 billion apples on top of a 1 m × 1 m tray[4] here on earth, and, as we shall see following, it is a much larger force/area than observed in practice. Why? Real crystals are finite in extent and are usually randomly oriented. Figure 6 shows a polished piece of copper as viewed under a microscope. Individual random crystals of copper are quite obvious.

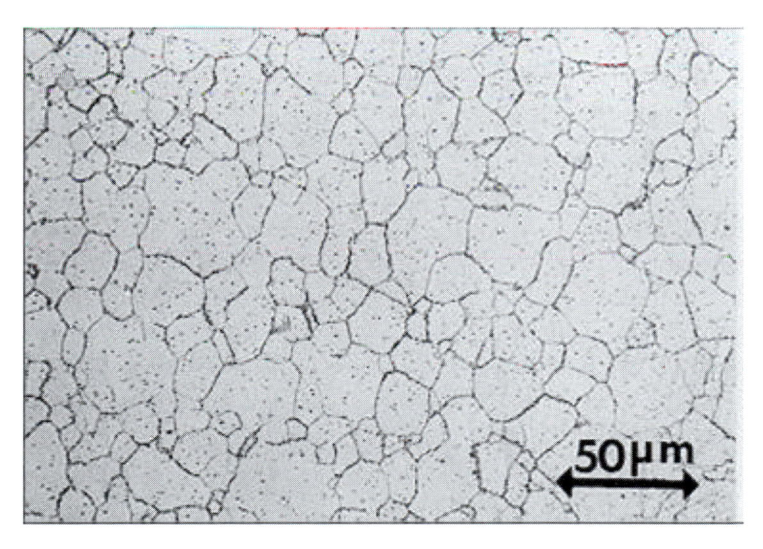

Figure 6

Polished Copper Showing Individual Crystals (Courtesy of the Copper Development Association, Inc.)

[4]The pressure of one atmosphere is just 100,000 Pa so we have to apply a million atmospheres tension to break our sample in the text.

Failure may occur between crystals at their relatively weak **"grain boundaries"** rather than within the crystals resulting in lower forces to separate them than the previous mathematical model.

In addition, not all materials are crystalline but have a completely random structure, called **"amorphous."** Many materials may also consist of mixed phases such as crystals in a matrix of amorphous material.

The other materials class we will deal with is polymers. The word *polymers* basically means "many molecular pieces". Polymers are repeating chains of small molecular assemblages. Generally, and always in this book, the term *polymer* will assume that these links are *organic chemicals.*[5] For example, the common polymer polyethylene consists of long chains of ethylene[6] molecules written as -[CH_2 - CH_2]- strung together, each dash representing the net attraction of electrons between two adjacent carbon atomic nuclei. Each carbon atom also has two off-axis hydrogen atoms and one on-axis C - C bond. Roughly speaking, the strength of a C - C bond is about 10% that of an iron-to-iron bond, so you might think that our model apparently implies a force/area of ~10 GPa as the upper limit to break a typical polymer bond. However, this is not true (Figure 7).[7]

Figure 7

Regular Array of Polyethylene (There is -CH_2- at the apex of each intersection.)

[5]Meaning based on carbon, hydrogen, and a few other atoms such as nitrogen, oxygen, and sulphur.

[6]The chemical structure of the gas ethylene is $CH_2 = CH_2$ with the "=" standing for a "double bond" between the carbon atoms. If this double bond is broken and rejoined to an adjacent carbon, it makes the repeating polyethylene structure found in that solid material. (Milk bottles are commonly made of this material.)

[7]Regular arrays of a polymer are also called crystals. They can coexist with a matrix of an amorphous polymer, in good analogy to the structure of metals. Crystals in polymers are usually the right size to efficiently scatter light, so you know that a white, opaque pure polymer such as Teflon™ is crystalline and a clear one such as "PET" or polyethylene terephthalate (used in soft drink bottles) is amorphous.

What you need is not the force between individual carbon atoms but that between adjacent polymer chains. While there are many kinds of possible molecular interactions, for the case in hand of polyethylene, the kinds of forces are lumped under the name "van der Waals" forces. They are quite weak, about 0.1% of the strength of the metal bonds, ~0.1 GPa. The upper limit for the strength of a polymeric material might therefore be of this magnitude.

We will now look at the application of metals and of polymers to the question of car bumpers. Modern cars use composite polymer bumpers, while older cars and some specialized cars still use metals. Why this is and what principles are needed to understand this application of materials is the subject of the rest of this chapter.

Defining Materials Requirements

Consider the choice of material for a car's bumper. Defining materials requirements for a bumper begins by understanding what a bumper is and what a bumper does. Take a look at a car: A bumper is a structure often integrated into the main chassis; typically it is 1–2 meters wide, 0.1–0.2 m high, and 0.02–0.04 m thick and is attached to the front or rear of the car about half a meter above the ground.

Figure 8 shows a "conceptual bumper" for which a material will be selected. A bumper does two main jobs:

1. It survives undamaged at a very low-speed (1–2 m/s) collision.[8]
2. It affords modest protection for the car, though sustaining major damage to itself, in a moderate-speed (3–5 m/s) collision.

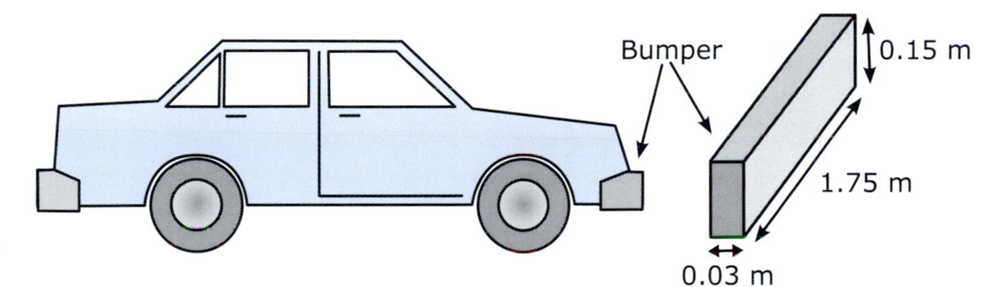

Figure 8

Conceptual Bumpers

What does an engineer mean by terms such as *collision*, *survive undamaged*, and *protection for the car*? Translation of such language

[8]Federal regulations call for protection against damage to the vehicle in a 2½ mph (1.18 m/s) collision; while this might seem to be a modest requirement, remember that modern cars are heavy!

into engineering variables, units, and numbers begins with the concept of energy conversion.

A collision is a rapid process of energy conversion. A moving car is carrying TKE. A collision rapidly converts that TKE into **"elastic"** and **"plastic"** energy[9] of materials. Elastic energy is a form of stored potential energy, while plastic energy is converted, via internal collisions of the atoms in the material, into heat. In a properly designed bumper, these two processes of energy conversion will occur in the bumper—not in the car proper and certainly not in you.

In a really low-speed collision, the energy conversion should leave the bumper looking as it did before the collision. In a moderate-speed collision, the bumper may be damaged or destroyed, but it should absorb and release the energy in a way that it leaves the car and the passengers undamaged. (In a high-speed collision,[10] other methods of protecting the passengers are needed, as discussed in Chapter 12.)

A wide range of bumper designs can achieve these energy conversion objectives. To select a design from these possibilities, an engineer first picks dimensions for a bumper. For simplicity, consider a bumper to be a rectangular solid object $1.75 \times 0.15 \times 0.03$ meters (width, height, and thickness, respectively). This is of the right order of magnitude. For a given level of protection, the engineer will then seek the lightest material (consistent with cost!) that will enable a bumper of these dimensions to meet the design requirements.

Why does the engineer seek the lightest material? In the days of metal bumpers, a bumper provided a significant part (about 5%) of the mass of the car. The higher the mass of the car, the lower the gas mileage (Chapter 6, Figure 13). Conversely, anything that reduces mass will increase gas mileage. So all other things equal, a car with lighter bumpers will have the higher gas mileage than a car with heavier bumpers. So our optimization problem will be choosing the material that will give the lightest bumper of given dimensions that will do a bumper's job.

Solving that optimization problem begins by using energy conservation to translate the design requirements from words into variables, units, and numbers.

EXAMPLE 1

Estimate the total energy that a 1.00×10^3 kg car has to absorb in a 2½ mph (1.18 m/s) collision with a wall. If all of this is absorbed in the bumper, what is the "specific" energy absorbed (i.e., per m^3 of bumper volume) given its dimensions of $1.75\ m \times 0.15\ m \times 0.03\ m$?

[9]We'll properly define these terms later, but for the moment realize that *plastic* in this context is not a word meaning "polymers" and that *elastic* means to spring back like an elastic band.
[10]The Insurance Institute for Highway Safety tests passenger cars up to 35 mph (15.6 m/s); see http://www.iihs.org/default.htm.

Need: Energy absorbed ____ in J.

Know: Speed of car, dimensions of bumper.

How: Conservation of energy; division of total energy by total volume (volume $= 1.75 \times 0.15 \times 0.03 = 0.00788$ m^3).

Solve: The bumper must absorb the total TKE from a 1.18 m/s collision:

That TKE is $\frac{1}{2} mv^2 = \frac{1}{2} \times 1.00 \times 10^3 \times 1.18^2$ [kg][m/s]$^2 = \mathbf{7.00 \times 10^2}$ **J.**

$\therefore$ **Specific energy** absorbed by bumper = total energy absorbed/ volume of bumper $= 7.00 \times 10^2 \times 10^{-6}/0.00788$ [J][1/m^3][MJ/J] $=\mathbf{0.089}$ **MJ/m^3.** (Megajoules are a convenient unit in this subject.)

Notice if we wanted 5.0 mph protection, the total energy absorbed goes to 2,800 J and the specific energy absorbed goes to 0.36 MJ/m^3 and at 35 mph (15.6 m/s) the corresponding numbers are 0.12 MJ and 1.6×10^2 MJ/m^3. ■

Is there a candidate bumper material that can absorb these amounts of energy elastically and so rebound to its original dimensions? Or will there be sustained damage to the bumper or, worse, beyond it? For us, the choice of a bumper material boils down to a choice between a polymer and a metal. A polymer offers light weight at some sacrifice of strength. A metal offers high strength but with a penalty of higher weight. Which shall the engineer choose? Answering these questions begins by considering the types of materials that nature and human ingenuity offer. Making that choice requires applying a new tool: the **stress-strain diagram.**

The stress-strain diagram is a tool for using measurements of two new variables, **stress** and **strain,** to quantify the terms **modulus of elasticity, elastic limit, plasticity, yield strength,** and **toughness.** Two basic experiments[11] are shown in Figure 9. Materials are either systematically stretched under tension, as in Figure 9a, or put into **compression,** as in Figure 9b.

Add more mass M (and force, Mg) in either of the situations and the wire stretches or the plate compresses accordingly. In this way one can plot a diagram of the stretch in L or compression in T vs. the applied load. The response is considered **elastic** if the material returns to its initial dimensions when the load is removed. Figure 10 is a composite plot of four elastic experiments. In tension, the double-length rod (or wire, etc.), Figure 10b, stretched in proportion to its load twice as much as did the shorter one, Figure 10a. The compression in Figure 10d is half that

[11]A third mechanical test is to twist the material and measure its torsional material properties.

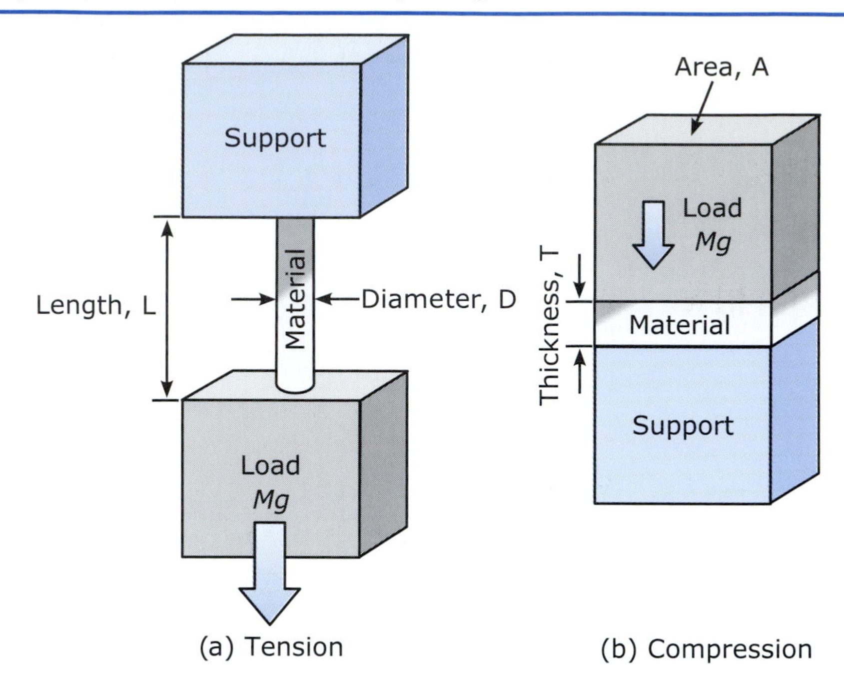

Figure 9

Basic Material Tests

in Figure 10c and is also proportional to its load. This is an expression of **"Hooke's Law".**[12]

$$(L - L_0) \propto L_0 \frac{Mg}{A} \tag{1}$$

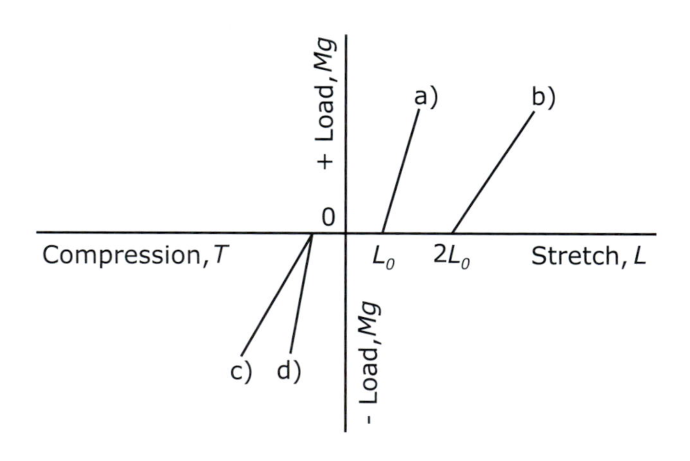

Figure 10

Elastic Response of a Material to Tension and Compression: a) Tension, initial length L_0, diameter D; b) Tension, initial length $2L_0$, diameter D; c) Compression, thickness T, load area A; and d) Compression, thickness T, load area $2A$

[12]Robert Hooke was a talented rival of Isaac Newton.

Notice that we have used a convention that tension is a positive load, stretching is a positive response to it, and the opposite is true for compression and the contraction response to it.

But why do we have to plot these data as clumsily as in Figure 10? Equation 1 is our clue that we can be more inclusive as well as more general. We write it as:

$$\frac{Mg}{A} \propto \frac{(L - L_0)}{L_0} \quad \text{or} \quad \frac{Mg}{A} = E\frac{(L - L_0)}{L_0} \text{ (thus defining } E) \tag{2}$$

The fractional (or for the percentage, × 100) stretching or compression—that is, the change in length divided by the original length, $(L - L_0)/L_0$—is called the **strain.** If the material stretches in tension, the strain is positive, and if the material is compressed, strain is conventionally negative. We often use the Greek epsilon ε to denote strain. Note that strain is a dimensionless number, since it is the quotient of two quantities with the same dimension, normally in SI units of meters/meters but just as well in feet/feet. The force causing the strain, divided by the applicable cross sectional area, is called the **stress,** denoted by the Greek sigma $\boldsymbol{\sigma}$**.** Again, a tensile stress is positive, and conventionally a compressing stress is negative. Stress has the dimensions of N/m^2 (i.e., Pa). The constant in equation (2) is called **Young's**[13] or the **Elastic modulus** and given the symbol $\boldsymbol{E}$**.** Equation (2) can thus be rewritten:

$$\sigma = E\varepsilon \tag{3}$$

Equation 3 is the way that engineers use Hooke's law, and it should be committed to memory in this form once it is understood.

EXAMPLE 2

When a mass of 1,000 kg of steel is carefully balanced on another piece of steel of width 1.00 m, length 0.15 m, and thickness 0.03 m, and sits on a base of steel, the length of the middle piece contracts by 0.00001 m.

(a) Is the middle piece of steel in tension or compression?
(b) What is the stress on the middle piece of steel?
(c) What is the strain in the vertical direction of the middle piece?

Need: Stress ____ N/m^2, strain ____ fraction.

Know: Mass of the object compressing the piece = 1,000 kg. Initial thickness of sample is 0.15 m and it suffers a contraction of -1.0×10^{-5} m in this direction when the load is applied.

[13]For historical reasons, this slope is also called Young's modulus, after the early nineteenth century scientist and scholar Thomas Young who first suggested this approach to understanding materials.

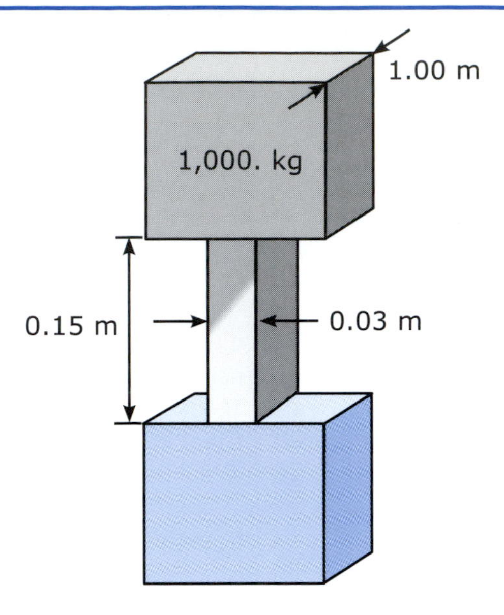

How: By definition, stress = force/area, and force = Mg; the supporting area is $1.0 \times 0.03\ m^2 = 0.03\ m^2$. By definition, strain = (change in length)/(initial/length).

Solve: (a) The piece is in compression.

(b) **Stress, σ** $= -1000. \times 9.81/0.03$ [kg][m/s^2][1/m^2] $=$ $\mathbf{-3.3 \times 10^5\ N/m^2}$ $=$ $\mathbf{-3.3 \times 10^5}$ **Pa.** The negative sign indicates compression.

(c) By definition, **strain** is an fractional contraction or $\varepsilon = -1.0 \times 10^{-5}/0.15$ [m]/[m] $=$ $\mathbf{-6.7 \times 10^{-5}}$ (a pure number; – sign indicates compression) ■

Suppose one made a large number of measurements like those described in Example 2 but with differing loads. Suppose one then plotted the results on a graph, with strain as the horizontal axis and stress as the vertical axis. For each axis, the negative direction would be compression, and the positive direction would be expansion. The resulting graph is called a **stress-strain diagram.** As will be shown now, it captures in one diagram three of the most important properties of a material: elasticity, strength, and toughness.

The stress-strain diagram of an idealized steel[14] plotted according to equation 3 looks like Figure 11. It also defines some new terms: **yield stress** (or **yield strength**), **yield strain,** and **plastic deformation.**

In the first instance, we have reduced the number of curves we need when we change the geometry of the material, or the applied load, to just

[14]Basically iron + some additives + some processing.

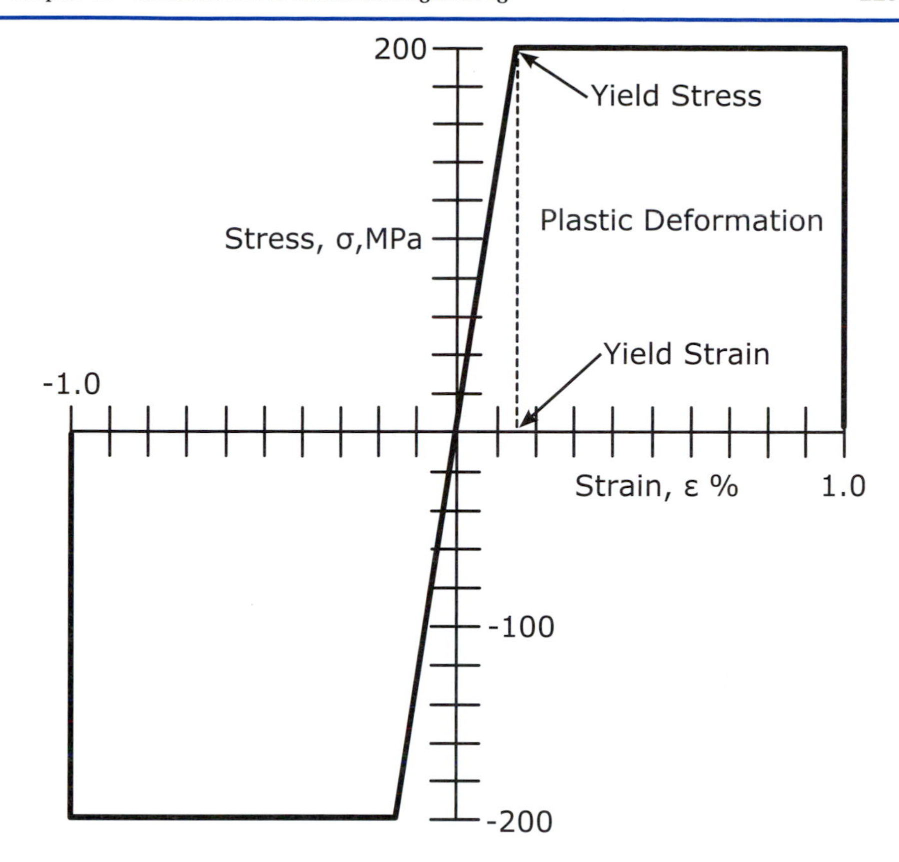

Figure 11

Simplified Stress-Strain Curves for an Idealized steel

two: one curve for tension and one for compression. In doing so, we have taken advantage of Hooke's law and collapsed the data into a more general and convenient form. It will no longer matter in this format if we change the sample's size or if we change the load on the sample by increasing the load mass or by reducing its cross-sectional area, and we don't need to worry about the absolute strains changing with the sample length. All the separate curves will collapse to a single curve.

The **yield stress** is the maximum stress at the limit of elastic behavior in which the specimen returns to its initial length when the load is removed. In this case it is 200. **M**Pa for this particular steel. The **yield strain** is the corresponding strain. Plastic deformation causes a *permanent* change in the properties of the steel when it passes beyond its elastic limits. Stretching here is called **plastic deformation.** For this idealized steel compressive properties exactly mirror the tensile ones.

Compare the yield stress of 200 **M**Pa to the theoretical failure value for iron, which was about 100 **G**Pa. In other words, our sample of steel is beginning to fail at a stress about 200 MPa or a factor of about 500 lower than its theoretical maximum. Presumably the failure is initiated at the crystalline boundaries rather than by pulling apart the rows of atoms of iron in the steel.

The elastic modulus E defined in the equation 3 is just the slope of the stress-strain curve. In this diagram, it calculates to be $(\sigma/\varepsilon) = 200$ MPa/ $0.0015 = 130$ GPa. Another variable of interest is **toughness.** In ordinary speech, people sometimes confuse it with (yield) stress/strength. To an engineer, however, strength and toughness have very different meanings. A material can be strong without being especially tough. One example is a diamond. It can be subjected to great force and yet return to its original shape, so it is therefore very strong. Yet, a diamond can be easily shattered and therefore is not very tough. A material can also be tough without being strong. An example is polycarbonate, a kind of plastic familiar as football helmets, spectacle lens, and a host of other uses. It is relatively easily dented and therefore not very strong. Yet, it absorbs a great deal of energy per unit mass without shattering and is therefore very tough.

In this introductory treatment, the stress-strain curve for strains greater than the yield strain will be a horizontal line parallel to the strain axis. In actual stress-strain diagrams, the shape of that portion of the curve is more complicated. This region is called the **"plastic"** region and in the simple linear treatment we are showing, it is called a **"perfectly plastic"** region. This overall model is called **"perfectly elastic, perfectly plastic."** It's a useful simplification of how real materials behave.

The horizontal line in the perfectly plastic region extends out to a maximum strain. At that strain the material "fails." The total strain at failure is as the sum of the elastic strain and as plastic strain. The important point is that a wire stretched beyond its maximum strain breaks. A plate compressed to its maximum strain shatters or spreads. The **toughness** of a material is defined as *the area under the portion of the stress-strain curve that extends from the origin to the point of maximum strain.*[15] Because this area (shaded in Figure 12) is the result of stress,

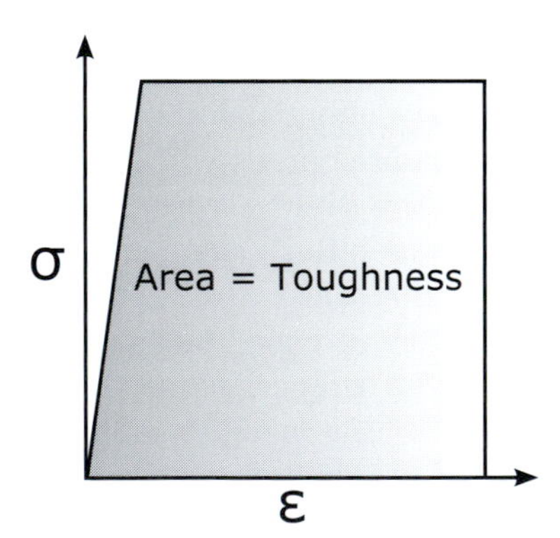

Figure 12

Toughness

[15]Actually there are several different definitions of *toughness* used by different authors, in particular the ability of a material to withstand sudden impacts. Under certain circumstances this may, in fact, measure the same thing as our definition.

measured in multiples of Pa or N/m^2, and of strain, measured in m/m, toughness has the units of N m/m^3, or J/m^3 (that is, energy per unit volume). It is usually measured in a dynamic experiment involving a strike by a heavy hammer. Toughness is therefore the preferential criterion to be used to predict material failure if a sudden load is applied.

If you go back to Figure 10 and think about the fact that you are moving a force (say F in newtons) through a distance in (say e in meters) as you stretch the material, the work done is, as always, force × distance = $F \times e$ [N][m].

$$\text{Work} = F \times e = \frac{F}{A} \times (e \times A) = \text{stress} \times \frac{e}{L_0}(L_0 \times A)$$

$$\text{or Work} = \text{stress} \times \text{strain} \times \text{volume}$$

where A is the cross-sectional area perpendicular to the applied stress in the sample and L_0 is the sample length. Toughness, defined as the work done until failure, represents the ability of the material to absorb energy. It is a prime variable for a large subset of materials research.

EXAMPLE 3

Given the stress-strain diagram for steel in Figure 11, determine (a) its modulus under compression, (b) its yield strength under tension, (c) its toughness under tension to yield, and (d) its toughness under tension to 1.0% deformation. Use appropriate SI units.

Need: E = ____GPa under compression, σ = ____MPa at yield, and toughness at yield and 1.0% deformation = ____ MJ/m^3.

Know: Stress/strain curve in Figure 11; toughness is area beneath the stress/strain curve.

How: Hooke's law $\sigma = E\varepsilon$, + stress/strain curves.

Solve: (a) Need slope in compression region for σ/ε curve.

E = -2.0×10^2 MPa/-0.0015 = **130 GPa** (to graph-reading accuracy)

(b) Under tension, **yield** is **~2.0×10^2 MPa** (to graph-reading accuracy).

(c) Toughness under tension:
At yield, toughness = triangular area, ½ × 0.0015 [m/m] × 2.0×10^2 [MN/m^2] = **0.15 MJ/m^3.**

(d) **At 1% strain,** add rectangular area: (0.01 – 0.0015) × 200. = 1.7 MJ/m^3.
∴**Toughness** = 0.1 + 1.7 = **1.8 MJ/m^3.** ■

The no-damage elastic portion of the curve is only a small fraction of the total protection afforded the car's structure. You can also see that plastic deformation, which will manifest itself in permanent deformation or crushing,[16] affords a much better sink to dissipate energy than does the purely elastic behavior. Of course, whether a particular bumper design can distribute this energy deposition uniformly and avoid exaggerated *local* deformation is up to the skill of the engineer.

We're now in a position to see if a steel bumper satisfies our first constraint. What does the stress-strain diagram tell us about its ability to absorb the energy of a collision yet return to its original shape?

EXAMPLE 4

Can the car's bumper from Example 1 absorb the TKE of the 2.5 mph collision assumed there? Or will it plastically deform? What if instead it is traveling at 5.0 mph? What is the weight of the bumper if its density is 7,850 kg/m^3? Assume the mechanical behavior is described by the stress-strain diagram of Figure 11.

Need: Energy to be absorbed in bumper during collision = ____ MJ/m^3 compared to energy of collision. Also, weight of bumper = ____ kg.

Know: From Example 1: TKE = 7.00×10^2 J at 2.5 mph; specific energy absorbed = 0.089 MJ/m^3; steel bumper volume = 0.00788 m^3; density of steel, ρ = 7,850 kg/m^3. At 5.0 mph, TKE = 2,800 J and the specific energy absorbed goes to 0.36 MJ/m^3.

From Example 3: elastic toughness = 0.15 MJ/m^3 and plastic toughness 1.8 MJ/m^3 at 1.0% strain.

How: Compare specific energy to be absorbed to capability of material to absorb it.

Solve: Assume the entire surface of the bumper contacts the wall simultaneously (i.e., the force on the bumper is uniform over its entire area).

At 2.5 mph, the specific energy to be absorbed = 0.089 MJ/m^3 of steel. Steel can absorb up to 0.15 M J/m^3, so it will elastically absorb this amount of energy—again if uniformly applied. If so, the bumper will bounce back to its original shape.

At 5.0 mph, the specific energy to be absorbed goes to 0.36 MJ/m^3 of steel. $\therefore$ The bumper will plastically deform since 0.36 MJ/m^3 > 0.15 MJ/m^3 but will survive since 1.8 MJ/m^3 > 0.36 MJ/m^3.

The weight penalty for this bumper is $\rho \times$ volume = 0.00788 $\times$ 7,850 [m^3][kg/m^3] = 62. kg or the weight of a physically fit student! ■

[16] No doubt with an accompanying repair bill that will disappoint the vehicle's owner!

Materials Selection

The previous section showed how the stress-strain diagram provides the information needed to determine if a material satisfies the design requirements. The question now becomes: Is there a material of less weight than steel that also satisfies the design requirements? To be specific, let us consider a polymer, a material that is much less dense than steel. Can it compete with steel in the bumper application? Again, our tool is the stress-strain diagram (see Figure 13 and Table 1).[17]

Notice it is not as "strong" as steel because its yield strength is considerably less, but it can be stretched much further while elastically returning to its original strength. Let's repeat Example 4 using this material for the bumper.

Table 1

Properties of a Polycarbonate

Elastic Yield, MPa	Strain% at Yield	Density, kg/m3
55.	30.	1,300.

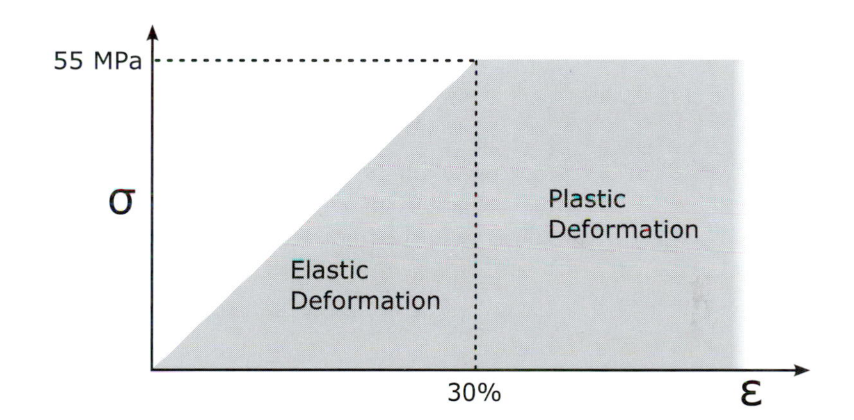

Figure 13

σ, ε Diagram for Polycarbonate

EXAMPLE 5

Can the car's bumper from Example 1 absorb the TKE of the low-speed 2.5 mph collision assumed there, or will it plastically deform? What if it is traveling at 5.0 mph, and what is the weight of this bumper? Again assume the compressive stress-strain diagram is the same as the tensile.

[17]Actually there are many kinds of polycarbonates; the properties of the one shown in Table 1 and those of a very common one made from a monomer abbreviated "BPA". It is used in crash helmets, water bottles, CDs/DVDs, and car bumpers. Another version of polycarbonate is used as a lens material for spectacles.

Need: Energy to be absorbed in bumper after collision = ____ MJ/m^3 compared to energy of collision. Also, weight of bumper.

Know: From Example 4: TKE = 7.00×10^2 J; specific energy absorbed = 0.089 MJ/m^3; bumper volume = 0.00788 m^3; density of polymer, $\rho = 1{,}300\ kg/m^3$. At 5.0 mph, TKE = 2,800. J and the specific energy absorbed goes to 0.36 MJ/m^3.

How: Compare specific energy to be absorbed to capability of material to absorb it. Need to calculate the latter from the σ, ε diagram.

Solve: Calculate the toughness by looking at Figure 13. It is given at yield by the triangular area, $\frac{1}{2} \times 0.30$ [m/m] $\times$ 55. [MN/m^2] = 8.3 MJ/m^3.

Even at 5.0 mph, the specific energy to be absorbed is 0.36 MJ/m^3—which is $\ll$8.3 MJ/m^3. Thus, this bumper can survive elastically.

The weight penalty for this bumper is $\rho \times$ volume = 0.00788 $\times$ 1,300 [m^3][kg/m^3] = 10. kg or much less than the weight of any student! ■

So, although polymer is not as "strong" as steel, its ability to compress further without breaking gives it adequate toughness to do the job. In addition, polymer is substantially less dense than steel. So *polymer is the winner*. In the last two decades, polymers have almost entirely replaced steel as the material from which automobile bumpers are made. Of course, there is still plenty of scope for the materials specialist to design the preferred configuration of the bumper. For example, the metal bumper may be manufactured with springs to absorb the impact. There is surely enough material in our steel bumper that we could use half of it in the form of coil springs, as in Figure 14.

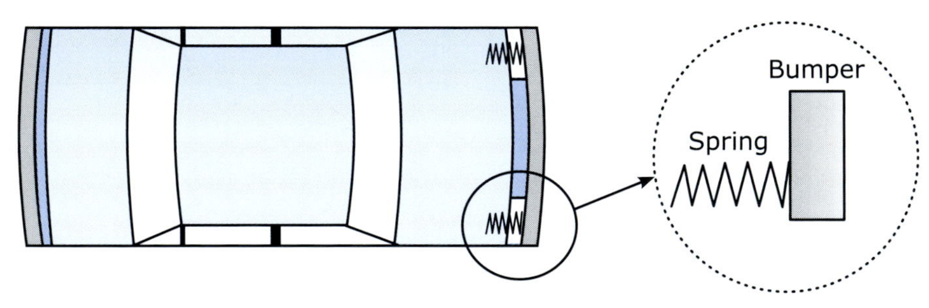

Figure 14

Mechanical Design also Influences Choice of Materials

This does not change the innate material characteristic of the steel.[18] Rather, by making a steel spring, it can be stretched or compressed much

[18]However, in compressing a coil spring there is some twisting and there is another modulus that enters the picture: the so-called modulus of rigidity (or torsion). Hooke's Law is also obeyed for these twisting motions.

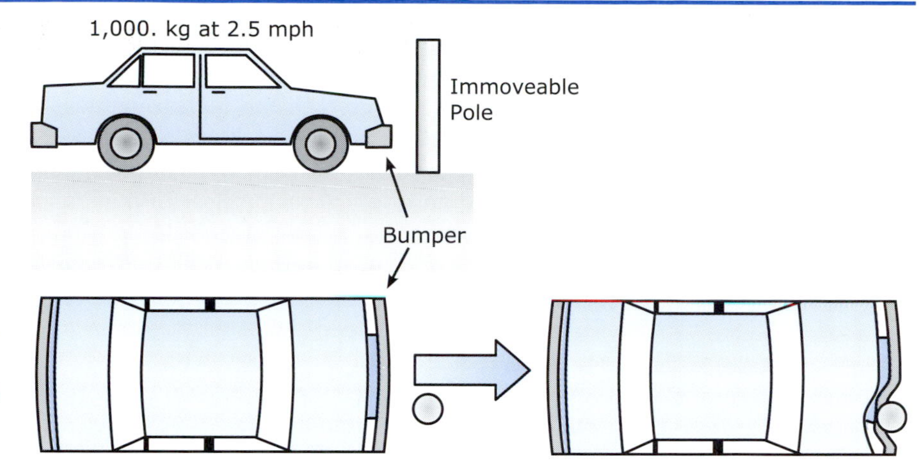

Figure 15

Not All Crashes Are Head On — The Material May Yield Locally in Low-Speed Impacts

further than an unshaped piece of the same material. It's as if we traded the material for one with a lower modulus (less steep curve) and a much larger strain to yield. Furthermore, there is plenty of scope to design for the randomness of crashes—that is, how the vehicles will interact with an immoveable object such as a pole as indicated in Figure 15.

This chapter is just as a glimpse into the most basic considerations that would go into a material design. Of course, materials engineers have become very clever at rearranging the macroscopic properties of materials as they need them. In the car bumper case, the polymer may be backed with a second one that is in the form of a foam—the advantage being its strain at yield is mediated by the thousands of gas-filled sacs within its structure.

Properties of Modern Materials

This discussion of selecting a material for a bumper gives a feel for the engineer's problem of materials selection. However, elasticity, strength, and toughness, though very important, are not the whole story. In the decision whether to use silicon or germanium to make transistors, the electronic and thermal properties of materials play a key role. In the selection of copper versus fiberoptic glass, information-carrying capacity becomes crucial. In the selection of steel versus polymers for refrigerators, considerations of appearance, manufacturability, and corrosion resistance become important. In the selection of a material for a hip joint transplant, compatibility with the human body becomes an essential materials requirement.

In addressing these varied requirements, twenty-first-century materials go far beyond the traditional categories of metals, polymers, and the other classes of materials shown in Figure 1. A wide range of "composite" materials now combine the properties of those original categories. Some are created by embedding fibers of one material within a matrix of a second material. This makes it possible to combine, for example, the high strength of a thin carbon fiber with the high toughness of a polymer matrix. Other new composite materials are literally tailored atom-by-atom. This makes it possible, for example, to provide materials with precise combinations of electronic and optical properties. In one sense, however, the emergence of these new composites is actually a matter of going "back to the future." The original materials used by humanity some 10,000 years ago, such as wood, stone, and animal skins, are complicated natural composites. Figure 1, as developed by Ashby, sums up this long-term history of humanity's materials use. Even with the new challenges of the twenty-first century, however, the traditional properties of elasticity, strength, and toughness will continue to play a central role in materials selection. As they do, the stress-strain diagram, the best representation of these properties, remains a crucial tool for the twenty-first century engineer.

SUMMARY

Materials engineers select materials. They do so by defining a constrained optimization problem. One property, such as performance or weight or cost, is to be optimized (that is, maximized or minimized), subject to meeting a set of constraints (design requirements).

Materials selection begins by quantifying those design requirements. It proceeds by searching the major classes of candidate materials, metals, polymers, composites, and so on for candidates capable of meeting those design requirements. The candidates are then subjected to further screening, based on such characteristics as the **strain** they exhibit in response to **stress.** This particular characteristic is plotted on a graph called the **stress-strain diagram.** This stress-strain, or σ ε diagram can be then used to determine key properties of the material: **modulus of elasticity, yield strength, plastic deformation, and toughness.** These properties, expressed in the appropriate units, provide the basis (the "know") for solving this particular constrained optimization problem of materials selection.

EXERCISES

The figure at the top of the next page depicts the situations described in Exercises 1–6.

1. A mass of $M = 1.0$ kg is hung from a circular wire of diameter 0.20 mm as shown in (A). What is the stress in the wire? (**A: 0.31 GP.**)

2. If the wire in (A) is stretched from 1.00 m to 1.01 m in length, what is the strain of the wire?

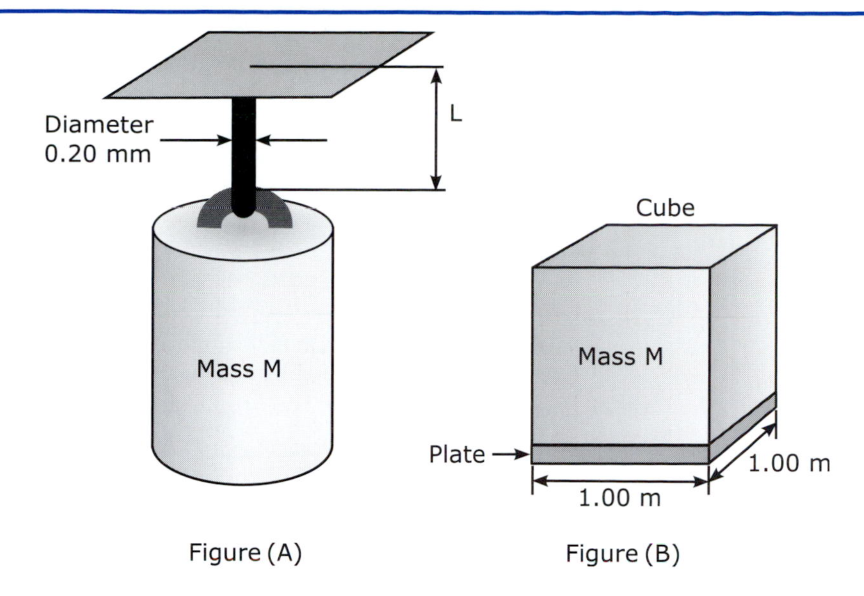

Figure (A) Figure (B)

3. In (B), when a block of metal 1.00 m on a side is placed on a metal plate 1.00 m on a side, the stress on the plate is 1.00×10^3 N/m^2. What is the mass of the metal cube? (**A: 102 kg.**)

4. Suppose the plate is as described in Exercise 3 and figure B was 0.011 m thick before the cube was placed on it. Suppose that placing the cube on it causes a compressive strain of -0.015. How thick will the plate be after the cube is placed on it? (**A: 0.011 or unchanged to three significant figures.**)

5. Suppose the wire in (A) is perfectly elastic. When subjected to a stress of 1.00×10^4 Pa, it shows a strain of 1.00×10^{-5}. What is the elastic modulus (i.e., Young's modulus) of the wire?

6. A plate of elastic modulus 1.00 GPa is subjected to a compressive stress of -1.00×10^3 Pa as in (B). What is the strain on the plate? (**A: -1.00×10^6.**)

   For Exercises 7–9, assume that a silicone rubber[19] has this stress-strain diagram.

7. What is the yield strength under compression of the silicone?

8. A flat saucer made of the preceding polymer has an initial thickness of 0.0050 m. A ceramic coffee cup of diameter 0.10 m and mass 0.15 kg is placed on a plate made of the same polymer as just indicated. What is the final thickness of the plate beneath the cup, assuming that the

[19]Silicone rubbers are very flexible; their structure consists of polysiloxanes of formula -)Si-O- in which the Si atom also has two CH_3 chemical groups per atom (but that are not shown). They are off the main chain. Incidentally, you must distinguish *silicon* (a brittle element used in electronics) from *silicone*, the soft rubber described here.

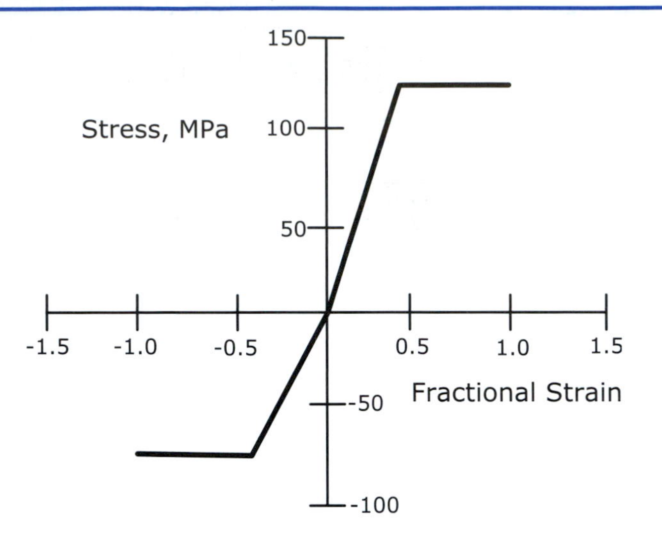

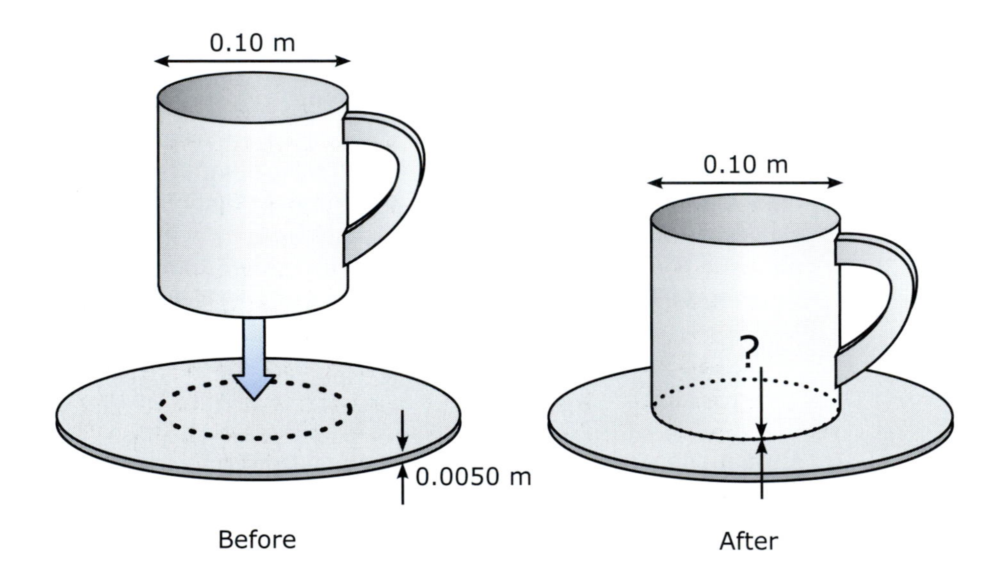

force of the cup acts directly downward and is not spread horizontally by the saucer?

9. What is the maximum number of such coffee cups that can be stacked vertically on the saucer and not cause a permanent dent in the plate? (**A: 4.3×10^5 cups**—a tough balancing act to follow!)

10. A coat hanger (like the one in the following figure) is made from polyvinyl chloride.[20] The "neck" of the coat hanger is 0.01 m in diameter and initially is 0.10 m long. A coat hung on the coat hanger causes the length of the neck to increase by 1.00×10^{-5} m. What is the mass of the coat? The stress-strain diagram is also provided. (**A: 1.3 kg.**)

11. How many such coats as per exercise 10 must be hung below the neck of the coat hanger to cause the neck to remain stretched after the coats are removed?

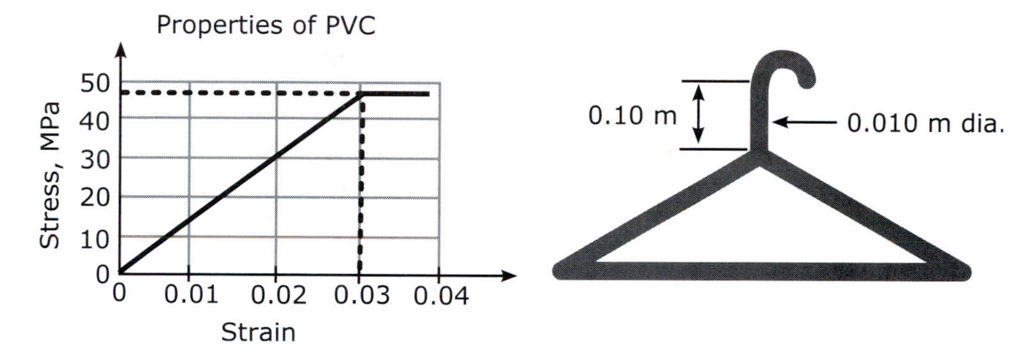

For Exercises 12–15, imagine the idealized situation as shown in (A) for an artillery shell striking an armor plate made of an (imaginary) metal called "armories." (B) is the stress-strain diagram of armories. Assume the material diagram is strained to failure.

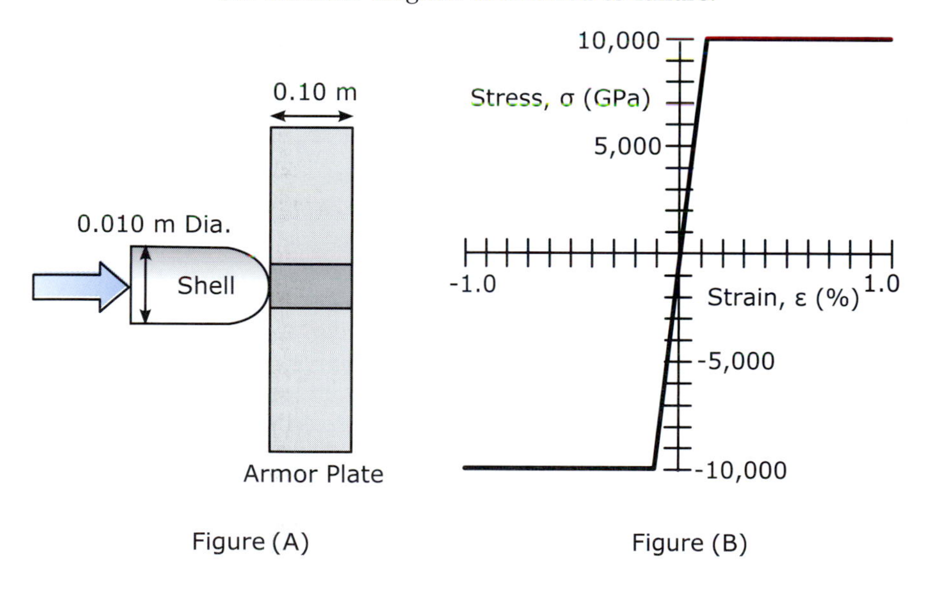

Figure (A) Figure (B)

[20]PVC is a very inexpensive polymer of the monomer $CH_2 = CHCl$ with repeating unit $-CH_2(CHCl)-$.

12. Suppose the shell has mass 1.00 kg and is traveling at 3.0×10^2 m/s. How much TKE does it carry?

13. Assume that the energy transferred in the collision between the shell and the armor plate in exercise 12 affects only the shaded area of the armor plate beneath contact with the flat tip of the shell (a circle of diameter 0.0010 m) and does not spread out to affect the rest of the plate. What is the energy density delivered within that shaded volume by the shell? (**A: 5.7×10^9 J/m^3.**)

14. Which of the following will happen as a result of the collision in the previous example? Support your answer with numbers.

    (a) The shell will bounce off without denting the armor plate.
    (b) The armor plate will be damaged but will protect the region beyond it from the collision.
    (c) The shell will destroy the armor plate and retain kinetic energy with which to harm the region beyond the plate.

    (**A: a**—but it is uncomfortably close to yielding and to a permanent set to the armor; an armored tank crew will at least get quite a headache.)

15. In exercise 14, what is the highest speed the artillery shell can have yet still bounce off the armor plate without penetrating it?

Exercises 16–19 involve the situation depicted below. Consider a "micrometeorite" to be a piece of mineral that is approximately a sphere of diameter $1. \times 10^{-6}$ m and density 2.00×10^3 kg/m^3. It travels through outer space at a speed of about 5.0×10^3 m/s relative to a spacecraft. Your job as an engineer is to provide a micrometeorite shield for the spacecraft. Assume that if the micrometeorite strikes the shield, it affects only a volume of the shield 1.0×10^{-6} m in diameter and extending through the entire thickness of the shield.

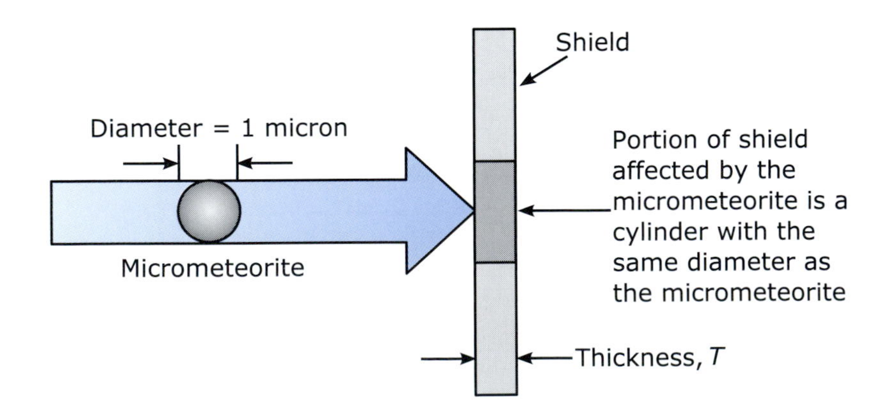

16. Using the stress-strain diagram for steel presented in this chapter, Figure 11, determine the minimum thickness a steel micrometeorite shield would have to be to protect the spacecraft from destruction (even though the shield itself might be dented, cracked or even destroyed in the process).

17. Using the stress-strain properties for a polymer (Figure 13 and Table 1), determine whether a sheet of this polymer 0.10 m thick could serve as a micrometeorite shield if this time the shield must survive a micrometeorite strike without being permanently dented or damaged. Assume the properties of the polymer are symmetric in tension and in compression. (**A: Yes,** it will survive unscathed.)

18. Suppose one was required to use a micrometeorite shield no more than 0.01 meters thick. What would be the required toughness of the material from which that shield was made if the shield must survive a micrometeorite strike without being permanently dented or damaged?

19. Suppose a shield exactly 0.01 m thick of the material in exercise 18 exactly met the requirement of surviving without denting or damage at a strain of –0.10, yet any thinner layer would not survive. What are the yield stress and modulus of the material? (**A:** σ_{Yield} **= –33 MPa,** E **= 0.33 GPa.**)

20. Your company wants to enter a new market by reverse-engineering a popular folding kitchen step stool whose patent has recently expired. Your analysis shows that by simplifying the design and making all the components from injection-molded PVC plastic you can produce a similar product at a substantially reduced cost.

However, when you make a plastic prototype and test its performance you find that it is not as strong and does not work as smoothly as your competitor's original stool. Your boss is anxious to get your design into production because she has promised the company president a new high-profit item by the end of this quarter. What do you do?

(a) Release the design. Nearly everything is made from plastic today, and people don't expect plastic items to work well. You get what you pay for in the commercial market.
(b) Release the design but put a warning label on it, limiting its use to people weighing less than 150 pounds.
(c) Quickly try to find a different, stronger plastic that can still be injection molded and adjust the production cost estimate upward.
(d) Tell your boss that your tests show the final product to be substandard and ask if she wants to put the company's reputation at risk. If she presses you to release your design, make an appointment to meet with the company president.

21. As quality control engineer for your company, you must approve all material shipments from your suppliers. Part of this job involves testing random samples from each delivery and making sure they meet your company's specifications. Your tests of a new shipment of carbon steel rods produced yield strengths 10% below specification. When you contact the supplier, they claim their tests show the yield strength for this shipment is within specifications. What do you do?

    (a) Reject the shipment and get on with your other work.
    (b) Retest samples of this shipment to see if new data will meet the specifications.
    (c) Accept the shipment, since the supplier probably has better test equipment and has been reliable in the past.
    (d) Ask your boss for advice.

Chapter 12

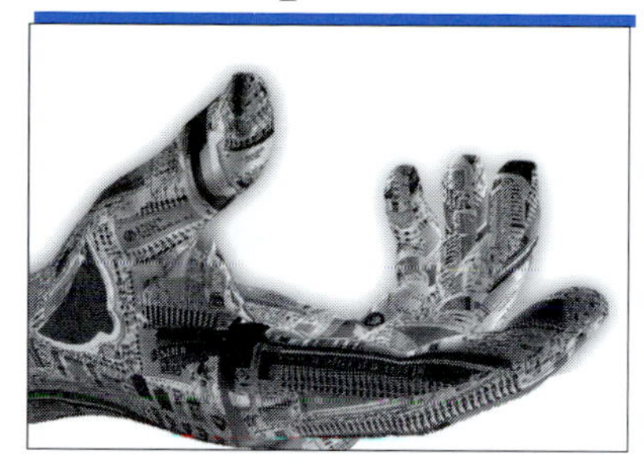

© iStockphoto.com/Antonis Papantoniou

Introduction to Bioengineering

Introduction

How do seat belts save lives? This is just one of the questions answered by the area of engineering called bioengineering. Bioengineering is the application of the methods of engineering to living things and, in particular, to humans. Aspects of the performance of living organisms can be understood using engineering techniques. Areas covered by bioengineers range from genetic engineering and biotechnology through biomechanics and biomaterials to signal processing and communications.

Through biocommunication engineering, engineers are developing new communication systems that might enable paralyzed people to communicate directly with computers using brain waves. Through bioinformation engineering, they are exploring the remarkable properties of the human brain in pattern recognition and as a learning computer. Through biomimetics engineers are trying to mimic living systems to create efficient designs. These areas range far beyond the example in this chapter, but their basic theme is similar.

By engineering analysis of the situations in which living matter might be placed and by characterizing in engineering terms the remarkable properties of living matter, knowledge can be gained that improves safety and protects health. That understanding can be used to make life safer and healthier. Among the tasks undertaken by bioengineers is the design of safety devices, ranging from football helmets to seat belts to air bags; the development of prosthetic devices for use in the human body, such as artificial hip joints; the application of powerful methods for imaging the human body, such as computed axial tomography (or CAT scanning) and magnetic resonance imaging (or MRI); and the analysis and mitigation of possible harmful health effects on humans subject to extreme environments, such as the deep sea and outer space.

This chapter presents some introductory engineering models for applying bioengineering to enhance safety. A focus will be on the use of seat belts as a safety device.

In this chapter you will be introduced to simple descriptions of human anatomy and of the effects of large forces on hard and soft human tissues. You will also learn (1) why collisions can kill, (2) how to make a first approximation of the likelihood of damage during collisions to the human body using a **fracture criterion,** (3) how to predict the injury potential of a possible accident using a criterion that could be called **"stress-speed-stopping-distance-area" (SSSA),** and (4) how to apply two other criteria for the effect of deceleration on the human body (the **30.*g* limit** and the **Gadd Severity Impact parameter**).

There are just two areas of human anatomy we will investigate: The first is to understand how serious blunt force trauma can affect the operation of the brain and other neurological tissue, and the second is to understand how bone protects soft tissues.

Biological Implications of Injuries to the Head

In many automobile accidents, the victims suffer severe head and neck injuries. Some of the accidents directly cause brain trauma, and others cause neck injuries. The kind of accident caused by severe overextension of the neck and associated tissues is called *whiplash.* Head and neck injuries are all too common in accidents where you suffer *high g decelerations.* What this means is that the victim absorbs very high decelerations on impact with another vehicle or with a stationary object. Still other victims are injured *within* the vehicle when they contact hostile interior components of the passenger compartment such as the dashboard or the windshield.

We will first take a brief look at human anatomy from the neck up to understand what forces of these magnitudes can do to neurological function. Figure 1 shows the essentials of the skull, the brain, the spinal column, and the spinal cord.

The skull protects the brain, which floats in a fluid-like layer. The base of the brain connects to the spinal cord through the spinal column, which provides protection for the spinal cord. The spinal column consists of individual bony vertebrae that surround the all-important spinal cord.

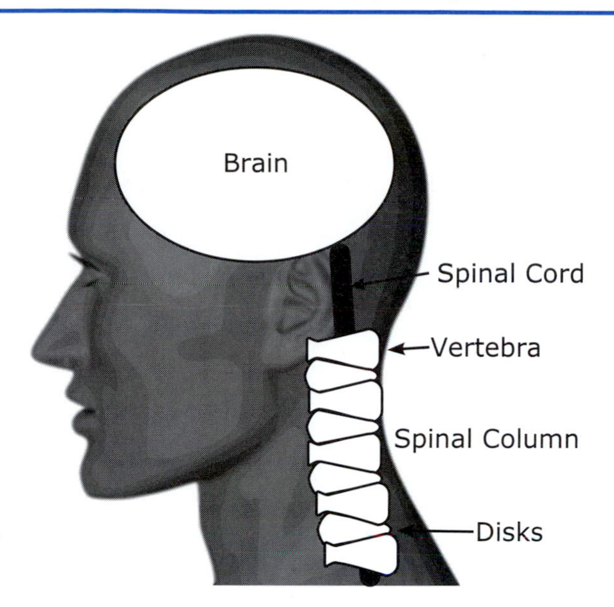

Figure 1

Part of the Human Nervous System

The spinal cord is the essential "wiring" that takes instructions from the brain to the various autonomic and voluntary bodily functions. The vertebrae are separated by cartilaginous matter known as "disks" to provide flexibility and motion. Injury to the disks can result in severe pain and, in the case of the spinal cord, in paralysis. Injury to the spinal cord[1] leads to very severe bodily malfunctions; injury to the brain causes a number of physical deteriorations or death. The brain is protected within the hard shell of the skull in which the brain floats in a fluid-like environment. The system, as efficient as it is, can suffer a number of possible injuries during high g decelerations (meaning decelerations of many times that due to gravity).

Consider the injuries that are illustrated in Figure 2: The brain will move relative to the skull in its fluid-like layer when experiencing high g's. In **hyperflexion** the skull will move forward relative to the brain, causing damage to the occipital lobe (the back of the brain) and in **hyperextension** it will move in the opposite direction relative to the brain and damage its frontal lobe. Further, damage can also occur to the basal brain, a potentially devastating injury, since it may interrupt the nerve connections to the spinal cord. Additionally there can be fractures to the vertebra and also extrusion or rupture of the protective disks.

[1] *Superman* movie star Christopher Reeve suffered such an injury in an equestrian fall and remained paralyzed for the rest of his life while bravely and effectively advocating for those suffering with spinal injuries.

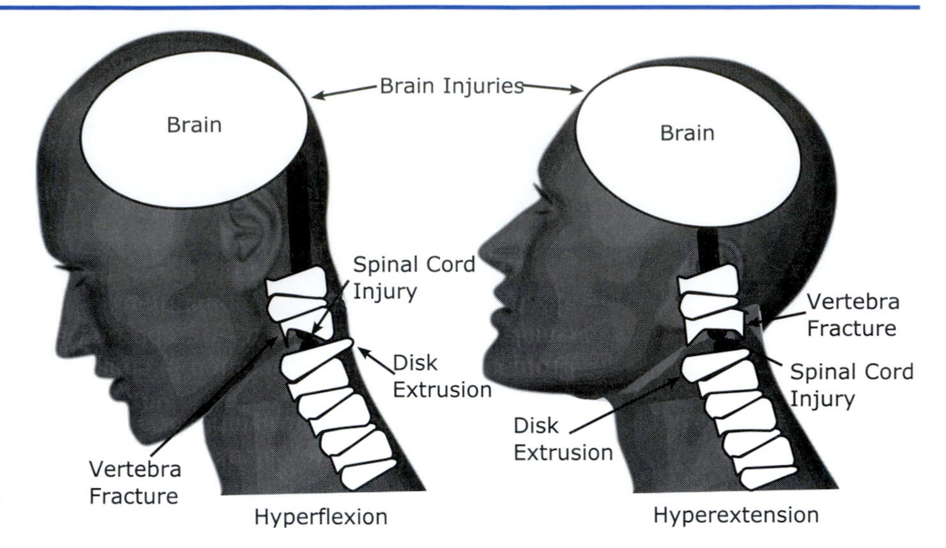

Figure 2

Brain Injuries Due to Extreme Whiplash

Why Collisions Can Kill

What is it about collisions that can kill or severely injure people? One answer is easy: Sudden stops can be harmful to your health! The challenge of this chapter is to express that answer using engineering variables. By using the right numbers and units for the values of those variables, criteria can be identified that distinguish a potentially fatal crash from one from which you walk away.

One key variable in the bioengineering analysis of automobile safety is deceleration. Just as the "zero-to-sixty"[2] time of acceleration is a measure of a car's potential for performance, the "sixty-to-zero" time of deceleration is a first measure of an accident's potential for injury. But while the zero-to-sixty time is typically measured in seconds, the sixty-to-zero time may be measured in milliseconds—that is, in thousandths of a second.

Rapid deceleration causes injury because of its direct relation to force. As discussed in the section on Newton's Second Law (Chapter 2), force is directly proportional to positive or negative acceleration (i.e., deceleration). And, as discussed in Chapter 11, force divided by area is stress. As further discussed in that chapter, the stress-strain curve gives the yield strength of a material. Suppose that the biologic material is flesh, bone, or neurons (nerve cells found in the brain, the spinal column, and local nerves). If its yield strength is exceeded, serious injury or death can follow.

[2]Legacy term for the time to go from a complete standstill to 60 mph.

An engineering analysis of collision injury combines the effects of acceleration with the properties of materials. So a first step in characterizing collisions is to determine the deceleration that occurs. We already have a tool for visualizing this: the $v - t$ graph.

EXAMPLE 1

A car traveling at 30. miles per hour (mph) runs into a sturdy stone wall. Assume the car is a totally rigid body that neither compresses nor crumples during the collision. The wall "gives" a distance of $D_S = 0.03\,\text{m}$ in the direction of the collision as the car is being brought to a halt. Assuming constant deceleration, calculate that deceleration.

Need: Deceleration = ____ m/s^2.

Know–How: First sketch the situation to clarify what is occurring at the impact. Then use the $v - t$ graph to envisage the collision. The slope of $v - t$ graph is acceleration.

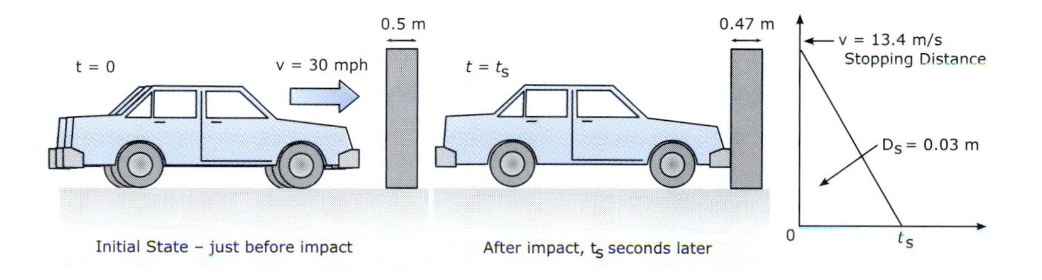

Solve: We first need to calculate the stopping time, t_S, to decelerate from 30. mph (13.4 m/s) to stationary.

From Chapter 10, if deceleration is constant, we know that distance equals the area under the $v - t$ graph or $D_S = 0.03\,[\text{m}] = \frac{1}{2} \times 13.4\,[\text{m/s}] \times t_S\,[\text{s}]$ where t_S is the stopping time.

$\therefore\ t_S = 0.06/13.4 = 0.0045\,\text{s}$.

Also the **deceleration** rate is the slope, $\triangle v/\triangle t = (0 - 13.4)/(0.0045 - 0) = 2980 = \mathbf{-3{,}000\ m/s^2}$. ■

What does this deceleration mean? How high is it? One good way to understand its damaging potential is to compare it to the acceleration due to gravity. This non-dimensional ratio is 306.

The use of acceleration in terms of g can be quite helpful. Referring to Chapter 2, what is the weight of a mass m on earth? It is mg; ditto, what is the force on a mass accelerating by an amount a? It is ma. But let us simultaneously multiply and divide that force by g—that is,

$F = ma = mg \times a/g =$ **the body's weight × number of g.**

In the preceding example, this means your head seems to weigh 306× its usual weight, ditto your legs, arms. An average human head has a mass of 4.5 to 5. kg so effectively it will experience force of ~4.75 [kg] × 9.8 [m/s^2] × 306 = 14,000 N or 3,200 lbf! Your whole body will experience a proportionate force.

You can also calculate the forces involved by kinetic considerations as opposed to the basically kinematic considerations above. The force × distance to stop is equal to the kinetic energy that is to be dissipated. Let's assume the driver is belted into the car and therefore decelerates at the same rate as the car.

EXAMPLE 2

The car and its belted driver of Example 1 suffers the same constant deceleration of 306 g or 3,000 m/s^2. What is the force he experiences during the collision if his mass is 75 kg? Assume the car is a totally rigid body that neither compresses nor crumples during the collision.

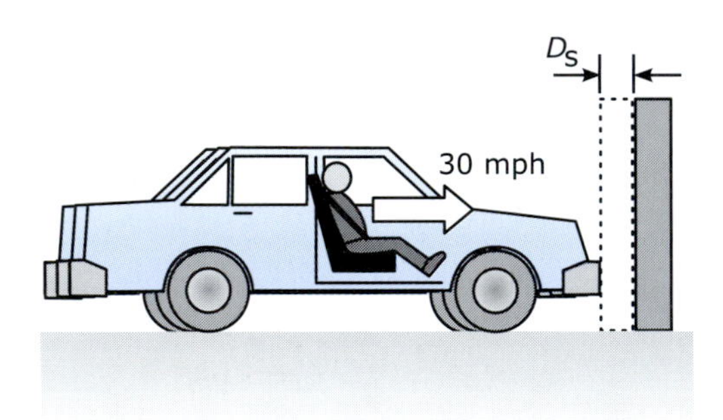

Need: Force stopping 75-kg man on sudden impact in SI force = _____ N.

Know: The driver's weight is 75 [kg] × 9.8 [m/s^2] = 735 N and deceleration rate is 306 g.

How: $F = ma = mg \times a/g$.

Solve: $\boldsymbol{F}$ = 735 × 306 = **2.24 × 10^5 N** (or ~50,000 lbf!). ■

The Fracture Criterion

Under what conditions does a force cause living material to fail? (That is, to crack, or break into pieces, or lose its ability to contain or protect fluid or soft structures, or to disintegrate.) As we saw in Chapter 11, a material breaks when the stress on it exceeds its strength. Stress and strength are variables with units (N/m^2) that can be calculated from a description of a situation and the material's stress strain diagram.

One aspect of the situation needed is the force exerted, which has been found in Example 2. Another aspect is the area over which that force is applied. This area is needed in order to calculate the stress on the material in question. Figure 3 shows an elementary picture of bone.

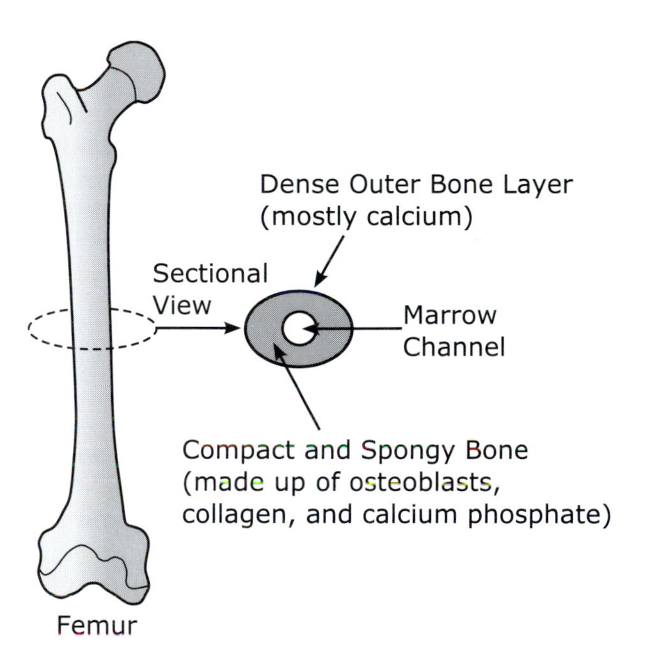

Figure 3

Structure of Body's Long Bones

Bone is a natural composite material that consists of a porous framework made of the mineral calcium phosphate, interspersed with fibers of the polymeric material "collagen." The calcium phosphate gives bone its stiffness, and the collagen provides flexibility. Figure 3 gives a view of the internal structure of the body's all-important long bones that have a hard outer layer, spongy interior layers with axial channels built from bone cells known as osteoblasts, and a central core of marrow that is responsible for renewing blood supply and other important cells. Bone is not uniform in its properties in different directions, a condition known as "anisotropy."

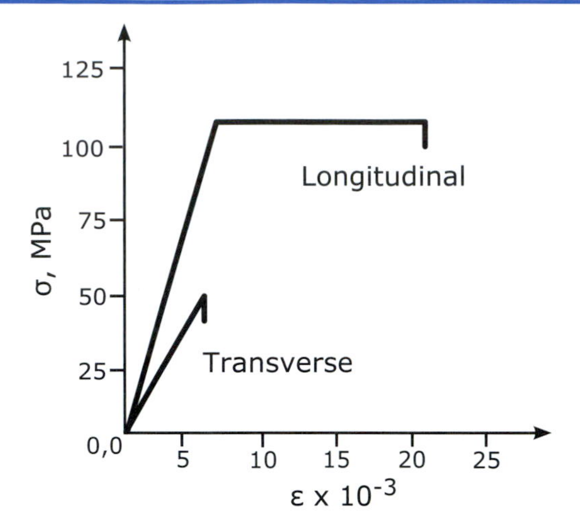

Figure 4

Assumed Bone Mechanical Properties

Table 1 Approximate Mechanical Properties of Human Bone

Direction	Elastic yield, MPa	Young's modulus, GPa	Strain at yield, %	Strain at fracture %	Toughness, at yield/fracture, MJ/m^3
Longitudinal	110	16.	0.7	2.	0.4/1.8
Transverse	60.	9.	0.7	0.7	0.2/0.2

This means, among other properties, that it has quite different mechanical properties along its axis and perpendicular to the axis.[3]

The properties of bone can be approximated by a stress-strain diagram[4] as in Figure 4. Notice bone is more rigid and flexible along (that is in the longitudinal direction) the bone's length than crosswise (that is in its transverse direction). These properties are due to the dense outer bone layer being tough and with the compact and spongy interior bone providing surprisingly high flexibility. Notice our model is a simplification of real bone behavior and is modeled as a perfectly elastic and perfectly plastic material. The corresponding data are as per Table 1.

Bone, like other materials, may fail in several different ways:

1. If bone is subjected to a local stress greater than its yield strength, it will take on a permanent deformation.

[3]Wood has similar anisotropy; you can split wood rather easily along the grain, but it is hard to break cross grain. The cellular patterns in wood are responsible for this useful behavior (for trees as well as for people).

[4]Adapted from A. H. Burstein and T. M. Wright, *Fundamentals of Orthopœdic Biomechanics* (Williams and Wilkins, Baltimore, MD, 1994), p. 116.

2. If that deformation exceeds the strain at fracture, the bone will break.
3. If the total energy deposited exceeds the toughness at yield × volume of affected bone, the bone will take a permanent set.
4. If the total energy deposited exceeds the toughness at fracture × volume of affected bone, the bone will break.

See Chapter 11 for further considerations along these lines. In general, bone failure modes 1 and 2 most strictly apply to the situation in which the load is applied slowly. For a suddenly applied load, it is usually the toughness criterion that applies. (Indeed, toughness is measured by the strike of a heavy hammer in a sudden blow.)

Most bones that fail in vehicular accidents are due to transverse stress rather than longitudinal stress[5] (for example, the ribs hitting the steering wheel if their owner is not seat-belted or air-bag protected). However, whiplash injuries, in which the neck is subject to longitudinal stress, may fail in that direction rather than in the transverse direction. In general, you would prefer not to permanently displace bone; therefore, it is the elastic yield strain that is generally the more conservative situation for the victim of a crash. Bone fracture is a more severe situation.

In another version of the accident as described in Examples 1 or 2, the driver is unbelted, so the skull hits the dashboard and dents. How does the stress on the skull compare with the elastic yield point? That depends on the area of the skull over which the force is applied. For a given force, the smaller the area, the higher the localized stress.

EXAMPLE 3

This is an accident similar to that in Examples 1 and 2, except the driver is not belted. Assume the driver's skull dents the dashboard to a depth of 0.03 m. The area of contact between the driver's forehead and the dashboard is 3. cm × 3. cm ($9. \times 10^{-4}\,\text{m}^2$), and the properties of bone are those given in Table 1. Will the collision *fracture* the skull?

Need: Yield strength of skull exceeded = _____ (yes/no).

Know: Force at impact is 2.24×10^5 N (since the head suffers the same deceleration of 306 g given $D_S = 0.03$ m).

How: Stress is force per unit area and from Table 1, the yield strength of the bone is 60. MPa.

Solve: Compare $60 \times 10^6\,[\text{N/m}^2]$ with the applied stress on impact. That stress is $2.24 \times 10^5\,[\text{N}]/(9. \times 10^{-4})\,[\text{m}^2] = 2.5 \times 10^8\,\text{N/m}^2$.

Yes. The bone will fracture, and the brain will be seriously damaged. Unfortunately, this is probably one fatally injured driver. ■

[5]Think of snapping a dead tree limb across your knee. The break occurs on the side away from your knee as fulcrum. That layer of branch is being *stretched* and it fails mostly in tension.

The Stress-Speed-Stopping Distance-Area Criterion

Let's now generalize the lessons of that example. Collisions kill or injure because they involve a high rate of deceleration. That deceleration causes contact with some physical material, and the result is experienced as a force. The smaller the area over which that force is applied, the higher the stress on the bone or tissue. The higher the localized stress, the greater the likelihood that the material will fail in that region.

We can express that insight as a relationship among variables. Consider first the generalized form of the $v-t$ diagram of a collision with a constant deceleration rate. Suppose a body of mass m is subjected to a deceleration of a and stops in a distance D_S. Then the force[6] experienced by the body = $mv^2/2D_S$. Now suppose that this force is experienced as the contact of an area A of the head or body with a surface. Then the stress can be calculated as follows:

$$\text{Stress, } \sigma = mv^2/(2AD_S)$$

This useful relationship may be called the **stress-speed-stopping distance-area (SSSA) criterion.** It states that, technically, it's not enough to say that "speed kills." What kills is the combination of high speeds, short stopping distances, *and* small contact areas!

Strategies for reducing the severity of collision follow from these insights. By decreasing speed, increasing the stopping distance, and increasing the area of application of the force, the effects of the collision on living materials can be reduced.

The presence of the v^2 term in the numerator of the SSSA indicates that decreasing speed is a highly effective way of decreasing collision severity. Cutting speed in half will, everything else remaining equal, reduce the stress of a collision to one-fourth of its previous value.

How might the other two terms in the SSSA be brought into play? Application of the area term is simple in concept, though more difficult in practice. The larger the area of contact between body and surroundings during a collision, the smaller the stress. This is one motivation behind the air bag. It is a big gas-filled cushion that can increase the area of contact by a large factor. Of course, that cushion must be deployed rapidly enough to be effective in a collision. It must also *not* be deployed unless a collision has just occurred. Only by conquering these two technical challenges (deployment in a few milliseconds when appropriate, nondeployment at all other times) did engineers turn the air bag into a valuable safety tool.

However, the area portion of the SSSA criterion is only part of the air bag story. It is an even smaller part of the story of an even more important

[6]Derived from these steps: $a=v/t$; $F=ma=mv/t$, and $D_S = ½\ vt$.

Figure 5

Seat Belts and Crumple Zones

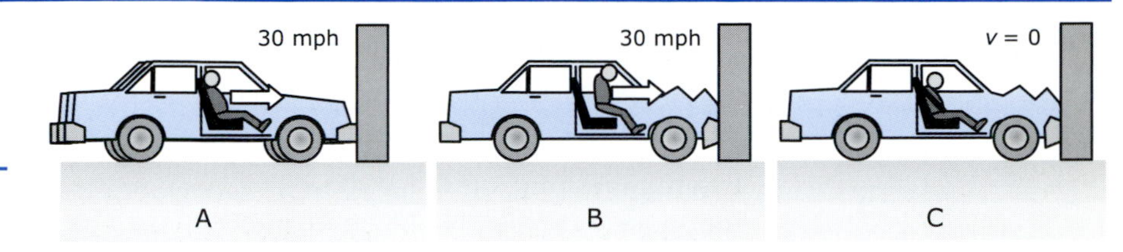

safety tool: the seat belt. A seat belt does indeed increase the area over which force is applied (compared to the forehead) and as importantly reduces the probability of an impact of the head with the dashboard. But that increase would not by itself account for the great life-saving and injury-reducing potential of seat belts. A more important part of that potential involves that other term in the denominator, the stopping distance, D_S.

A more practically important contribution of seat belts is to *attach the driver to a rigid internal passenger shell while the rest of the car shortens by crumpling.* The high effectiveness of the seat belt as a safety device has only been achieved in combination with the design of a car that significantly crumples during a collision. Let's now assume that the car has such a "*crumple zone.*" That is illustrated in Figure 5.

Picture A shows the situation of driver, car, and wall during a 30. mph collision. The car rapidly crumples to a stop several milliseconds later, as shown in picture B. But the driver is still moving forward at the speed of 30. miles an hour (as described by Newton's Law of Motion). It is the collision of the driver's head and the windshield that brings the driver to a stop in a short distance on the order of a few 1/100 of a meter. By contrast, when the driver is wearing a fastened seat belt, the body stops with the car, as shown in picture C.

In the latter case, the driver gets the full benefit of the crumple of the car, which is on the order of a half to one meter. This tenfold increase in stopping distance results, according to the SSSA criterion, in a reduction in stress to one-tenth of its previous value. This can be the difference between life and death.

EXAMPLE 4

A car traveling at 30. miles per hour (13.4 m/s) runs into a sturdy stone wall. Assume the car's crumple zone results in a stopping distance of 0.60 m. Assume that the 75 kg driver is wearing a fastened seat belt. Assume that the area of contact of seat belt and body is 4.0 cm × 30. cm (0.012m^2). Determine the maximum stress on the driver's body.

Need: Maximum stress on the driver's body _____ N/m^2.

Know–How: We could simply repeat the analysis of the previous examples, but the SSSA criterion provides a shortcut.

Solve: Stress, $\sigma = mv^2/2aD_S$.

Substitute into the SSSA formula and $\sigma = 75. \times 13.4^2/(2 \times 0.012 \times 0.60)$ [kg] [m/s]^2[1/m^2][1/m] = **9.4 × 10^5 N/m^2 ≈ 1 MPa** (≪ elastic yield of bone, Table 1).

The number of g's is $9.4 \times 10^5 \times 0.012/(75. \times 9.8)$ [N/m^2][m^2][1/kg][s^2/m] = 15. (Previously it was 306, so survivability has been greatly enhanced.) ■

To sum up: here is an answer to our original question, expressed in engineering variables and units. The principal way seat belts save lives is by attaching the driver securely to the inner shell of the car, enabling the driver to take advantage of the car's crumple zone of about 0.5–1 m. During a collision from 30. miles an hour, that strategy restricts deceleration to less than about 150 m/s^2, resulting in stresses on the body that are less than the 1 MPa and almost surely less than required to break bones.

Criteria for Predicting Effects of Potential Accidents

The preceding sections have established the point that the effects of deceleration on human bone and on tissue offer a major safety issue. We have just indicated one way to address that issue by translating deceleration into force, and force into stress. A comparison of maximum stress experienced by the body with the compressive strength of bone provides a first criterion for the prediction of accident severity.

The effects of acceleration and deceleration on the human body go far beyond the potential to break bones. These effects range from the danger of blackouts of pilots experiencing very high acceleration to the bone loss and heart arrhythmia problems experienced by astronauts exposed for a long time to the micro-gravitational forces of space flight.

We will continue to focus on the effects of high accelerations. Just how many g's can the human body stand? As a first approximation, engineers drew on a wide range of experiences, such as those of pilots and accident victims, to arrive at an initial criterion: The human body should not be subjected to *more than 30 g's*. At 30 g's and above, the damaging effects of acceleration or deceleration on the human body can range from loss of consciousness to ruptured blood vessels to concussion to the breaking of bone to trauma or death. This criterion still serves as an *initial* rule of thumb for the design of safety devices.

EXAMPLE 5

Does the experience of a driver without a seat belt who experiences a 30 mile per hour collision and is stopped in 0.1 m by a padded dashboard exceed the 30 g criterion? Does the role of a seat belt in taking advantage of the crumple zone to increase collision distance to 0.6 meter meet the 30 g criterion?

Need: Deceleration without seat belt = __ (less than/equal/more than 30. g).
Deceleration with seat belt = __ (less than/equal/more than 30. g).

Know-How: $F \times D_S = (1/2)mv^2$ and $\therefore F = (1/2)mv^2/mgD_S = (1/2)v^2/gD_S$.

Solve: Doing the arithmetic:

Deceleration without seat belt is 92 g—that is, **more than 30 g.**
Deceleration with seat belt is 15 g—that is, **less than 30 g.**

Consistent with our previous analysis, the combination of seat belt and crumple zone has reduced the driver's deceleration from a probably fatal 92 g to a probably survivable 15 g. ■

However, experiences under extreme conditions, such as high-speed flight, soon revealed that the 30. g criterion was seriously incomplete. In some cases, humans have survived accelerations far above 30 g. In other cases, accelerations significantly below 30 g caused serious injury.

The missing element in the simple 30 g criterion is **time.** Deceleration has many effects on human tissue, ranging from destruction at one extreme to barely noticeable restriction of blood flow at the other. Each of these effects has a characteristic time interval needed to take effect. Very high decelerations, measured in hundreds of g's, can be survived *if* the exposure is short enough. Deceleration at moderate g levels, on the other hand, can prove fatal if the exposure is long enough (see Figure 6).

In Figure 6, the axes are expressed in terms of the $\mathbf{log_{10}}$ of the variables. Not only can we get a scale covering orders of magnitude by this neat trick, but also we can infer something important about the relationship. In fact, it is linear, not between the actual variables g and t_S but between $\log g$ and $\log t_S$.

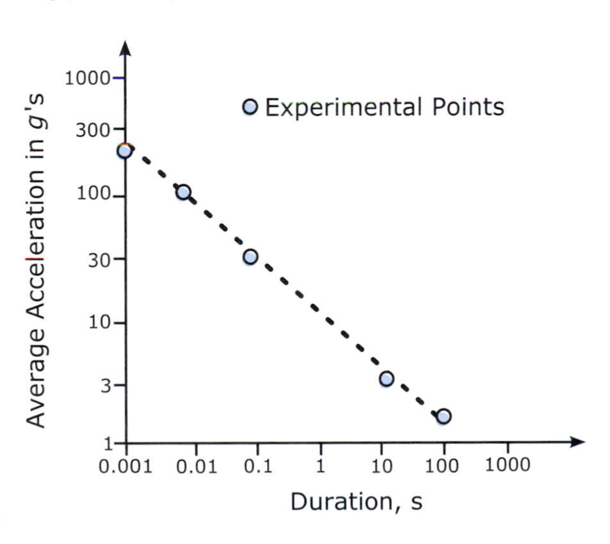

Figure 6

Injury Criterion Based on g and t (S.A. Berger et al. 1996. *Introduction to Bioengineering*, Oxford University Press.)

The dotted line represents a *rough boundary* between accelerations likely to cause serious injury or death (above and to the right of the line) and survivable accelerations (below and to the left of the line). Notice that the previous "30 g" criterion is the point on that line corresponding to a duration of 0.1 seconds (the right order of magnitude for the duration of an automobile accident).

Figure 6 can also be expressed as a formula. A straight line on a log-log plot corresponds to the fact the number on the x-axis is raised to a certain power. In this case, the formula for the line dividing serious injury from survivability is:

$$a = (0.002\ t_S)^{-0.4}$$

with a expressed in the nondimensional units of g and with t_S expressed in seconds. That is, the slope of the log/log graph is −0.4. This is *not* a basic scientific law. Rather it is an empirical relationship summing up the net effects of many physical and biological properties. It is, however, a useful guide. It is made more convenient by being rearranged into the form:

$$a^{2.5}\ t_S = 500$$

The quantity on the left is defined as the **Gadd Severity Index (GSI).**[7]

$$\mathbf{GSI} = \boldsymbol{a^{2.5} t_S} \text{ (with } t_S \text{ in seconds and } a \text{ in } g)$$

The GSI can be treated as a numerical index, not an engineering variable. Note that, for introductory purposes, this is a simplified version of the GSI, assuming either constant deceleration or that a meaningful average deceleration can be measured. First, calculate the GSI, and then compare the result to 500. If the GSI is greater then or equal to 500, there is a serious danger of injury or death.

EXAMPLE 6

Calculate the GSI for a driver without a seat belt who experiences a constant deceleration in a 30. mile per hour collision and is stopped in 0.05 m by the dashboard.

Need: GSI = _______ (a number).

Know–How: By an analysis similar to Example 5.

$D_S = 0.05 \text{ m} = \frac{1}{2}\, vt_S = \frac{1}{2} \times 13.4 \times t_S$ [m/s][s].

$\therefore\ t_S = .05/6.7 = 0.0074$ s.

$\therefore\ a = (0 - 13.4)/(0.0074 - 0) = -1810 \text{ m/s}^2\ \therefore\ a = 190\, g$

Solve: GSI $= a^{2.5}\ t_S = 190^{2.5} \times 0.0074 = 3{,}700$. This suggests that severe injury or death is likely. ■

[7]The Gadd Severity Index was introduced in C. W. Gadd, in 1961. "Criteria for Injury Potential." National Research Council Publication #977 (Washington National Academy of Sciences) pp. 141–145.

EXAMPLE 7 Calculate the GSI for a driver with a fastened seat belt who experiences a constant deceleration in a 30. mile per hour collision and is stopped by a 1.0 m crumple zone.

Need: GSI = _______ (a number).

Know–How: By an analysis similar to Example 6.

$$D_S = 1.0\ m = \tfrac{1}{2} \times 13.4 \times t_S$$
$$t_S = 1.0/6.7 = 0.15\ \text{s}$$
$$a = -13.4/.15 = -89.\ \text{m/s}^2$$
$$\therefore a \text{ in } g\text{'s} = 9.\,g$$

Solve: **GSI** $= a^{2.5}\, t_S = 9.^{2.5} \times 0.15 =$ **36.** This suggests that severe injury or death is very unlikely. ■

SUMMARY

Bioengineering applies our previously learned methods of engineering to living bodies, organs, and systems. This chapter illustrated just one aspect of bioengineering, the use of biomechanics and engineering analysis of biomaterials to improve safety. In this chapter you learned why collisions can kill, how to make a first approximation of the likelihood of collision damage to the human body using a fracture criterion, how to predict the injury potential of a possible accident using the **stress-speed-stopping distance-area (SSSA)** criterion, how to apply two criteria for the effect of deceleration on the human body (the $30\,g$ limit and the **Gadd Severity Impact** parameter) and how to analyze bioengineering problems.

EXERCISES

Exercises 1 and 2 concern the following situation: A car is traveling 30. mph and hits a wall. The car has a crumple zone of zero, and the passenger is not wearing a seat belt. The passenger's head hits the windshield and is stopped in the distance of 0.10 m. The skull mass is 5.0 kg. The area of contact of the head and the windshield is $0.010\ \text{m}^2$. Assume direct contact and ignore the time it takes the passenger to reach the windshield.

1. Provide a graph of $v - t$ of the collision of the skull and the windshield, and then graph the force experienced by the skull as a function of time. (**A: 4,500 N.**)
2. If the compressive strength of bone is $3.0 \times 10^6\ \text{N/m}^2$, will the collision in the previous exercise break the skull?

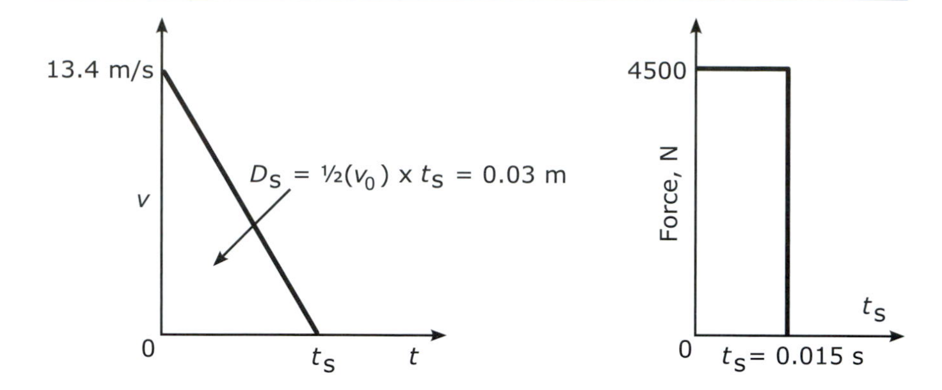

Exercises 3 and 4 concern an experiment in the 1950s when Air Force Colonel John Paul Stapp volunteered to ride a rocket sled to test the resistance of the human body to "g forces." The sled accelerated from 0 to 625 miles per hour in 5.0 seconds. Then the sled hit a water brake and decelerated in 3.0 seconds to a standstill. Assume that Stapp was rigidly strapped into the sled and that he had a mass of 75 kg.

3. Prepare $v - t$ and $F - t$ graphs of Stapp's trip, and compute the "g forces" he experienced in the course of acceleration and deceleration. **(A: 5.7 g–9.5 g.)**

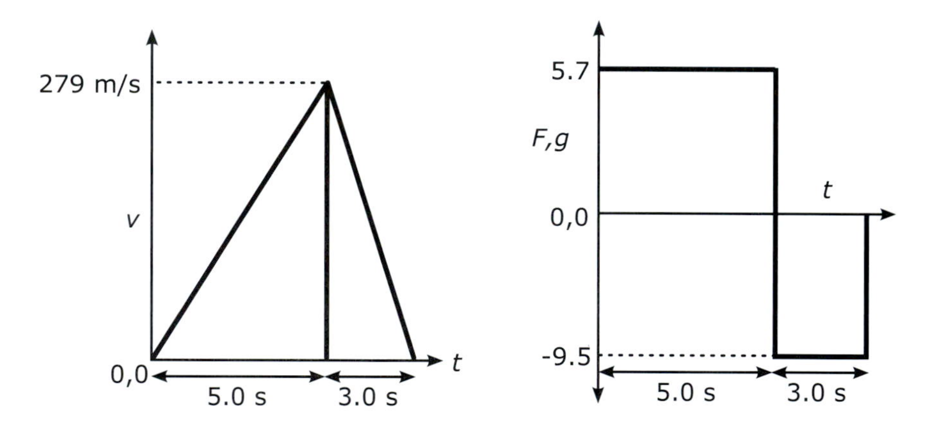

4. Using the force vs. time graph (Figure 6) for human resistance to "g forces," predict whether Stapp suffered serious injury in the course of his record-breaking trip.
5. A tall person sits down on a sofa to watch TV. Assume that the center of gravity of the person falls 1.0 m with constant gravitational acceleration in the course of sitting down. The sofa compresses by 0.05 m. Assume constant deceleration. Determine the g forces experienced by the person in the course of this sitting down. (**A: 20. g**—so be kinder to couch potatoes!)

Exercises 6 and 7 concern an infant's rear-facing safety seat as illustrated here.

6. A rear-facing child safety seat holds a child of mass 12 kg rigidly within the interior of a car. The area of contact between the seat and the child is $0.10\,\text{m}^2$. The car undergoes a 30. mph collision. The car's crumple zone causes the distance traveled by the rigid interior to be 1.0 m. Give the stress experienced by the child's body in terms of a fraction of the breaking strength of bone, assuming an infant's bones break at a stress of $10.\,\text{MN/m}^2$. (**A:** As a fraction of breaking stress, $\mathbf{1.08 \times 10^{-3}}$, bones should hold and the infant should be safe.)
7. A rear-facing child safety seat holds a child of mass 25 kg rigidly within the rigid interior of a car. The area of contact between the seat and the child is $0.10\,\text{m}^2$. The car undergoes a 30. mph collision. The car has no crumple zone, but a harness attached to the car seat stops it uniformly within a distance of 0.30 m. According to the Gadd severity index, will the child sustain serious injury or death?
8. Consider a parachute as a safety device. When the parachute opens, the previously freely falling person has typically reached a speed of about 50 m/s. The parachute slows to a terminal speed of about 10 m/s in 1.3 s. Approximating this set of motions by a constant deceleration, what is the maximum g experienced by the parachutist? (**A:** $\mathbf{-3.1}\boldsymbol{g}$**.**)
9. In the previous problem, the force exerted by the parachute is spread by a harness in contact with $0.50\,\text{m}^2$ of the parachutist, and the parachutist has a mass of 75 kg. What is the force per unit area (stress) experienced by the person during the deceleration?
10. The parachutist in the previous two exercises hits the ground (still wearing the parachute!) and is stopped in a distance of 0.10 m. If this final deceleration is constant, calculate the Gadd Severity Impact of the landing. (**A: GSI = 370.**)
11. A 75 kg person jumping from a 1.00×10^3 m cliff will reach a terminal speed of 50. m/s and uses a 1.00×10^2 m bungee cord to slow the descent. The bungee cord exerts a force F proportional to its extension, where F (in newtons) = (5.0 N/m) × (extension in m) and is designed to extend by 5.00×10^2 m in the course of bringing the user to a stop just above the ground. Is the maximum deceleration in g's experienced by the falling

person more or less than the maximum deceleration experienced by a parachutist undertaking the same leap (excluding landing forces)?

12. The air bag is designed to inflate very quickly and to subsequently compress if the driver hits it. A collision uniformly stops a car from 30. mph to 0.0 and then triggers the air bag. The driver is not seat belted and so hits the inflated air bag. This acts as a local "crumple zone" and consequently compresses by 0.20 meters as his head is brought to rest. According to the Gadd Severity Index, will the driver suffer serious injury? Assume constant deceleration of the driver's head after hitting the air bag. (**A: GSI = 420,** and no, the driver should not suffer serious injury or death.)
13. Two 100 kg football players wearing regulation helmets collide helmet to helmet while each is moving directly at each other at 10.0 m/s and come to a near instantaneous (<1 millisecond) stop. The area of contact is 0.01 m^2, and the helmets are each designed to provide a crumple zone of 0.025 m. What is the maximum stress exerted on each player? (**A: 4.0×10^7 N/m^2.**)
14. A designer of football helmets has two options for increasing the safety of helmets but for economic reasons can implement only one. One option is to double the area of contact that will be experienced in a helmet-to-helmet collision. The other is to double the crumple distance experienced in a helmet-to-helmet collision. Which will be more effective in reducing the maximum stress? (**Hint:** Try the previous exercise first.)
15. A soccer player "heads" a wet 0.50 kg soccer ball by striking with his forehead a ball moving toward him. Assume the player initially moves his head forward to meet the ball at 5.0 m/s and the head stops after the ball compresses by 0.05 m during impact. Assume the deceleration of the head is constant during impact. Compute and comment on the calculated Gadd Severity Impact of heading a soccer ball under these conditions.
16. A car strikes a wall traveling 30. mph. The driver's cervical spine (basically the neck) first stretches forward relative to the rest of the body by 0.01 m and then recoils backward by 0.02 m as shown below. Assume the spine can be modeled by a material with a modulus E = 10. GPa and a strength of 1.00×10^2 MPa. Will the maximum stress on the cervical spine during this "whiplash" portion of the accident exceed the strength of the spine? Assume a 0.15 m length of the cervical spine. (**A:** Stress on cervical spine is **greater than** its tensile strength.)

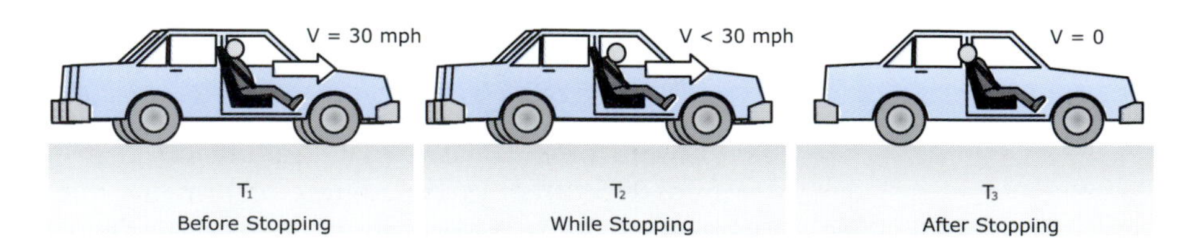

17. Which do you think has been more effective in reducing fatalities on American highways, seatbelts or airbags? Give an engineering reason for your answer, containing variables, numbers, and units. (**Hint:** Recall the SSSA formula previously developed and consider what other safety element is designed into a modern automobile.) Then go on the Web and see if you were right.
18. As a bioengineer at the Crash Safety Test Facility of a major automobile company, you are asked to provide more data for the Gadd Severity Index (Figure 6). Your boss suggests using live animals, dogs and cats from the local pound, in hard impact tests and then inspecting them for injury. You know their injuries will be severe or fatal, and using dogs and cats seems cruel. What do you do?

    (a) Nothing. Live animals are used regularly in product testing, and besides, they will probably be killed in the pound anyway.
    (b) Suggest using dead animals from the pound, since their impact injuries probably don't depend on whether or not they are alive.
    (c) Suggest using human cadavers since you really want data on humans anyway.
    (d) Suggest developing an instrumented human mannequin for these tests.

19. You are now a supervisor in the bioengineering department of a major motorcycle helmet manufacturer. Your engineers are testing motorcycle helmets manufactured by a variety of your competitors. Motorcycle helmets contain an inner liner that crushes upon impact to decrease the deceleration of the head on impact. This liner material is very expensive and can only be used once (i.e., once the helmet sustains a single impact it must be replaced). Your company has developed an inexpensive liner that will withstand multiple impacts but is less effective on the initial impact than any of your competitors. The vice president for Sales is anxious to get this new helmet on the market and is threatening to fire you if you do not release it to the manufacturing division. What do you do?

    (a) Since your company has invested a lot of money in the development of this helmet, you should release it, and besides, if you don't, someone else will.
    (b) Recommend continued testing until your company's product is at least as good as the worst competitor's product.
    (c) Contact your company's legal department to warn them of a potential product liability problem and ask for their advice.
    (d) Go over the vice president's head and explain the problem to the company's president.

20. During World War II, Nazi Germany conducted human medical experimentation on large numbers of people held in its concentration camps. Because many German aircraft were shot down over the North Sea,

they wanted to determine the survival time of pilots downed in the cold waters before they died of hypothermia (exposure to cold temperatures). German U-boat personnel faced similar problems.

In 1942, prisoners at the concentration camp in Dachau were exposed to hypothermia and hypoxia experiments designed to help Luftwaffe pilots. The research involved putting prisoners in a tank of ice water for hours (and others were forced to stand naked for hours at subfreezing temperatures) often causing death.

Research in the pursuit of national interests using available human subjects is the ultimate example of questionable bioengineering. Since the Nazi scientific data were carefully recorded, this produces a dilemma that continues to confront researchers. As a bioengineer today, should you use these data in the design of any product (such as cold-weather clothing or hypothermia apparatus for open heart surgery)?

(a) Since these experiments had government support and were of national interest at the time, they should be considered valid and available for scientific use now.
(b) You should use these data, since similar scientific experiments have been conducted in other countries during periods in which national security was threatened, and these data are not questioned today. Even the United States conducted plutonium experiments on unsuspecting and supposedly terminally ill patients (some of whom survived to old age!).
(c) This is just history and should have no bearing on the value or subsequent use of the data obtained.
(d) Do not use the data.

Chapter 13

© iStockphoto.com/Josh Rodriguez

Introduction to Chemical Engineering

Introduction

Why don't you pull your car up to an oil well and fill it up? You probably don't have an oil well in your backyard, but what if you did? Could the oil that comes directly out of the ground run your car?

This chapter uses those questions to introduce the area of chemical engineering. The short answer to the third question is no. Your car would sputter to a stop if the tank was filled with oil from an oil well. In response to that negative answer, chemical engineers design oil refineries that convert crude oil from the well into gasoline at the pump (and many other useful and related compounds). Gasoline is simply one component of crude oil that has been tailored to keep your car running smoothly and efficiently.

Chemical engineers also work with catalysts and with chemical reactors that depend on catalysts to achieve their goals. Catalysts are compounds that speed chemical reactions and promote one or more desirable products of a chemical reaction. In oil refining, **catalytic reactors** are used to make automotive fuels or heating oils according to the seasonal demands. In addition, chemical engineers design car **exhaust** and **catalytic converters** that ensure that cars meet pollution standards. They also develop processes for making polymers, invent ways of improving the delivery of medicines or of starving the delivery of nutrients to cancer cells, develop pollution cleanup processes, make modern electronic materials, and do a wide range of other jobs.

The principal tool in this chapter is one you have already seen: **stoichiometry**—the conservation of matter measured in appropriate units. Here it is used in connection with selected **control boundaries** chosen to emphasize input and output streams of various processes. In this chapter we focus mostly on physical processes so that molecular identity is conserved. To exemplify the power of this methodology, you will (1) address the challenge of **oil refining;** (2) learn about **flowsheets,** a tool that is the essence of traditional chemical engineering; (3) use **process analysis** (related to a larger conceptual methodology known collectively as **unit operations**) and use it to model basic chemical engineering processes; (4) model a highly simplified process of **distillation** and use this model to predict the output of the various products of an oil refinery; (5) model how refinery output can be improved by such processes as **flash distillation,** and **thermal and catalytic cracking;** and (6) apply **stoichiometry** to analyze another important chemical engineering innovation, an automobile's **catalytic converter.**

People have been doing chemical engineering ever since the prehistoric discovery of the psychoactive properties of fermented grains and grapes. However, the modern profession of chemical engineering dates only from the late nineteenth century. At that time, groups of chemists and engineers worked jointly to design chemical plants. They decided that they had enough in common to create a new discipline, which was formalized by a like-minded group of faculty at MIT who built on the experiences of a previous generation of British chemists and engineers.

Oil Refining

The new discipline soon had a major challenge from the rapid growth of the automobile industry. Millions of cars had to be regularly, reliably, and safely supplied with fuel. We have seen that chemical energy is typically supplied to automobiles mostly in the form of hydrocarbons (Chapter 5). These molecules, made of hydrogen and carbon atoms, combine with oxygen to yield carbon dioxide, water, and energy. The most convenient sources of hydrocarbons lie concentrated in deposits under the ground, the remains of animals and plant matter that lived hundreds of millions of years ago. The name for these hydrocarbon deposits, crude oil, encompasses a wide range of substances. At one extreme, these deposits contain just light gases containing only a few carbon and hydrogen atoms per molecule—for example, the molecule composed of just one carbon atom joined to four hydrogen atoms is called methane, or "natural gas." At the other extreme, crude oil contains a thick viscous sludge[1] with its molecules containing dozens of carbon and hydrogen atoms per molecule.

[1]Some crude oils are even solids at room temperature.

Crude oil from the ground would fill your tank with an impractical mixture of molecules. The heavy or viscous ones would clog up the fuel system. They would not ignite in the combustion chamber. Crude oil also contains significant amounts of foul-smelling, smog-producing, sulphur compounds. The lightest hydrocarbons in crude oil evaporate easily and explode readily.[2]

A better choice is the Goldilocks solution: Fill the tank with molecules that are not too big, not too small, but just right. For automotive engines this means molecules each possessing about 5–10 carbon atoms and 10–25 hydrogen atoms. This class of molecules includes the compounds collectively called "gasoline" in the United States and many equivalent names elsewhere.[3] The ideal molecule[4] for providing chemical energy to the Otto Cycle, the one used as a standard, is isooctane, whose composition is C_8H_{18}. Figure 1 gives an idea of just three components of the hundreds of compounds present in crude oil that are present in widely different amounts in crude oils from different geological sources.

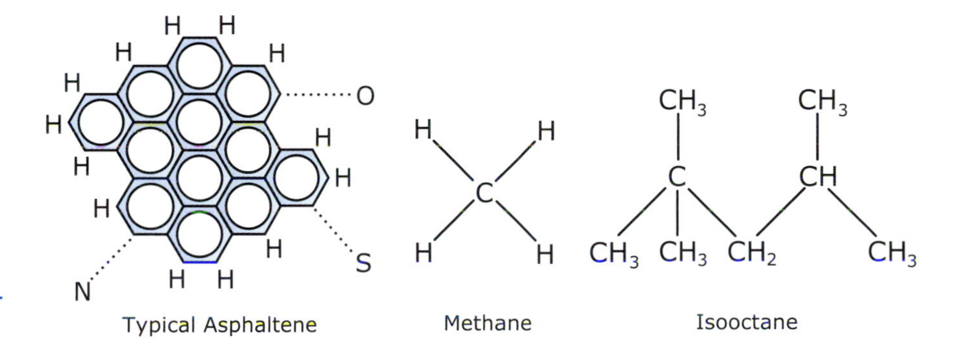

Figure 1

Some Molecules Present in Crude Oil

The composition of crude oil is widely variable. Many of you will also study chemistry, and the formulae in this diagram will become clear. Meanwhile, here is an outline of what you will need to understand in this chapter.

Asphaltene (essentially "tar"), is a "condensed structure" consisting of fused hexagonal *aromatic*[5] (or benzene) rings. The way you read this structure is that carbon forms hexagonal rings with itself that can connect to identical rings. At each interior point there are just carbon atoms. At exterior

[2]Although vehicles can also be adapted to run on natural gas, they need a high-pressure fuel supply system to do so.
[3]http://www.casanovasadventures.com/catalog/camp/p1040.htm.
[4]Isooctane is the industrial standard for gasoline, and the fuel you buy for your car is rated in "octane number," a measure of how well the fuel will burn in your engine without causing internal damage and loss of engine power.
[5]"Aromatic" because many members of this grouping have a pleasant odor. Don't be fooled! They are significantly carcinogenic!

points there are usually hydrogen atoms connected to carbons at each point, but we have also allowed "dotted" structures that include a few other kinds of atoms. Within the rings, the bonds are a hybrid of *single and double bonds*. Single ("covalent") bonds, conventionally represented by "—", consist of two electrons shared between adjacent atoms; double covalent bonds share two electrons from each atom for a total of four. Double bonds are conventionally represented by the symbol "=." The atoms of carbon arrange themselves so that they have a total of four bonds on the average everywhere in the molecule. The circle within each hexagon represents the nonlocalization of the electrons that form these bonds.

In Figure 1, asphaltene is modeled by the simple chemical formula $C_{38}H_{16}$. This formula ignores the "heteroatoms" of oxygen, nitrogen, and sulphur and the structures that support them. The molecular mass of $C_{38}H_{16}$ is 472; however, real asphaltenes present in crude oil may have molecular masses of several thousands so that the asphaltene formula should only be taken as a representative *portion* of an actual asphaltene, and the actual molecule might be a multiple of that shown. In addition, the dotted lines in the formula mean there is more structure that we have ignored but that, in refinery practice, is vital to include. Even though "heteroatoms" are a small portion of crude oil, they are the source of much of the pollution potential and must be removed.

Methane, CH_4, the principal component of **natural gas.** It has a tetragonal structure with a centrally located carbon atom and again illustrates the carbon atom's capacity to form four chemical bonds. It dissolves in crude oil fairly readily and constitutes a significant fraction by weight of many crude oils.

The **isooctane** structure, is a semistructural representation of a 3D structure. Clearly the formula C_8H_{18} does not convey all of this information. In isooctane each carbon atom still bonds in four ways—in this case, either to a hydrogen atom or to another carbon atom.

There are other classes of molecules in crude oil that are useful. Linear hydrocarbons, including methane and isooctane, are known as paraffins (or aliphatic hydrocarbons) and have many uses. Pentane, molecular formula C_5H_{12}, is a volatile hydrocarbon used as a solvent for oil-based paint and other products. Its structure is shown in Figure 2 in schematic form with the dots representing the relative position of the carbon atoms.

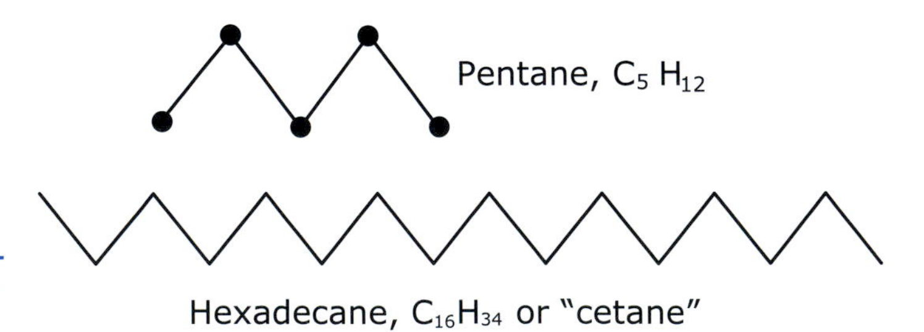

Figure 2

Structure of Aliphatic or Paraffinic Hydrocarbons

We can make our molecules even more conceptually schematic by drawing just the backbone bonds and omitting the dots representing the carbon atoms at the apices of the chain while mentally attaching the necessary hydrogen atoms. Hexadecane with 16 carbon atoms and 34 hydrogen atoms is thus illustrated in Figure 2. This molecule is also known as "cetane." Cetane is to diesel engines as octane is to gasoline engines. This large linear[6] (meaning no offshoot carbon atoms from the main chain as in isooctane) hydrocarbon is an ideal fuel for diesel engines. It requires the high temperatures that are characteristic of the diesel engine to fully vaporize it and it burns at the requisite rate, which are necessary steps for efficient combustion. Conversely, the isooctane molecule is an idea fuel for gasoline engines because it vaporizes and burns readily at the lower temperature conditions seen in gasoline engines.

Oil Refineries

Oil refineries separate the components of crude oil into their constituent parts to produce hydrocarbons that are more suitable for use as solvents, heating oils, and so forth.

EXAMPLE 1 An oil deposit contains crude oil that is a mixture of, by **mols,** 7.0% C_5H_{12}, 33.% C_8H_{18}, 52.% $C_{16}H_{34}$, and 8.0% $C_{38}H_{16}$. What is the percentage of each substance by mass?[7]

Need: By mass (or wt), C_5H_{12} = ____%, C_8H_{18} = ____%, $C_{16}H_{34}$ = ____%, and $C_{38}H_{16}$ = ____%.

Know–How: Imagine a total of 100 kmols. Divide them among the various species.

7.0 kmols C_5H_{12}, 33. kmols C_8H_{18}, 52. kmols $C_{16}H_{34}$, and 8.0 kmols $C_{38}H_{16}$.

Determine mass of each species by multiplying number of [kmols] by the species molecular mass in [kg/kmols].

That is, $7.0 \times (60 + 12)$ kg C_5H_{12}, $33. \times (96 + 18)$ kg C_8H_{18}, $52. \times (192 + 34)$ kg $C_{16}H_{34}$, and $8.0 \times (456 + 16)$ kg $C_{38}H_{16}$.

Solve: By mass $C_5H_{12} = 2.5\%$, $C_8H_{18} = 19.0\%$, $C_{16}H_{34} = 59.4\%$ & $C_{38}H_{16} = 19.1\%$ (use a tabular format to solve this problem—it's especially helpful to use a spreadsheet).

[6]This is a somewhat misleading usage, since the longer the molecule, the more it can twist into knots; "linear" refers to the ideal stretched state as drawn.
[7]We can equally use "wt%" as "mass%" if confined to Earth's gravity.

		mol	molecular mass	mass, kg	% by mass	Atoms	molecular mass
Pentane	C_5H_{12}	7	72	504	2.5%	C	12.0
Isooctane	C_8H_{18}	33	114	3762	19.0%	H	1.00
Cetane	$C_{16}H_{34}$	52	226	11752	59.4%		
Asphaltene	$C_{38}H_{16}$	8	472	3776	19.1%		
Totals:		100		19794	100.0%		

Notice that this particular crude oil has ~ 20% by mass of aromatics in the form of asphaltenes (tars); this component could be used directly for road paving, but it is usually more valuable if upgraded by adding extra hydrogen (which can be indirectly produced within the refinery). Usually "light crudes" are favored by refineries because they naturally contain higher percentage of components desired for automotive fuels.

From the mid-nineteenth century, mixtures of hydrocarbons, such as the one described above, were pumped out of the ground into wooden barrels. (The 42 U.S. gallons or 35 Imperial gallon barrel is implied in the expression "barrels of oil per day" and is the commercial volume unit for oil.)[8] The original wooden barrels were shipped by rail to oil refineries. There, a maze of pipes, pumps, and pots were used to separate the oil into its components. In preautomobile days, the most valuable component was kerosene (a mixture of hydrocarbons basically similar to cetane) for use in lamps. Each of these oil refineries was individually and probably idiosyncratically designed, without much consideration of engineering fundamentals.

Process Engineering

In the early decades of the twentieth century, a new idea emerged. Chemical engineers, such as Arthur D. Little, an independent consultant, and W.W. Lewis, a professor at MIT, looked at chemical plants in a new way. They viewed the plants, however complicated, as arrangements of simple elements. These elements, which served as a sort of alphabet for chemical plant construction, came to be known as **process steps.** It also turned out that while many of the process steps seemed quite different, they were really closely related. For example, industrial distillation (separation of a boiling mixture by differences in relative volatility of the individual components), absorption (bubbling one gas through a liquid solvent), stripping (ditto, but for extraction of a dissolved species), and liquid extraction (removal of an impurity in one liquid by another) are all described by the mathematics of "countercurrent" operations (i.e., two fluids moving in opposite

[8]A "standard" sized oil refinery is about 50,000 barrels/day (B/D) or about 275,000 kg/hr of mixed hydrocarbons. The largest refineries process several million B/D. The U.S. consumes about 20 million barrels per day of petroleum.

directions) and constitute a single **unit operation.** Engineers working in unit operations looked for similarities between different operations to understand their underlying common features.

A single process step relates a given input to a given output. It may internally deal with how quantities of mass and of energy are used in individual operations. For example, mass flow is conserved in all process steps in steady state. In some process steps so are the molecular species but not if the step includes a chemical reaction. The *total* amount of mass of material that enters a process must equal the total amount of mass that exits whether or not there have been chemical changes within the process step.

The conservation of a quantity, such as mass or energy, can be illustrated by a tool we previously used, the **control surface.** Consider a straightforward process operation, a heat exchanger that is to heat or cool a process flow stream.

EXAMPLE 2

Draw a process diagram of a **heat exchanger** with a mass flow rate M of material to be heated, and its associated control surface.

Need: A drawing illustrating the fact that a heat exchanger is a structure into which an amount of material M flows in through the control surface, and the same amount of material, M, flows out.

Know–How: Represent the heat exchanger by a circle and the control surface by a dotted oval line surrounding the circle. Use a zigzag line to represent a metal coil inside the heat exchanger that carries a heating fluid such as steam in and out. The process stream (M) is separated from the heating steam by the metal coil and the process stream simply absorbs heat from the steam inside the metal coil.

Solve: Clearly the process flow out of the heat exchanger is equal to that entering (provided there is no leak and provided the heat exchanger is always full). The same is true of the steam line irrespective of whether or not it changed phase from a vapor to a liquid.

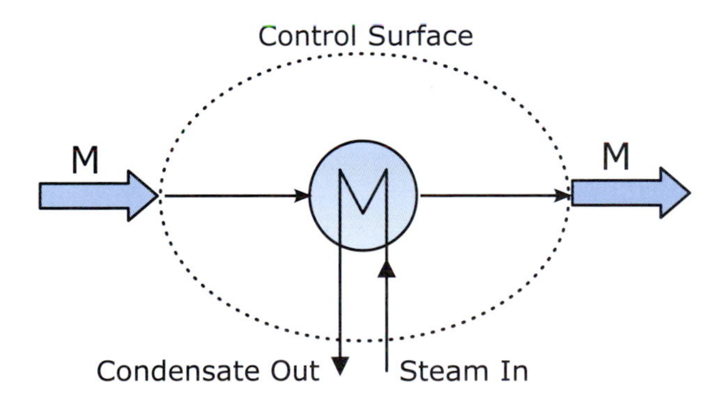

■

In all the examples in this chapter, the flow of mass into a unit operation will equal the flow out. However, between input and output, things will change, which, after all, is the purpose of a process element. For example, a **pump** will change the pressure of the input into a higher output pressure. In the process, the pressure from the pump causes the material to move in a certain direction. A heat exchanger will change the temperature of the input material to a different (higher or lower) temperature as will be illustrated in a **process diagram** by the symbol shown on the right in Figure 3. The arrows indicate the direction in which mass moves. The heat exchanger in this sketch is stylized with one fluid moving in the "tubeside" and the other through the "shellside." In Figure 3 the secondary fluid doing the heating or cooling is understood and not explicitly drawn (while in other cases both fluid streams may be shown explicitly). While we must mentally remember for such straightforward operations that control surfaces could be drawn, typically we will not draw them.

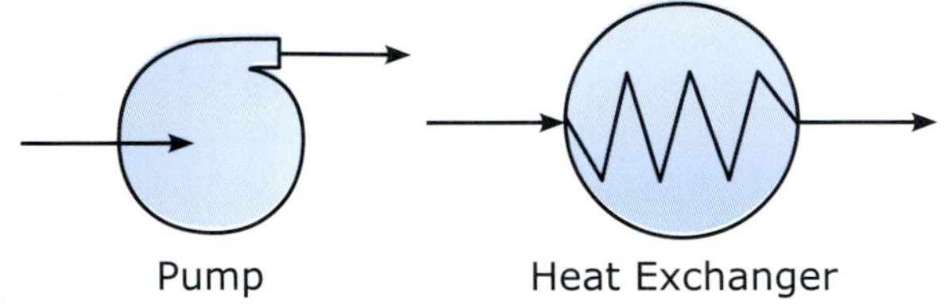

Figure 3

Common Process Elements

EXAMPLE 3

Crude oil at 50°C is heated to make it flow more easily. For this purpose it is pumped into a heat exchanger that is heated by condensing steam at 200°C. This time, the steam is on the outside of the metal coil, since it is easier to pump highly viscous crude oil directly through the tubeside than on the shellside of the heat exchanger. Draw a process diagram of this heating system.

Need: A process diagram.

Know–How: Simple process symbols for heat exchange and for pumps.

Solve:

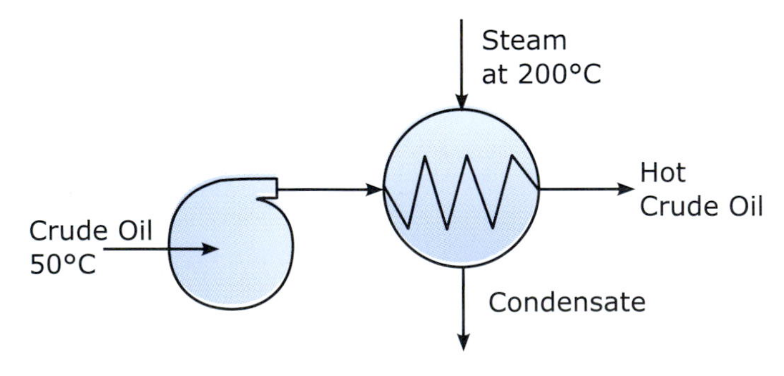

Notice that at this point we cannot determine the crude oil outlet temperature. To do this we would need to know (1) the amount of oil flowing, (2) its heat capacity, (3) the amount of steam used, and (4) how big the heat exchanger was to effect the heating of the oil. All we know for certain is that the "hot crude oil" is hotter than 50°C but cooler than 200°C. It is the purpose of process engineering to put numbers onto such states.

Distillation

Let's return to the problem of oil refining. Its purpose is separating and modifying components of crude oil. In our simplified treatment, each component is characterized as a chain of carbon and hydrogen atoms of a given mass. How can different molecules be separated from one another? The simplest view of a refinery is shown in Figure 4.

Figure 4

Gross Process Diagram of a Refinery

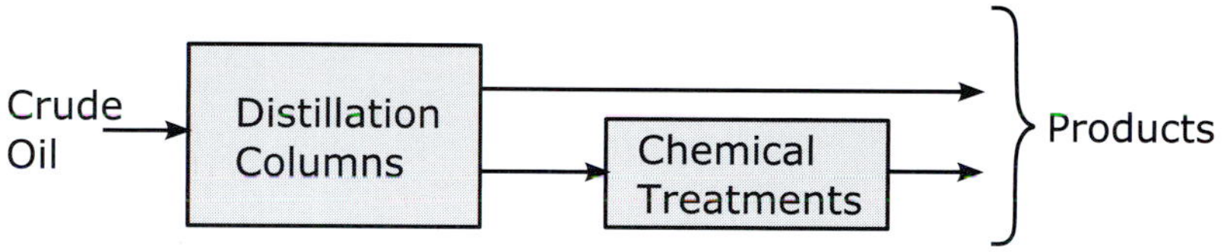

The first major process or (unit) operation in most refineries is a set of distillation columns. This is a purely *physical* separation process that relies on the different relative volatilities among the various components of crude oil. Generally, they are arranged in sequence with the most volatile components removed first (such as methane, pentane, isooctane, etc.) and eventually in the last distillation columns the least volatile compounds such as asphaltenes are removed. But because distillation is a physical treatment, *the various molecules retain their identity (i.e., there are no chemical reactions)*.

This is *not* true of chemical treatments in which the molecules are broken and reformed into other molecules. Some of the "heaviest" components can be *chemically* re-arranged in large reactors colloquially called **catalytic** or **cat crackers.**[9]

Distillation is conceptually a relatively simple unit operation. The key unit operation is the fact that condensed liquid (**condensate**) flows down the distillation column while vapor of a different chemical composition flows up (Figure 5).

[9]Jargon in which "cracking" means rearranging the molecules of a light fraction of crude to more useful molecules. "Catalytic" (sometimes shortened to "cat") means in the presence of a catalyst to speed the reaction and to increase selectivity toward desired products.

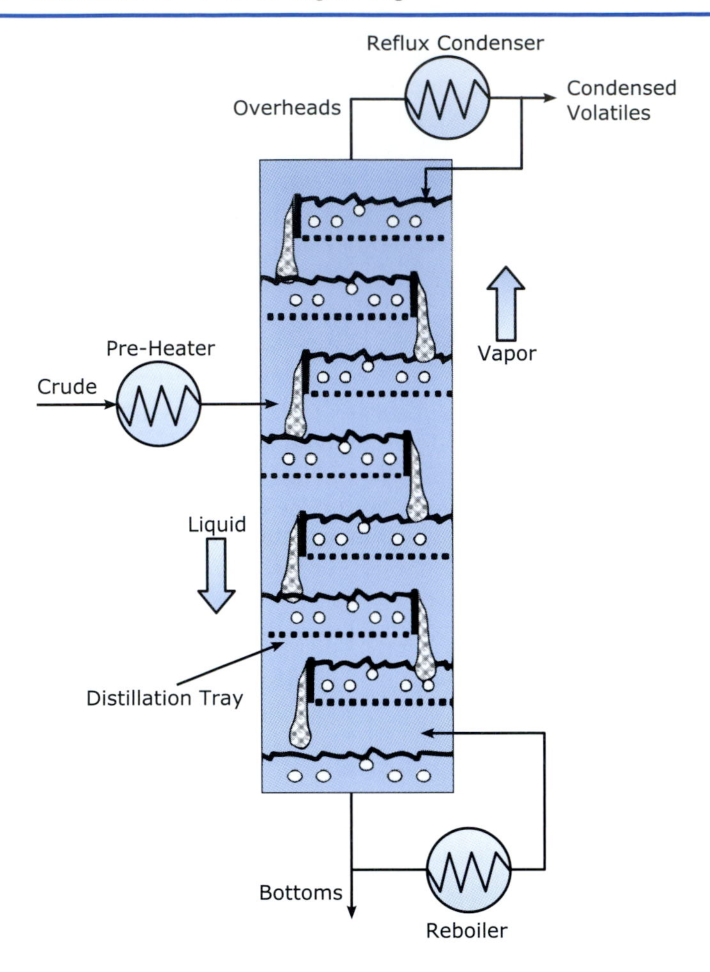

Figure 5

Schematic of a Typical Distillation Column

The driving force for distillation is the difference in composition between the liquid condensate at every location and the vapor with which it is in contact. A distillation column is the process made up of pipes, pumps, and heat exchangers, plus a vertical column containing trays. The trays are devices that temporarily hold back the liquid condensate, while contacting the downflowing liquid with upflowing vapor of a different chemical composition (hence the term "countercurrent" to describe it). The vapor contacts the liquid by flowing through an arrangement that does not allow passage of the liquid. For the feed, a heat exchanger adds enough heat to evaporate some of the crude oil into vapor. At some point in the column, the hot crude oil and some of its vapor are pumped in. The vaporized components go up the column and are washed by condensate flowing down in order to extract any remaining amounts of the less volatile components trapped within. The condensate is produced by condensing the vapor at the top of the column by a **"reflux condenser"** (just another heat exchanger, this one dedicated to cooling the vapor overheads to liquid). The more of the vapor that is condensed at the top of the column, the purer the "overheads" in the desired volatile material. The bottom part of the column operates similarly; condensate runs down the column and contacts upflowing hot vapor with the result that additional volatile components are extracted from it. This vapor is produced by heating the product "bottoms" in a **"reboiler"** (just another heat exchanger, this one dedicated to boiling a liquid). The more of the bottoms that are heated and vaporized, the more volatiles are stripped from the bottoms.

The overhead product at the top of the column is richer in the more volatile components, and the bottom product is richer in the less volatile components, than is the original crude oil. The detailed theory makes it clear that one can never reach 100% separation, but one can approach arbitrarily close by having many trays and larger reflux condensers and reboilers. Two process diagram symbols for a distillation unit are shown in Figure 6. The more detailed one is the more literal and includes the reflux condenser and the reboiler; however, often we do not need to know about

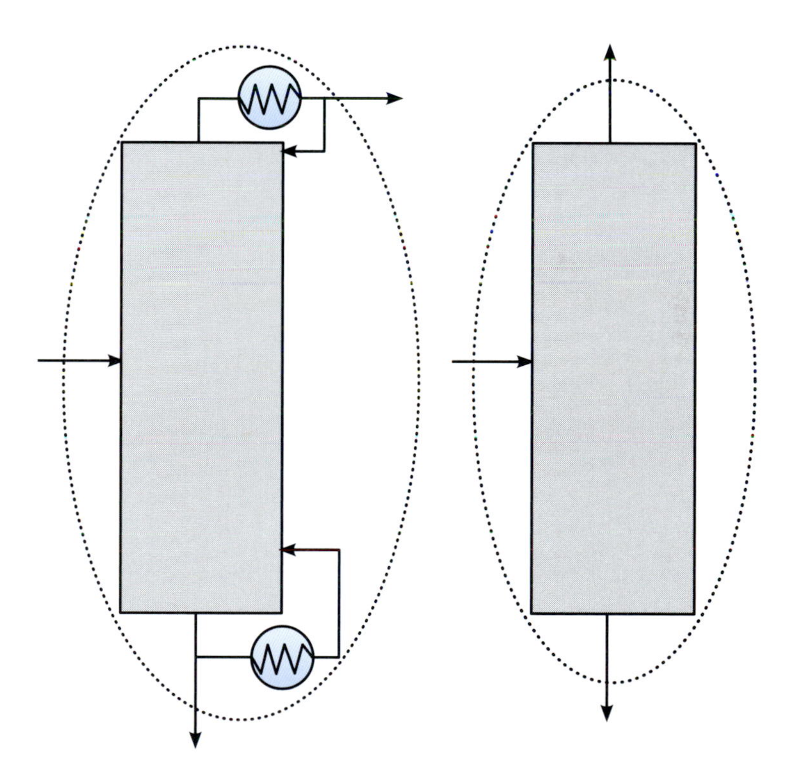

Figure 6

Process Diagrams of a Distillation Column

the internal flows, and we need just the inlet and outlet flow streams. Again, you should imagine the distillation column symbol as surrounded by a control surface "dotted line" though which are arrows denoting the flows. That control surface reminds us that the mass flow rate into a distillation unit must equal the mass flow rate out. For the two control surfaces in Figure 6 the internal flows are irrelevant and the two diagrams are completely equivalent.

EXAMPLE 4

A crude oil contains four components by mass: 10.% asphaltenes, 5.0% light gases, 40.% isooctane, and 45% cetane. An input stream of 1.00×10^4 kg/hr of this crude oil is fed to the distillation column in an oil refinery. Assume the distillation process separates out a top and a bottom stream. If the bottom consists of 100% asphaltenes, what are the flow rates by mass and what is the composition of all of the process lines?

Need: Composition and rate of flow of each component in each line.

Know: Principle of conservation of mass across control volume. Also, the distillation process is a physical one and therefore the type and quantities of molecules are preserved.

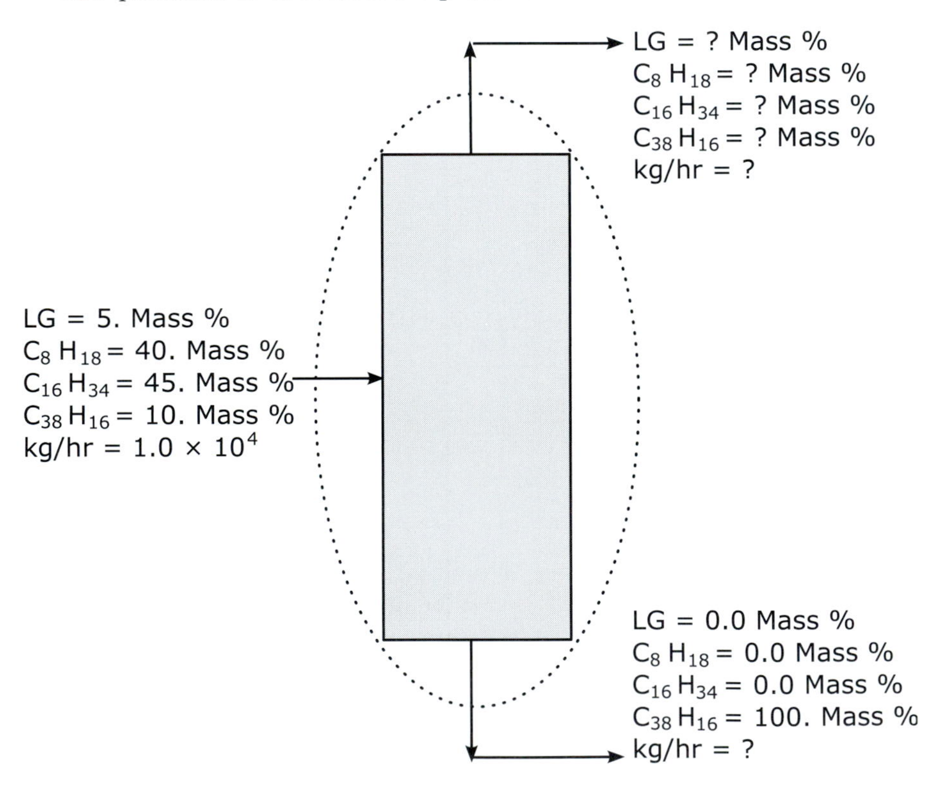

How: Use a process diagram showing inputs and outputs of each stream with a corresponding spreadsheet to plot the knowns and unknowns.

Solve: We immediately know the amount of each component in the feed; 5.0×10^2. kg/hr LG, 4.0×10^3 kg/hr isooctane, 4.5×10^3 lb/hr cetane, and 1.0×10^3 kg/hr asphaltenes.

Thus, the total overheads are $1.0 \times 10^4 - 1.0 \times 10^3 = 9.0 \times 10^3$ kg/hr and the composition can be deduced from the feedstock mass% (*e.g.*, mass% isooctane in overheads is $(4.0 \times 10^3$ [kg/hr])/(9.0×10^3 [kg/hr]) = 44.4 mass %). It is equally easy to complete the spreadsheet below (answers are in **bold**):

Stream	kg/hr	Mass % LG	kg/hr LG	Mass % isooctane	kg/hr isooctane
Feed	10,000	5.0%	500	40.0%	4,000
Bottoms	**1,000**	**0.0%**	**0**	**0.0%**	0
Overheads	**9,000**	**5.6%**	**500**	**44.4%**	**4,000**

Stream	Mass % cetane	kg/hr cetane	Mass % asphaltene	kg/hr asphaltene
Feed	45.0%	4,500	10.0%	1,000
Bottoms	**0.0%**	**0**	**100.0%**	**1,000**
Overheads	**50.0%**	**4,500**	**0.0%**	**0**

Notice the overhead steam is not pure; it consists of three components in the ratios of LG: isooctane: cetane = 5.6: 44.4: 50. by mass%. Since these components are potentially valuable products, they would be further refined in standard refinery practice. A second distillation column would normally be added to effect further separation.

EXAMPLE 5

Define a distillation column to separate light gases and isooctane from cetane given the initial feed composition of light gas to isooctane to cetane is 5.6: 44.4: 50. by mass. Assume that the light gases and isooctane go overhead together; further assume that you can achieve 100% separation of overheads from bottoms.

Need: Composition of the two distillation streams.

Know: Method of Example 4, which is the same kind of problem except that we need a new control surface for our mass balance:

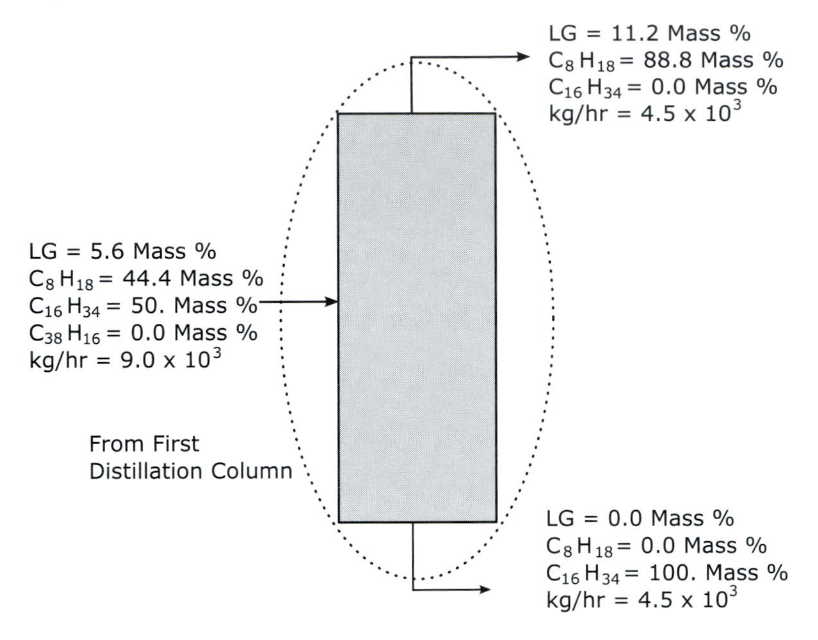

How–Solve: Same as Example 4. Notice the incremental approach to the process design of this chemical plant. By decomposing it to its elements, the overall system can be easily designed. This is a surprisingly powerful method. More complex systems have recycle streams that increase the arithmetic but not the principles.

	A	B	C	D	E	F
7		Input	LG	C_8H_{18}	$C_{16}H_{34}$	$C_{38}H_{16}$
8	% by mass	100.0%	5.6%	44.4%	50.0%	0.0%
9	kg/hr	9.00E+03	5.04E+02	4.00E+03	4.50E+03	0.00E+00
10	O/H, kg/hr	4.50E+03	5.04E+02	4.00E+03	0.00E+00	0.00E+00
11	O/H, %	100.0	11.2	88.8	0.0	0.0
12						
13	Bottoms, kg/hr	4.50E+03	0.00E+00	0.00E+00	4.50E+03	0.00E+00
14	Bottoms, %	100.0	0.0	0.0	100.0	0.0

■

The second distillation column has accomplished more of what we are seeking: separation of the cetane fraction (for diesel fuel and kerosene) from the two lighter fractions. But how do we separate the light gases from the isooctane needed for gasoline?

How Refinery Output Can Be Improved

With the arrival of the automotive age, gasoline (hydrocarbon molecules with 6–10 carbon atoms per molecule) became a very profitable product for oil refineries. Automotive engineers also learned that automobiles ran best when that the composition of gasoline was close to the chemical structure of isooctane. So chemical engineers looked for ways to improve the isooctane output of oil refineries.

A method of extracting isooctane from the light gases in the overhead stream is **flash distillation.** Its process model is shown in Figure 7.

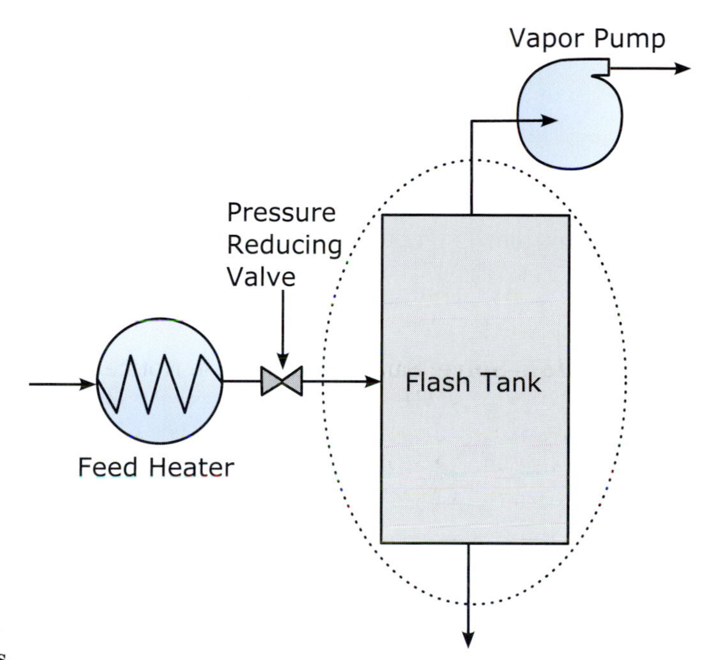

Figure 7

Flash Distillation Process

It operates by suddenly reducing the pressure on a hot mixed stream, which causes the more volatile components to "flash" into vapor, while the less volatile ones remain in the liquid state. The essential mechanical components are a vapor pump, a flash tank, and a valve that separates the flash tank from the higher pressure upstream fluid. To produce a sufficient inlet temperature there may be an upstream feed heater. Figure 7 also suggests a useful control surface (the dotted oval) for the analysis of a flash process; within it, no external heat is added or removed from the system. Since heat is needed to evaporate the lighter component, it gets the necessary energy by cooling the contents of the system so that the products leave the flash system at a lower temperature than when they went in.

EXAMPLE 6

A distillation column's overhead stream contains 4.5×10^3 kg/hr of light gases and isooctane with 11.2 mass % being light gases (5.0×10^2 kg/hr) and the remainder (4.0×10^3 kg/hr) being isooctane. A flash distillation unit is placed in this overhead flow stream. Its overhead stream output contains only light gases and 2. mass % isooctane (C_8H_{18}). Its bottom stream contains 98 mass % isooctane (C_8H_{18}), and the balance is light gas. How much isooctane, in kg/hr, can be extracted in the bottom stream of the flash distillation unit?

Need: Isooctane flash unit bottom flow stream mass flow rate = ___ kg/hr.

Know–How: In this example, some additional complications have been introduced—less than 100% separation has been achieved. Finally, we only have measurements of isooctane and have to compute the balance of light gases. The latter may be subject to round-off errors.

The total mass flows of overhead O and bottom B give the overall mass flow rate: $O + B = 4.5 \times 10^3$ kg/hr. Also, the isooctane in the overhead and bottom equals that entering the control boundary.

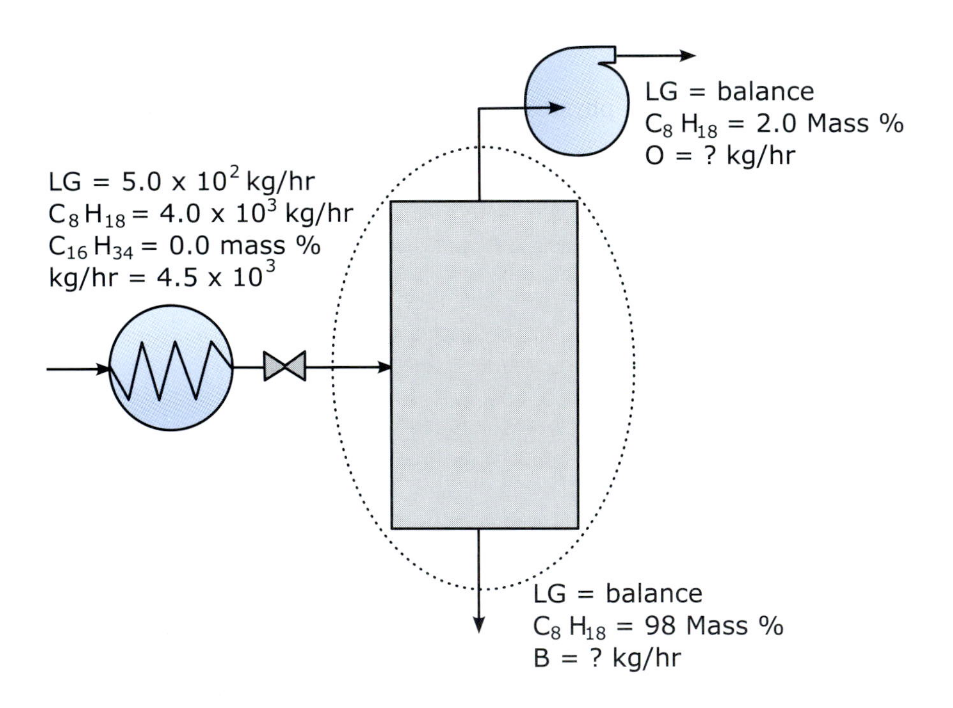

Solve: The overhead is 2 mass (or wt.) % isooctane and the balance is light gas and the bottom is 98 mass (or wt.) % isooctane.

We have to find the unknown mass flow rates O and B in kg/hr first. To do this we need *two independent* equations. The overall mass flow rate balance and the isooctane mass flow rate balance are:

Total mass flow rates:

$$O + B = 4.5 \times 10^3 \text{ kg/hr} \quad (1)$$

The isooctane mass flow rates:

$$0.020 \times O + 0.98 \times B = 4.0 \times 10^3 \text{ kg/hr} \quad (2)$$

Thus, $B = 4.073 \times 10^3 = 4.1 \times 10^3$ kg/hr (and $O = 4.27 \times 10^2 = 4.3 \times 10^2$ kg/hr) and the isooctane flow in the bottom is $0.98 \times B = 3.99 \times 10^3 \approx$ **4.0×10^3 kg/hr.** ■

Note the ratio of the concentration terms: $0.98/0.02 = 49$. These concentrations may be measured quantities with *associated errors of measurement.* In particular, the measured isooctane in the overheads could be subject to a small error—say, 10%, meaning it lies between 1.8 mole % and 2.2 mole % light gases. Notice that the subsequent calculations would propagate this error. Errors of measurement can be a serious limitation of stoichiometric methods. Typically, had the light gasses been independently measured it is unlikely the totals would be exactly 100% in either the overheads or the bottoms.

So a combination of distillation columns and flash separation further physically separated the isooctane from the input stream. But is there any way to increase the isooctane output even further? To do this, the physical process of distillation must be supplemented by chemistry. That is, molecules must be taken apart or put together.

One example of such chemistry is **thermal cracking.** It is the breaking apart of large molecules by heat to produce smaller molecules. It works because heat breaks and reforms new chemical bonds. Again, we will take a process approach. We will still use stoichiometric algebra to approximate the complicated chemistry of atomic bonding. We will do so by introducing a module called a **thermal cracker,** illustrated in Figure 8. It consists of a heavy duty pump (because the bottoms feed is likely to be quite viscous and will need to be pumped to the next process step) a preheat furnace that burns

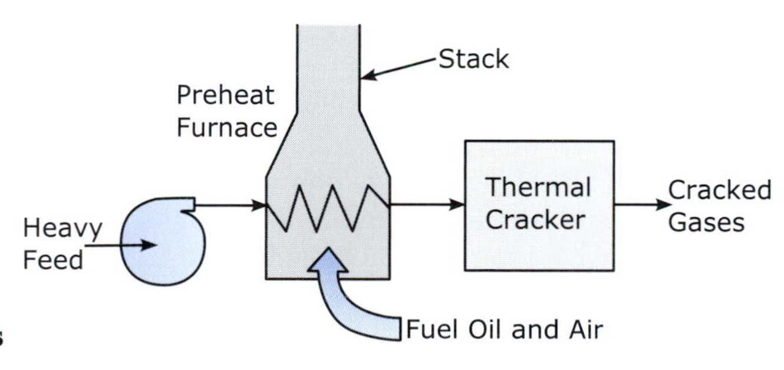

Figure 8

Thermal Cracking Process

otherwise low-quality oil as its fuel, a stack to vent the combustion products, and the thermal cracker itself, likely a large pressure-tight vessel that is designed for the high temperatures of the process. The thermal cracker can be attached to the bottom of a distillation column. Its input is usually the heavy component distilled from crude oil. Its operation is to chemically break the input molecules into smaller pieces. Since chemical bonds are broken, we can no longer use stoichiometric balances that preserve the molecules that are present in the feed. We will have to revert to the stoichiometry at the atomic level that we previously used in the combustion chapter.

EXAMPLE 7

The 1.0×10^3 kg/hr of bottoms of a crude oil primary distillation column is 100% asphaltenes with an empirical formula $C_{38}H_{16}$. Thermal cracking will produce some isooctane with some heavier components:

$$2C_{38}H_{16} = C_8H_{18} + C_{68}H_{14}$$

(Note the stoichiometric balance in this equation was done by the methods introduced in the combustion chapter.)

What is the total isooctane output mass flow rate of the refinery with the thermal cracker in use?

Need: Isooctane output mass flow rate = ____ kg/hr.

Know–How: The isooctane output will be the sum of the isooctane output of the distillation process and the isooctane output of the thermal cracker. The molecular mass of isooctane is 114.

Solve: Isooctane output mass flow rate $= \frac{1}{2} \times 1.0 \times 10^3 \times 114/(12. \times 38 + 16 \times 1.00) = 121$ kg/hr (notice we have used a stoichiometry that counted the atoms, not the molecules, of the process since the molecules were changed by the cracking chemistry). ■

The other isooctane stream that was flash distilled is about 4.0×10^3 kg/hr so we have added an additional 3% isooctane. Can we extract more? Notice as well that the downside of extracting more light molecules in a cracking reaction is to produce more of the heavier molecules. In Example 7 we designated these heavier molecules by the empirical formula "$C_{68}H_{14}$"; and it might be called "residual oil" or tar or simply asphaltene. But whatever it is called, this is just a designation for the average heavy molecules.

We can easily complete the crude oil refining process by adding a flash unit and a thermal cracker to the distillation column model, and then using a "solve the problem on the picture" approach as in the previous example. A schematic of a process design of a simple and small oil refinery is shown in Figure 9. Modern refineries will have many additional units in order to maximize the yields of desirable products.

Using heat to reassemble molecules is a fairly crude technique. However, chemistry provides us with a much more sophisticated and selective tool for molecular reassembly. This tool is **catalysis.**

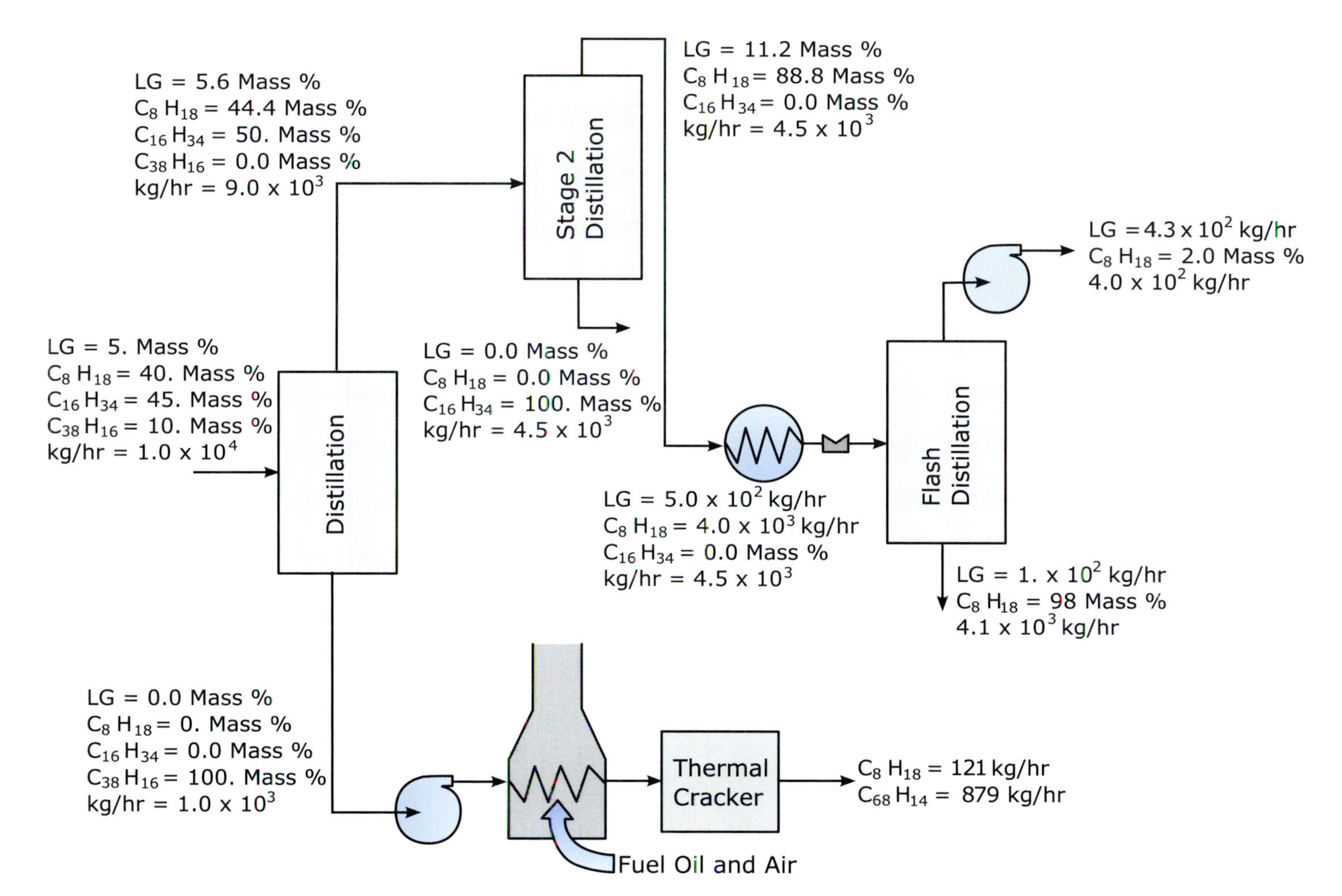

Figure 9 Flowsheet of a Crude Oil Refinery Producing Gasoline and Diesel Fuels

Catalytic Conversion

Chemistry mediated by catalysis is a way to make chemical reactions both faster *and* more selective by directing molecules to increase the likelihood of a desired reaction occurring. Both these attributes of a good catalyst are keys to efficient chemical processing for oil refining. A catalyst is a physical structure, often a porous ceramic material, on which molecules can react. In some cases the catalysts are extended solid matrices of material and in others they may be small particles that are made to circulate with reactive gases. The latter are known as "fluidized bed catalytic crackers," and they are heavily used in the oil industry.

On many catalysts there are small amounts of carefully selected metal atoms attached to the internal porous surfaces of the ceramic. Much of the chemistry occurs in conjunction with these metallic atoms as well as the ceramic substrate on which they are dispersed. Unlike the reacting molecules, the catalyst is chemically unchanged by the reaction, so it has a long lifetime. The chemical details are complicated. We will only hint at them by representing a catalyst by a sort of corrugated surface that can "catch" or "trap" molecules. Typically catalysts contain precisely arranged internal pores in which molecules of interest can be contained to undergo the required chemical reactions.

Catalytic conversion has two main uses in an oil refinery. It can be used to make the cracking process more effective. A catalytic cracking process is an essential part of modern oil refineries. It can also be used to combine lighter molecules into heavier ones. For example, light gas molecules can be combined into isooctane, further increasing the refinery's gasoline output. Other catalytic processes combine heavy gases with hydrogen[10] to make desirable light products.

Another essential use of catalytic conversion is part of an automobile's exhaust system. A car's catalytic converter selectively combines pollutant molecules such as unburned hydrocarbons and carbon monoxide gas in the engine's exhaust to form molecules less harmful to the environment than the original exhaust molecules. The catalyst in these systems is either one or more trace metals on the surface of small ceramic beads or in the form of a monolith (a bulk porous ceramic piece).

One of the functions of a catalytic converter is to remove the poisonous gas carbon monoxide (CO) that is emitted in small amounts from automobile engines. It coverts CO into less immediately harmful carbon dioxide (CO_2).

[10]Where does the hydrogen comes from? You can make it by burning hydrocarbons with steam over a catalyst based on the metal nickel, a process known as "reforming." This, too, is a common refinery process. The hydrogen so produced is then combined with highly condensed hydrocarbons that were deficient in hydrogen as compared to carbon. Hydrogenation can produce useable fuel hydrocarbons, either isooctane for gasoline (in summer) or, by slightly adjusting the conditions, cetane for diesel or for fuel oil in winter.

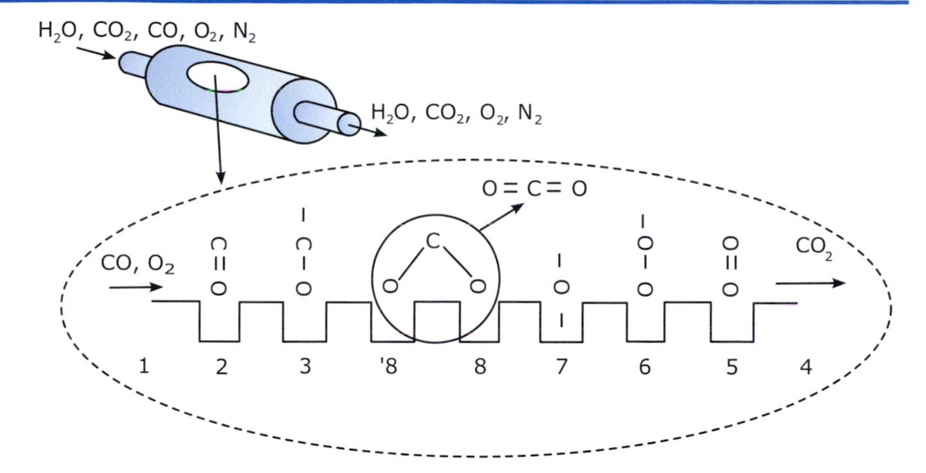

Figure 10

Catalytic Oxidation of CO to CO_2

The way the catalytic converter does this is indicated schematically in Figure 10.

The interior surfaces of the catalyst (shown as a series of small pits) temporarily trap the carbon monoxide and oxygen molecules (from excess air that is pumped into the exhaust gas). This makes them more likely to react with neighboring oxygen molecules than would be the case in the absence of the catalyst. A plausible reaction sequence is shown in Figure 10. Carbon monoxide (1) enters the catalyst chamber and is attracted to the catalytic site (2) by one weak bond, thus "liberating" the other bond (3). Meanwhile, molecular oxygen is also attracted to catalytic site (5) and opens its internal bonds (6) so that an oxygen atom attaches to a surface site (7). If that site is located near the CO site (8), the adsorbed CO molecule ('8) can attach to the site (8). This forms the stable molecule O=C=O (CO_2) and it desorbs to leave the system.

Automotive pollution standards are usually stated in terms of grams of pollutant emitted per mile of travel. Current standards restrict carbon monoxide emissions of automobiles to about one gram per mile. Example 8 gives a sense of how catalytic converters help meet those standards.

EXAMPLE 8

An automobile engine emits carbon monoxide (CO) into its exhaust at the rate of 0.08 g/s. The engine's exhaust is then run through a catalytic converter that combines 95% of the carbon monoxide molecules in the exhaust with oxygen molecules to produce carbon dioxide (CO_2) that continues out through the tailpipe. If the car is traveling at 30. miles per hour, what is the carbon monoxide emission of the car in g/mile?

Need: CO = ____ g/mile.

Know: Emission of CO in g/s and speed of car in mph.

How: Use the dimensional analysis method to relate [g/mile] = [g/s] × [s/mile].

Solve: CO = (1.00 – 0.05) ×0.08 = 0.004 g/s.

The speed of the vehicle is 30. mph = 30./3,600 [miles/hr][hr/s] or 0.0083 miles/s.

Therefore, the amount of **CO** produced = 0.004/0.0083 [g/s][s/mile] = **0.48 g/mile.** ■

SUMMARY

If we were to sum up the *traditional* profession of chemical engineering in a few words, those words would be "unit operations." These are the different process steps whose commonalities are used to analyze them with a set of consistent mathematical relationships. While all of the unit operation methods are too complicated to be developed here, the first step is often to create a flowsheet with the different process steps represented by blocks. Using these simple building blocks of process operations, chemical engineers have solved complicated problems, such as how to turn the mixture of hydrocarbons given to us by previous generations of biomatter into the gasoline that powers our cars.

The tool of process engineering can model a wide range of other systems, from the car's catalytic converter, to that most remarkable of chemical factories, the living cell. In the twentieth century, chemical engineers addressed the challenge of oil refining. They created unit operations. They used it to design, model, and improve the chemical factories that turn natural resources into useful products. You have sampled the capabilities of process engineering in tackling such problems as flowsheeting, with process steps such as piping, pumping, heat transfer; distillation, cracking and catalytic combination. Many of the processes that use these building blocks were developed without much concern for the long term sustainability of the resources that provide the mass and energy entering the control surface, or about the effects on the environment of the mass and energy leaving the control surface. Those challenges of long term sustainability and environmental impact are major tasks for the chemical engineers of the twenty-first century.

EXERCISES

1. A pipe carries a mixture of, by **mols** (which are proportional to the number of molecules), 15% C_5H_{12}, 25% C_8H_{18}, 50.% $C_{16}H_{34}$, and 10.% $C_{32}H_{40}$. What is the percentage of each substance by mass? **(A: Mass%: 0.58% C_5H_{12}, 15% C_8H_{18}, 61% $C_{16}H_{34}$, and 23% $C_{32}H_{40}$.)**
2. A pipe carries a mixture of, by **mass**, 15% C_5H_{12}, 25% C_8H_{18}, 50.% $C_{16}H_{34}$, and 10.% $C_{32}H_{40}$. What is the percentage of each substance by **mols** (which are proportional to the number of molecules)?

3. In a car's cooling system, a circulating pump forces water through the car's engine, where the water absorbs heat from the cylinders at an average temperature of about 600°C. The water leaves the engine at a temperature of about 110°C and passes through the car's radiator, where it transfers its heat to the atmosphere at 30°C. Draw a process diagram of the car's cooling system.
4. The glass pipe system shown has a liquid flowing in it at an unknown rate. The liquid is very corrosive and flow meters are correspondingly very expensive. Instead you propose to your boss to add a fluorescent dye via a small pump and to deliver 1.00 g/s of this "tracer" and activate it by UV light. By monitoring the dye fluorescence downstream, you determine its concentration is 0.050 wt.%. What is the unknown flow rate in kg/hr? (This is a useful method of indirectly measuring flows.) **(A: 7,500 kg/hr.)**

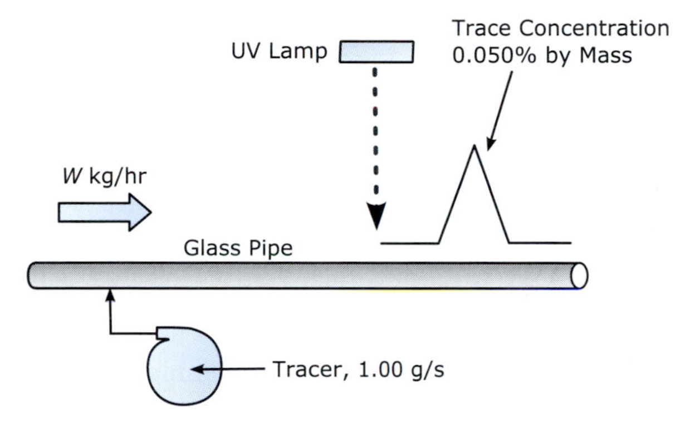

5. The glass pipe system shown has one measured flow rate and the remaining ones are not known. You add a fluorescent dye via a small pump to deliver 1.00 g/s of this "tracer" and activate it by UV light.

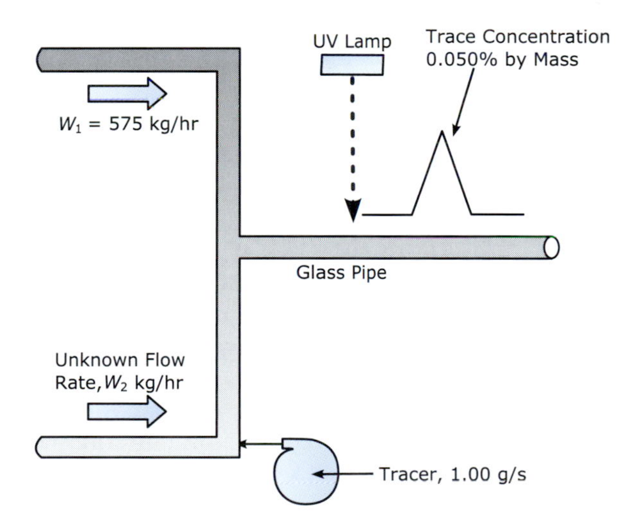

By monitoring the dye fluorescence in the combined stream you determine its concentration is 0.050 wt.%. What are the unknown flow rates in kg/hr?

6. In the pipe system shown here, one liquid flowing at an unknown rate W_1 contains an impurity of concentration C_1. When mixed with a pure stream of the same liquid flowing at a known rate W_2, the resulting impurity concentration is C_2. Show that: $W_1 = \frac{C_2}{C_1 - C_2} W_2$.

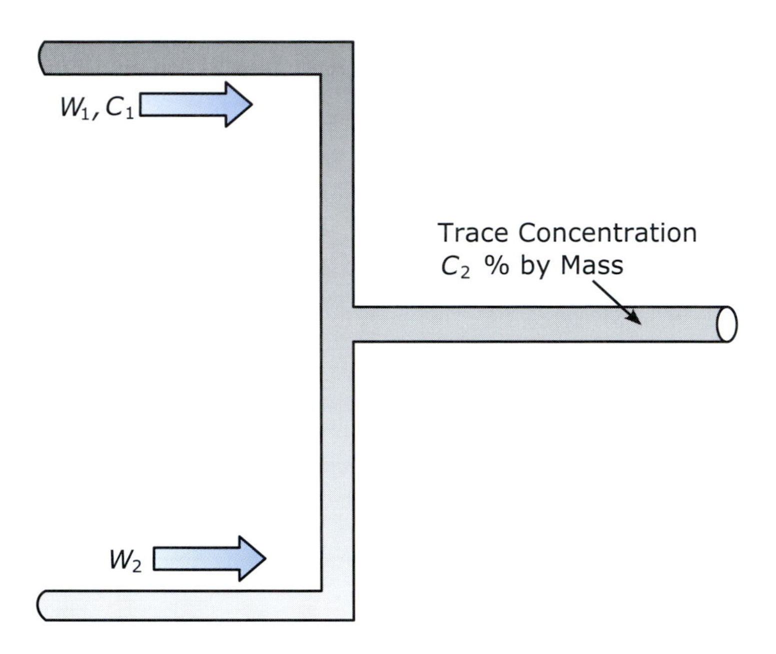

7. In the previous problem, $W_1 = 850.$ kg/hr and the impurity is present to the concentration of $C_1 = 2.50$ mass %. What is W_2 if (a) $C_2 = 2.00$ mass %, (b) 2.0 mass % (c) 2. mass %, and (d) 2 mass %? **(Selected A: (a) 210 kg/hr, two significant figures, (c) 200 kg/hr, one significant figure.)**
8. A crude oil contains just three components by mass: 15% asphaltenes, 1.0% light gases (containing about 1–5 carbon atoms + hydrogen) and the remainder light "distillate" that we will say is isooctane.

   1.0×10^4 kg/hr of this crude oil is fed to the distillation columns in an oil refinery. Assume the distillation process perfectly separates the stream of gas between a mixture of light gas and isooctane in the top stream and the bottoms of asphaltene.

   Draw a process diagram of this distillation column and indicate on the drawing the flow rates and composition by component of the output streams.

A:

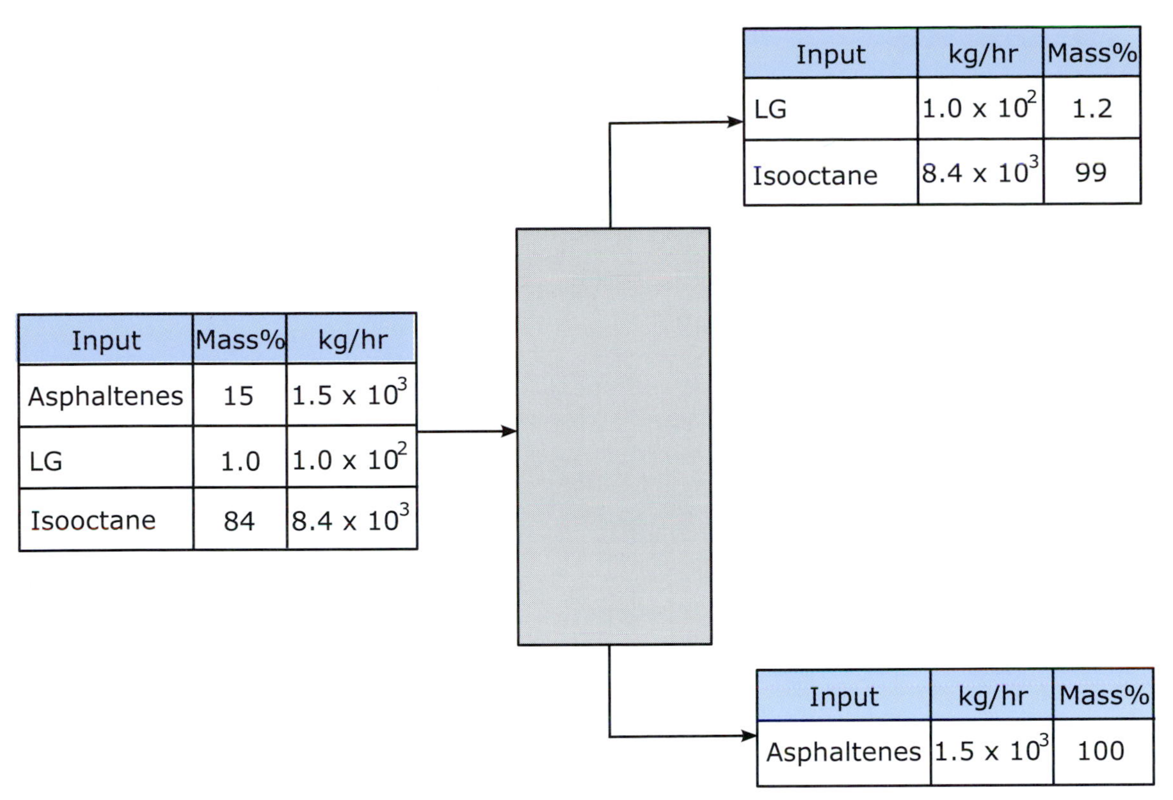

Input	Mass%	kg/hr
Asphaltenes	15	1.5×10^3
LG	1.0	1.0×10^2
Isooctane	84	8.4×10^3

Input	kg/hr	Mass%
LG	1.0×10^2	1.2
Isooctane	8.4×10^3	99

Input	kg/hr	Mass%
Asphaltenes	1.5×10^3	100

9. A crude oil contains just three components by mass: 15% asphaltenes, 1.0% light gases (containing about 1–5 carbon atoms + hydrogen), and the remainder light "distillate" that we will say is isooctane.

   1.0×10^4 kg/hr of this crude oil is fed to the distillation columns in an oil refinery. 5.0 mass % of the isooctane dissolves in the bottom stream. Determine the flows of the components in the overhead and bottom streams.

10. An oil refinery tank farm can deliver 1.0×10^4 kg/hr of a crude oil consisting of 15 mass % asphaltenes, 25 mass % isooctane, 50. mass % heavy oil, and 10. mass % light gases. It is put through a four-stage distillation column. The top or first stage separates out only light gases in the overheads. The second stage separates out 96 mass % isooctane, 2.0 mass % light gases, and 2.0 mass % heavy oil. The third stage separates out only heavy oil. The bottom stage separates out only asphaltene. Create a process diagram model of this refinery. Transfer the picture to a spreadsheet. Show the mass flows for each component at each of the four stages.

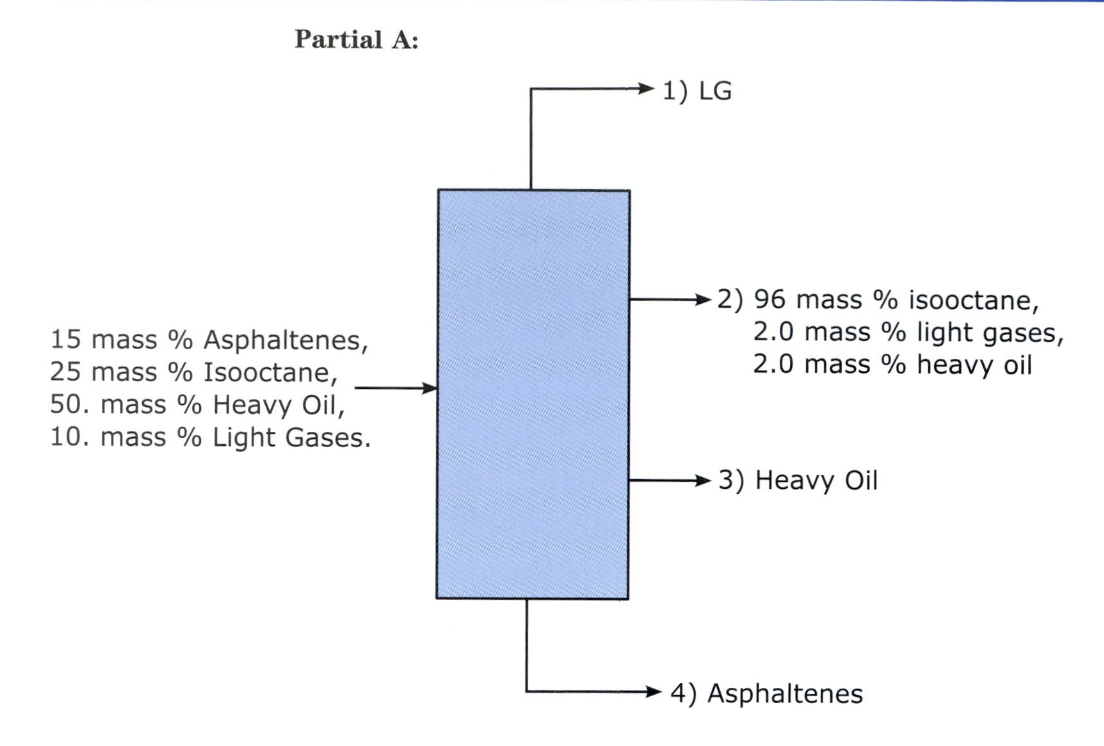

11. An 8,200 kg/hr stream of light gases and isooctane with 98.5% isooctane (C_8H_{18}) by mass and 1.5% light gases by mass is flashed. The bottoms contain 99 mass % isooctane and the overheads 4.0 mass % isooctane. What are the O/H, bottom mass flows in kg/hr? (**Hint:** You will have to solve *simultaneous* algebraic equations.) (**Partial A: Bottoms flow = 8,157 kg/hr.**)
12. An 8,200 kg/hr stream of light gases and isooctane with 98.5 **mol**% isooctane (C_8H_{18}) and 1.50 **mol** % light gases is flashed. The bottoms contain 99 **mol** % isooctane and the overheads 4.0 **mol** % isooctane. Assume average molecular mass of light gases is 65 kg/kmol. What are the O/H and bottom mass flows in **kg**/hr?
13. An oil refinery wants to increase its production of gasoline relative to its diesel oil. The product of one of its distillation columns contains an intermediate product that is 50% cetane ($C_{16}H_{34}$) and 50% (C_8H_{18}) isooctane by mass. This stream is sent through a "thermal cracker." The thermal cracker breaks up one-half of the cetane molecules. The resulting composition contains only two species: C_8H_{18} and C_2H_4 (ethene[11]). What is the *molar* composition of the product of the thermal cracker? (**A:** In mol%: **10.0% $C_{16}H_{34}$, 40.2% C_2H_4 & 49.9% C_8H_{18}.**)

[11]In earlier chemical nomenclature this molecule was called "ethylene." It is characterized by a double bond between the two carbon atoms. By distributing the hydrogen atoms equally on each carbon atom, the rule that every carbon has four bonds is maintained.

14. Many chemical plants recycle undesirable products to reprocess them to useful end products. In a process to make oil from coal, 25 mol % of the original carbon in the coal is recycled in the form of methane, CH_4, and 25 mol % is lost to the atmosphere in the form of CO_2. The process burns coal in a sub-stoichiometric quantity of air and adds steam to form a mixture of oil, here simplified to CH_2, as well as the CH_4 by-product and also the product losses as CO_2. What is the *molar* composition of the product of the thermal cracker? An overall process diagram is shown here:
(A: O_2/Coal = 13.5% [mol/mol], H_2O/Coal = 25% [mol/mol], CO_2/ Coal = 25% [mol/mol], & CH_2/Coal = 75% [mol/mol].)

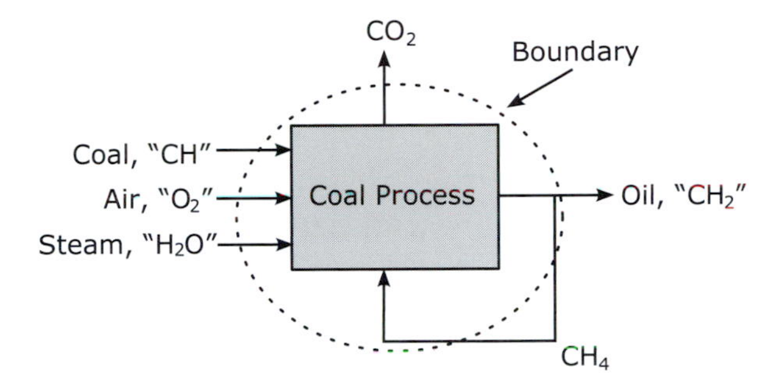

15. In the previous exercise, the operator of the coal plant realizes there is a (profitable) demand for natural gas (CH_4) in the winter and diverts all of the CH_4 to that product instead of recycling it. Calculate the molar amounts of oxygen, of steam consumed, of oil and natural gas produced, and CO_2 emitted relative to the amounts of coal consumed.
16. Catalytic converter: An automobile engine generating 30. kW of mechanical energy emits carbon monoxide (CO) into its exhaust at the rate of 0.033 g/s. The engine's exhaust is then run through a catalytic converter that combines 65% of the carbon monoxide molecules in the exhaust with excess oxygen to produce carbon dioxide (CO_2) which is then emitted through the tailpipe. If the car is traveling at 30. mph, what is the CO emission of the car in g/mile? **(A: Tailpipe emissions = 1.2 g/mile.)**
17. It is desired to reduce the carbon monoxide emission of the automobile in the previous exercise to 1.0 gm of CO per mile. Assume again a 30. mph speed, and an engine generating 30 kW of mechanical energy and emitting 0.033 g/s into the exhaust stream. What percentage of the CO molecules must the catalytic converter convert to CO_2 molecules in order to meet the 1.0 g/mile CO emission target?
18. It is December 1928 and your name is Thomas Midgley, Jr.;[12] you have just invented a new miracle refrigerant composed of

[12]See an abbreviated story of his life: http://www.uh.edu/engines/epi684.htm.

chlorinated flurocarbons (CFCs) that will make your company a lot of money.

At midnight on Christmas Eve you are visited by three spirits who show you your past, present, and future. In your future you discover that the chlorine in your miracle refrigerant will eventually destroy the Earth's ozone layer and put the entire human race at risk. What do you do?

(a) Claim the entire vision was due to a bit of undigested meat and forget it?
(b) Run to the window and ask a passerby to go buy a large goose?
(c) Destroy all records of your CFC work before anyone can use it?
(d) Start work developing nonchlorinated hydrocarbon refrigerants?

19. It is 2021 and you are a process engineer at a large oil refining company. The world is rapidly moving toward a hydrogen energy economy and your company has been trying to develop an efficient and cost-effective way to extract hydrogen from crude oil. You have found a low-cost process, but it produces a considerable amount of undesirable chemical by-product pollutants. However, during your work on this project you have also discovered an effective and inexpensive way to extract hydrogen directly from seawater. You realize that revealing this process would effectively eliminate the world demand for petroleum and would probably cause serious financial damage to your company. What do you do?

    (a) Quit your job and start your own hydrogen producing company.
    (b) Talk to your supervisors and reveal your process to them to see if they wish to pursue implementing it as part of their company.
    (c) Contact a patent lawyer not associated with your current employer and try to patent this potentially lucrative new process.
    (d) Without your employer's permission, publish an article in a well-read chemical or energy magazine revealing your process and giving it to the world free of charge.

Chapter 14

© iStockphoto.com/Michael Knight

Cars of the Future—What Will They Be Like?

The Car Culture

Since the beginning of the automobile era, over 100 years ago, Americans have had a love affair with their cars. Automobile styling reflected social standing and temper of their times. Aerodynamic styling, rocket-like tail fins, big toothy chrome grills, bright colors, and sporty interiors are all part of American automotive history.

Figure 1

1928 Antique automobile. © iStockphoto.com/ Richard Scherzinger

Figure 2

1957 Chevrolet rear taillight.
© iStockphoto.com/ Ed Hughes

Today we have entered an era of more conservative styling but with essential and inclusive electronics. Automotive power sources are shifting from the inefficient internal combustion engine to the more sophisticated and efficient "hybrid"[1] power source and perhaps eventually to a "fuel cell"[2] power source that converts chemical energy directly into electrical energy with no moving parts. This coincides with the complete electrification of modern automobiles, ultimately producing drive-by-wire systems that minimize mechanical components.

The early twenty-first century will see the conversion of the automobile from what started as primarily a mechanical device into a mechatronic device that is as dependent on electrical and computer components as it is on mechanical components.

[1]A "hybrid" vehicle is driven by a combination of electric motors and an internal combustion engine.
[2]"Fuel cells" are continuously fueled batteries that combine fuel and oxygen without actual combustion to produce electrical energy.

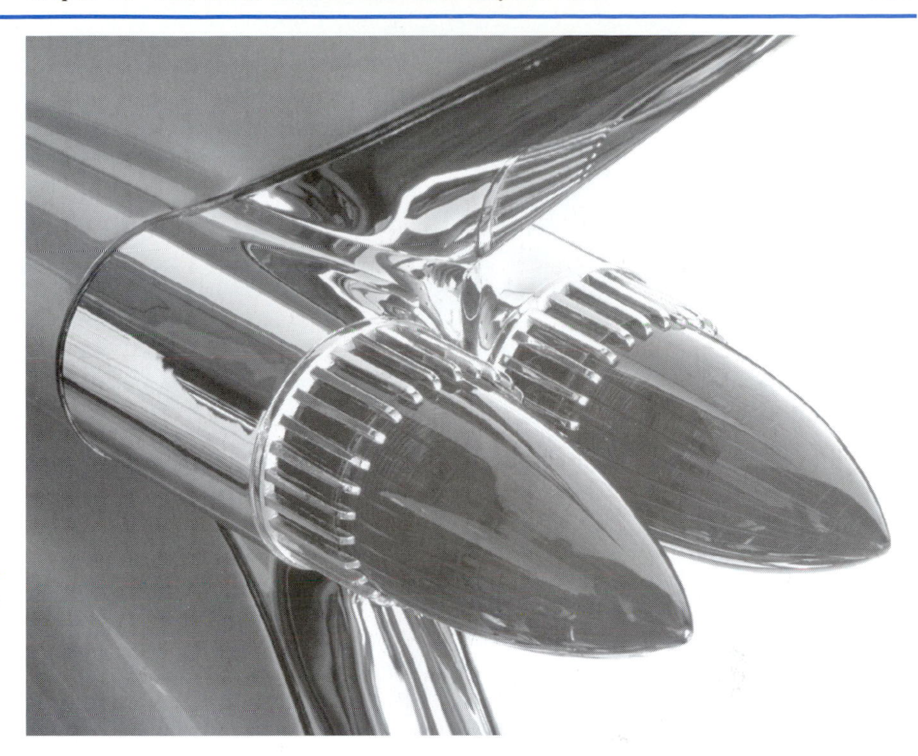

Figure 3

The tail-fin and taillights of a 1950's luxury automobile.
© iStockphoto.com/ David Kelly

Big Changes Ahead: Convergence

Early twenty-first-century automotive development will focus on merging mechanical, electrical, and computer systems to produce "intelligent" mechatronic devices that will augment human driving skills. The term "digital mobility" means using digital data from numerous vehicle sensors to control a vehicle's basic functions. All of these basic digital functions in future vehicles will be networked to improve energy efficiency, reduce emissions, and enhance occupant safety.

"Convergence" is the perfect word to describe what is currently happening to automotive technology. For example, the center of the dashboard normally housed the vehicle's radio and environmental (heater and air conditioning) controls. In the future the central dashboard position will become the vehicle's information center, containing navigation systems, digital music, voice recording, environmental controls, telephone, vehicle diagnostics, Internet connection, and so forth.

Consumers are demanding smarter and more integrated vehicles. The convergence of personal computer technology and the car's digital radio will produce vehicle entertainment centers containing gigabyte hard

Figure 4

GM's "Autonomy" concept car.
Courtesy of General Motors Corporation

drives, wireless Internet connection, memory cards, voice recognition, word processing ability, and many more unimagined delights.

Eventually, with full automation, in addition to entertainment, future consumers may want the ability to safely work from their car, just as they now work from home. Smart highway systems will take control of the vehicle and "drive" it safely to a destination selected in the global positioning system memory.

Until the 1970s, the only truly electronic technology in a car was its radio. Today, automotive electronics drive up to 90 percent of functional systems in vehicles, with more sensors and actuators continually being introduced.

Each new feature requires an electronic control unit, and the electrical power required to operate the controls of future vehicles is constantly increasing. Vehicle batteries have been upgraded from 6 volts in the early twentieth century to 12 volts in the late twentieth century and will be at 42 volts in the early twenty-first century to be able to supply the electrical power required by advancing vehicle design.

In recent years, a wide range of automotive innovations have been researched including electric and hybrid vehicles, radar braking systems, power train controls, vehicle self-diagnostics, interior motion sensing, and drive-by-wire systems.

Future automotive electronics will be driven by robust software development. This will integrate with faster, smaller, and smarter electronics, all connected and embedded. Software upgrades, done wirelessly without going to a dealership, will personalize a driver's vehicle interface and preferences. Cars of the future will have digital microprocessor-based controls on all of their systems, from power source to steering, braking, and navigation. Intelligent control systems will also interact with the nonvehicle

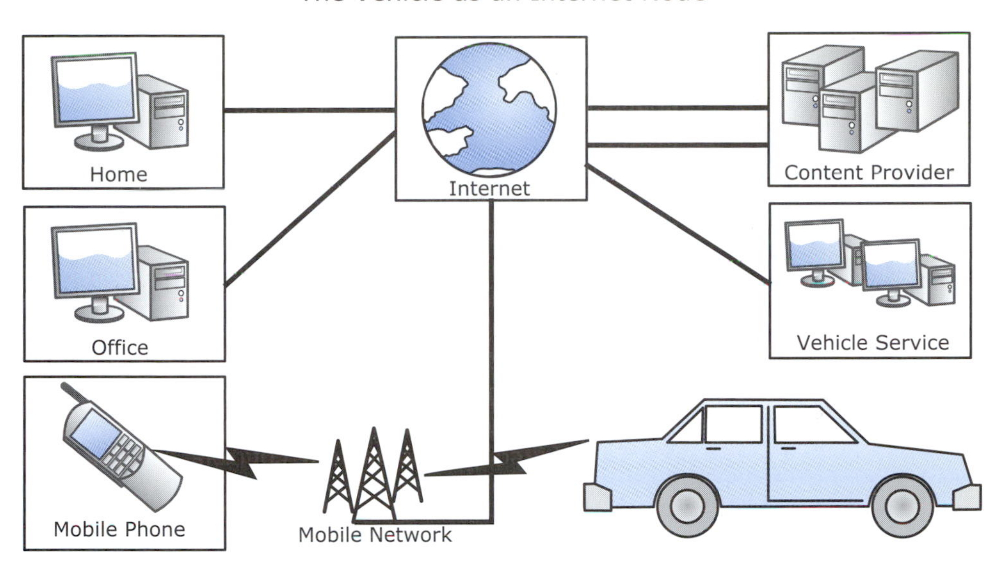

Figure 5 Automobiles will become internet nodes

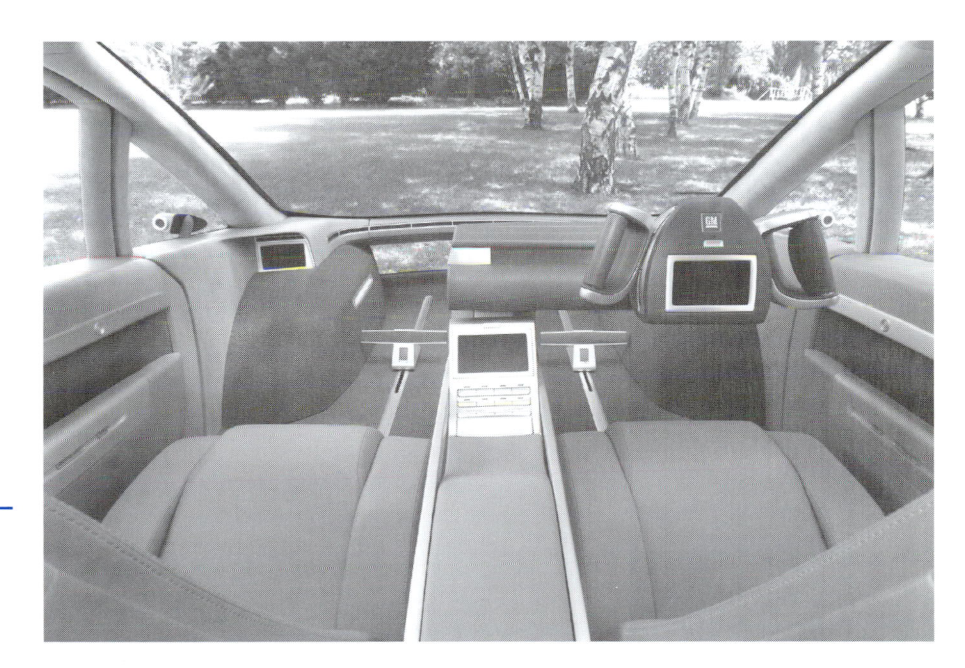

Figure 6

Dashboard of the GM Hi-Wire concept car. Courtesy of General Motors Corporation

systems such as PDAs, cell phones, the Internet, vehicle interior environment control, personal entertainment systems (games), and intervehicle networking capability for personal communication, traffic control, and accident avoidance.

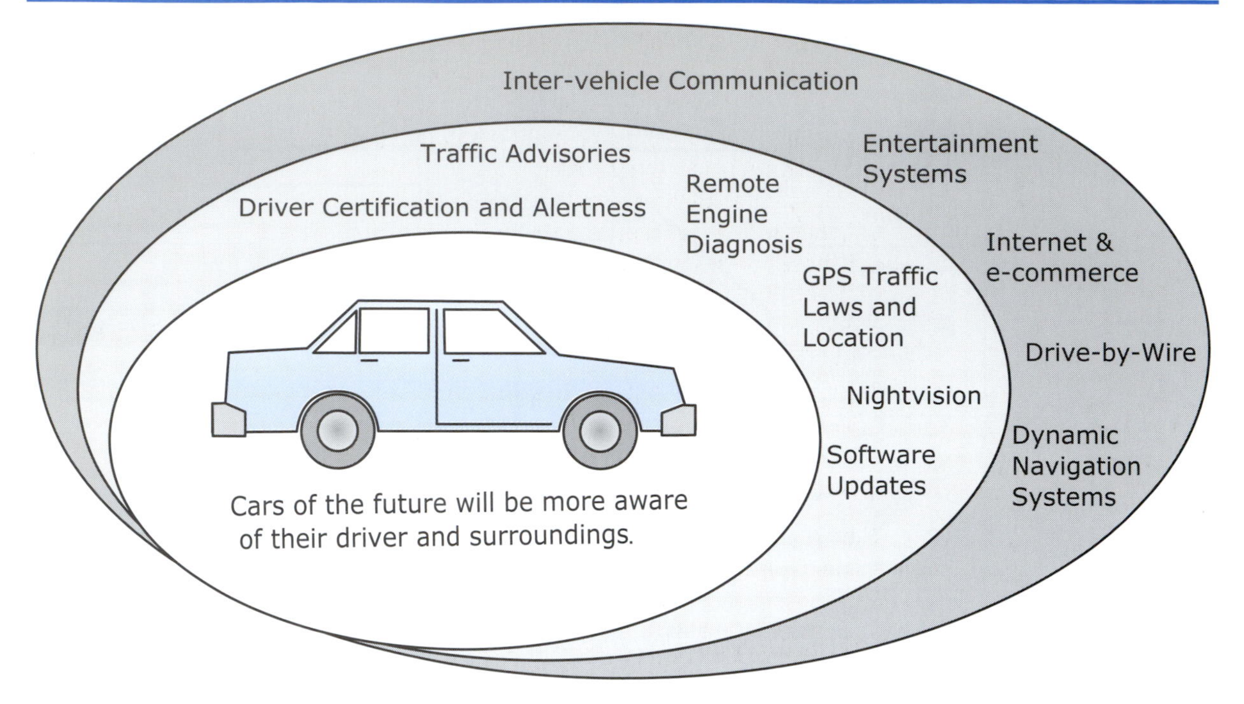

Figure 7 Future cars will be aware of their driver and surroundings

Automotive Electronics

Automotive electronics in the twenty-first century will undergo significant changes. Increased demands by consumers for improved automotive safety and convenience, new regulatory standards, competition from foreign suppliers, lean manufacturing, and continued advances in microelectronics and information technologies will all converge to produce radically different vehicle operating systems in the future.

Advances in semiconductor, sensor, and computer technologies are driving automotive electronics to new levels. Information, communications, and geographical navigation technologies are converging into a new field called telematics.[3] Future automotive electronic systems will include new or upgraded technology in many areas:

- Antilock braking systems (ABS)
- Traction control systems (TCS)

[3]*Telematics* is the English language version of the French word *telematique*—coined by Simon Nora and Alain Minc in the book *L'informatisation de la Société* (La Documentation Francaise, 1978); translated as *The Computerization of Society* (MIT Press, 1980). The word *telematics* means the blending of computers and telecommunications. Examples include the Internet and the Minitel system in France.

- Automatic stability control (ASC) systems
- Driver facial recognition systems for anti-theft protection
- Driver fatigue or intoxication recognition (producing vehicle shutdown)
- Airbags and similar occupant-restraint systems
- Electronic fuel injectors
- Engine management systems and other power train electronic systems
- On-board navigation and driver information systems
- Entertainment and cellular phone systems
- Electronic security systems
- Smart headlights (that provide maximum illumination for road and traffic conditions)
- Rear backup system with audio and visual response
- Traffic congestion avoidance systems
- Accident avoidance systems
- Rain-sensing windshield wipers

Future telematic systems will provide vehicles with a flexible, easy-to-use interface that allows occupants to use their PDA, cellular phone, and related technologies while driving safely. Telematic systems will also allow the vehicle to be continuously connected to the car manufacturer to provide services such as vehicle tracking, pay-per-use insurance, telediagnostics, and remote vehicle system updating.

Smart Automobile Safety Systems

Smart safety systems can prevent intoxicated individuals from starting cars, detect impending collisions and automatically avoid them, and support the driver's ability to control the vehicle safely in all situations. Typical situations requiring smart safety technologies include: vehicle incipient rollover, lane departure warnings, rear vision obstruction, object detection in front of the vehicle, and vehicle dynamic stability under various weather and road conditions.

Smart Tire Systems

Correct tire pressure will improve handling, decrease fuel consumption, and extend tire life. Intelligent tires will contain individual microchips that will work together to monitor tire pressure, temperature, and wear, reporting changes to the driver through the car's computer.

The embedded tire sensors will be passive and not rely on batteries for power. In some high-line vehicles, self-powered (from captured

Figure 8

Tire pressure sensor. Courtesy of Daimler-Chrysler Corporation

vibrational energy converted to electricity) pressure transmitters are already available.

Smart Exhaust Systems

Automotive exhaust systems are designed to meet legal noise limits and reduce mechanical noise. However, modern luxury and sports cars have a sound (music to some, but noise to others) that is characteristic of their brand and model. Automotive acoustics are now an essential part of a car's character, just as unmistakable as its appearance. Sound design includes targeting desired frequency ranges and suppressing the undesired frequencies. Since exhaust noise frequency depends on engine speed, the exhaust system will be "tuned" to emphasize various frequency bands to produce a pleasant, or a sporty, sound that is also audible inside the car. For other drivers "sound cancellation technology" will basically eliminate most of the external noise that reaches the driver so that the exhaust system noise will disappear.

Smart Drive-By-Wire Systems

Drive-by-wire cars of the future will replace nearly all of the twentieth-century mechanical systems for steering, braking, vehicle environmental control, and lights with computer-controlled electronic actuators. One of the most complex new systems will be steer-by-wire technology. The mechanical connection between the steering wheel and the front wheels will be replaced by electronically controlled actuators that will turn the front wheels and provide force feedback to the driver. A steer-by-wire system must be very robust, safe, and reliable. To replace conventional mechanical steering, steer-by-wire systems will be able to tolerate electrical failure in any subsystem. A fault-tolerant steer-by-wire system without any mechanical backup will be able to detect and correct system failures quickly and safely.

Smart Highway Systems

Our reliance on automobiles in a nation with a rapidly aging highway infrastructure compromises safety, produces congestion, and reduces vehicle and transportation efficiency. Automated Highway System (AHS) technologies offer solutions to a broad range of problems. An automated highway system will transmit control information to all vehicles present. It will control highway infrastructure, maximize safety and efficiency, and reduce congestion and transportation costs. This technology will create a local area network between nearby vehicle computers that will implement collision avoidance systems by controlling the vehicle's speed, electronic brakes, electronic steering, and any other systems that supplement or replace human driving judgment. In the last decades of the twentieth century, the number of streets and roads in America increased by about two percent, but the number of cars in America increased by more than 50 percent (and, on average, they each traveled 77 percent more miles). This has produced increased traffic congestion that affects all major metropolitan areas. Without a massive move away from personal transportation to public mass transportation, we will need to manage personal traffic flow more efficiently.

Smart highways are under development that will safely "platoon" vehicles by taking control of all the cars and moving them along, bumper-to-bumper, at an efficient constant speed while driver and occupants watch TV, chat, work on their computers, or just enjoy the scenery.

Smart Traffic Systems

"Smart traffic signals" and electronic roadside signs will wirelessly communicate with the vehicle's computer system and provide the driver with instant updates on traffic and weather conditions ahead. Future cars will have imbedded satellite-based global navigation systems that will communicate with the driver and occupants through the vehicle's computer.

Not only will the vehicle know where it is at, it will know the traffic laws appropriate to that location. For example, the vehicle will know the posted speed limit and warn the driver when it is exceeded. It will also log the location and duration of speed limit violation and have the capability to report this to the authorities when queried wirelessly by roadside monitoring stations. The vehicle will also know when to stop at intersections, and log the driver's failure to come to a complete stop when required. These logging and reporting capabilities are certainly technically feasible. Whether they are legally feasible is less certain. Does automatically querying a vehicle's internal log constitute an unconstitutional search?

Smart Advanced Power Systems: Variable Displacement and Variable Valve Engines

Several technologies have been developed to reduce fuel consumption in automobile internal combustion engines. One of the more important involves deactivating several of the engine's cylinders by closing both their intake and exhaust valves. Variable displacement engines will have to seamlessly alternate between deactivating cylinders when less power is needed and reactivating the cylinders when more power is needed. This optimizes fuel economy without sacrificing vehicle performance. With electronic throttle control the engine can deactivate or activate cylinders in 40 milliseconds.

Also, an electronic valve control (EVC) system will replace the engine's mechanical camshaft, controlling the intake and exhaust valves with electronic actuators for independent valve timing and displacement. The EVC system controls the opening and closing time and lift amount of each intake and exhaust valve with independent actuators on each valve. Changing from a mechanical camshaft driven valve to independently controlled actuator valves produces increased engine power, greater fuel economy, and more environmentally friendly emissions.

Fuel Cell Systems

The automobile industry is under a great deal of pressure to produce and market low emission vehicles. Fuel cells show great promise as a replacement for internal combustion engines in automobiles due to their high efficiency, low emissions, and quiet, continuous operation.

Fuel cells use electrochemistry rather than combustion to generate electricity from hydrogen. However, most commonly, they run on fuels like natural gas, methanol, or gasoline and use a fuel reformer to convert these hydrocarbon fuels into hydrogen and other components. A fuel cell then combines the hydrogen with oxygen from the air to generate electricity. The need to convert a fossil fuel to hydrogen is a serious inefficiency in the overall fuel cell cycle as well as a source of CO_2.

A fuel cell is similar to a battery in that electrochemical energy is used to produce electrical energy. Like batteries, fuel cells can be stacked in series to increase the voltage of the system. Unlike batteries, fuel cells never need to be recharged since their fuel is continuously delivered. Some fuel cells use premanufactured hydrogen from an external tank and oxygen from the atmosphere to produce electrical power. Fuel cells combine the best attributes of batteries (electrochemical energy conversion) and internal combustion engines (rapid refueling via an external fuel supply).

Fuel cell energy conversion efficiency[4] (the ratio of the electrical energy to as-supplied fuel energy) ranges from 45 to 70 percent compared to the 15 to 30 percent efficiency of an internal combustion engine. Fuel cells have high efficiency at low loads while internal combustion engines have

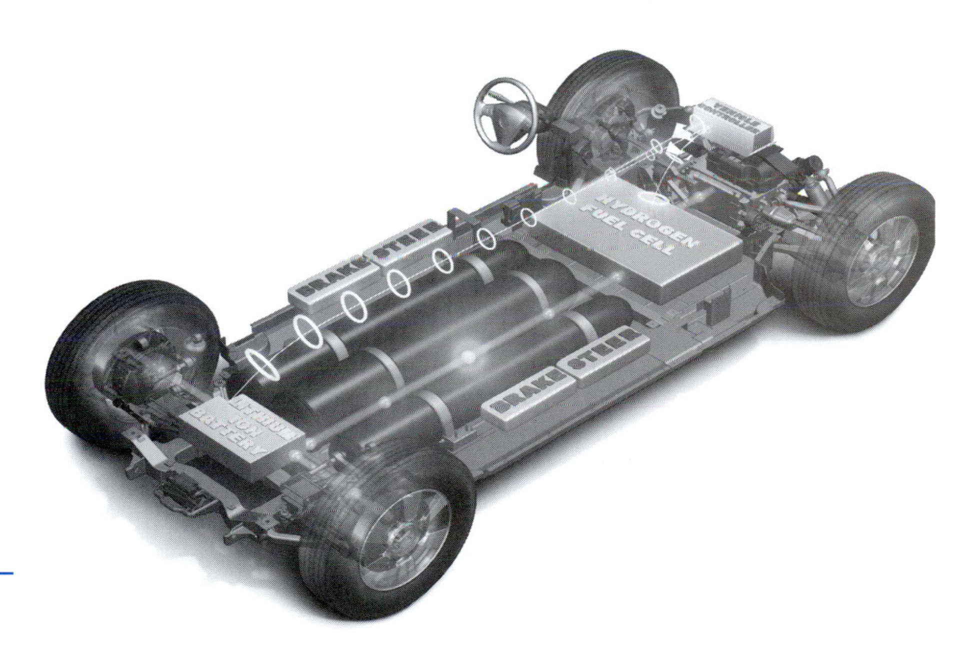

Figure 9

Fuel cells of the GM Sequel concept car

[4]These fuel cell efficiencies quoted are exclusive of significant inefficiencies realized during the manufacture of hydrogen. While this manifests itself as *additional* CO_2 releases, these releases are remote from the vehicle itself.

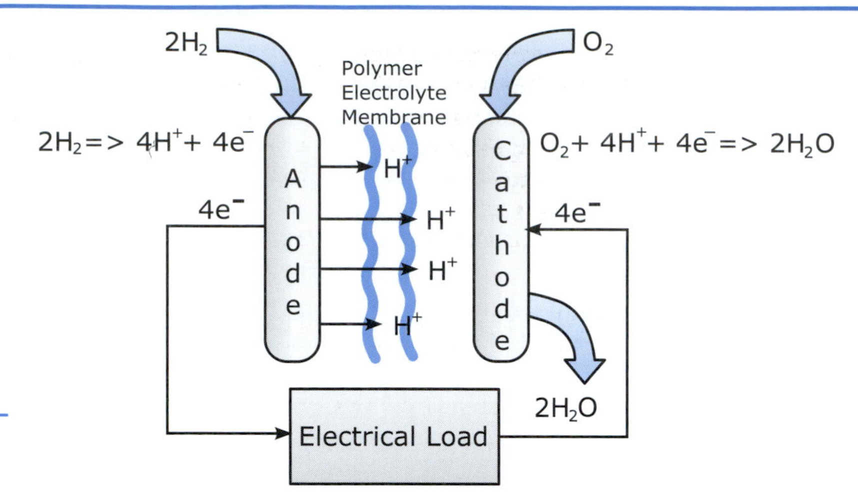

Figure 10

Fuel cell chemical reactions

high efficiency at high loads. There are several types of fuel cells, distinguished mostly by the chemistry, electrolyte, and fuel.

Enjoying the Ride—Smart Vehicle Entertainment Systems

Vehicle entertainment systems play a large role in the future development of automotive electronics. Since the first car radio was introduced in the mid 1930s, in-vehicle entertainment systems have driven the development of new consumer electronics in the automotive industry.

Automotive electronics evolved to include transistor radios, electronic tuning, cassette and CDs, multispeaker premium audio systems, rear seat DVD video, satellite digital audio broadcast, and MP3 radio. Future automotive electronics will showcase satellite digital TV, hard disc drive "music juke-boxes," and advanced WiFi distribution of music and video.

Digital entertainment has made the transition from the home to the automobile. Plasma screens, MP3 music, DVD movies, and sophisticated computer games have penetrated the automotive industry. Rear seat video, digital receivers, satellite reception, digital playbacks, and high-bandwidth wireless connectivity are all part of tomorrow's car.

The evolution of car radio is another example of the convergence of diverse technical fields: RF electronics, mobile wireless communications, the Internet, personal computers, consumer electronics, and automotive human-machine interfaces. Besides traditional AM/FM programming, radios today also play a variety of media: cassette, CD, MP3, DVD-A, and so on, in addition to 100 channels of satellite digital audio programs. In the future, radio (or its equivalent) will continue to be the entertainment center

Figure 11

Color displays are integrated in the backs of the driver and front passenger seat backrests. These are hooked up to the TV receiver and the DVD player.
From Daimler-Chrysler media.

Figure 12

Automotive video console.
© iStockphoto.com/
Roberta Casaliggi

of the vehicle, and consumers will expect to have access to personalized information anywhere and anytime.

To fulfill this vision, the next generation of radios will likely be based on a computing platform with sizeable built-in memory and running software-defined functions and features. It will also have built-in broadband two-way communications to simplify the consolidation and management of music and data files and to enable future dynamically reconfigurable applications. In a few short years rear video systems have gone from a "nice-to-have" feature to a "must have" feature. In a short period of time the market has migrated from a VHS player with a 5.8" LCD screen to a ceiling-mounted system that includes a 7" screen and DVD player. The future promises to be quite dynamic.

How Do We Get There?

The wireless industry is characterized by rapid, dramatic high-tech changes with a less than two-year cycle time and an equivalent lifecycle. The automotive electronics industry is working toward reducing the typical development cycle of two to three years down to one to two years but with a life cycle of 10 years or more. In addition to the electronic development benefits seen in the wireless industry, the automotive industry needs to place more emphasis on the quality and reliability of their designs because of inherent safety requirements.

New automotive technologies have traditionally been initiated in the aftermarket.[5] Since the development cycle of an aftermarket product can be as much as half of that of an equivalent original equipment part, it is the primary enabler of early technological development. Aftermarket development lead times can be less than one year. However, an electronic device that is offered as original equipment by a car maker will typically be subjected to rigid vehicle testing producing a development cycle of over two years.

The electronic and semiconductor content of automobiles continues to grow rapidly. In recent years we have seen the emergence of new categories of multimedia products such as satellite radio, rear seat video system, and global navigation products. As these products gained popularity through the aftermarket, automotive manufacturers faced the challenge of delivering them as standard equipment in their vehicles.

[5]*Aftermarket* is an umbrella term for the collective network of vendors who design and sell components that are intended to replace the stock manufacturer's parts in order to alter the appearance or performance of the vehicle. Alternatively, an aftermarket part can be a straight replacement for a stock item at a lower price with no intention to cause such a change.

Looking Forward

The future of automotive technology is clearly very challenging and highly interdisciplinary. With over 600 million automobiles in the world today, our dependence on petroleum for fuel is causing increasing financial, ecological, and international problems. Even if we improve vehicle efficiency (e.g., hybrids) and implement hydrogen-based fuel cell technology, it will take decades to reduce the environmental impact. By 2005 there were more than 210 million cars in America alone, and we add about 17 million more each year. (We currently have more cars than drivers.) Furthermore, with a population of over 1.3 billion (more than four times that of the U.S.), China is on the road to becoming the largest car market in the world. With annual car sales at 5 million in 2004, it is estimated that 50–60 million Chinese can now afford a car. By 2010 China is projected to be the world's largest automobile consumer. In China's bigger cities today the bicycle has already been replaced with streets jammed with cars. The population of India is also over 1 billion, and they, too, are rapidly developing a car culture.

How will the engineers and politicians in the United States and the world deal with this growing technological transportation problem? In Denmark, where rail and bike infrastructures are well developed, to reduce car ownership a car's registration fees exceeds its retail price. Germany has tripled its length of nationwide bike paths, and Swiss towns use their car-free status as a tourist attraction.

For every mile driven in a car, we use two to three times as much fuel as by mass transit, but this is a solution that has not been very appealing to Americans. Yet, it has worked well in other parts of the world. Western Europeans use public transit for about 10% of all urban trips, compared with Americans at only 2%.

When the automobile was introduced in 1900, horse drawn carriages were the main form of transportation, and there was an impending environmental dilemma then, too. As the U.S. population grew, so did the horse population. If this continued unchecked, the amount of horse manure produced would have been sufficient to blanket the entire island of Manhattan one mile deep. The automobile saved us from that! It is now up to you, our readers, to solve the current problems wrought by the automobile.

Part 2

Hands-on

Chapter 15

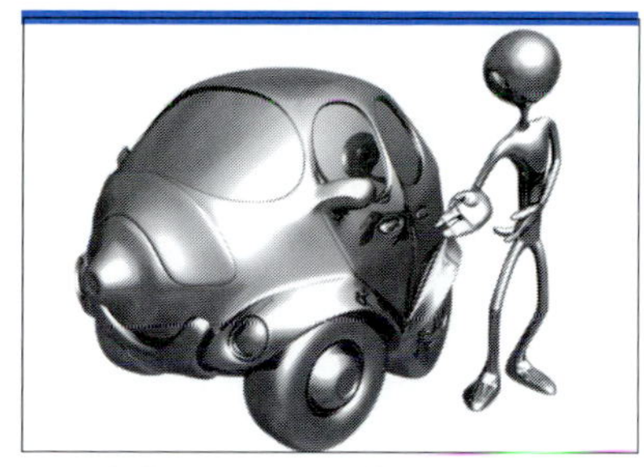

© iStockphoto.com/Scott Maxwell

Introduction to Engineering Design

This chapter describes **the nature of engineering design,** suggests some **benefits of a hands-on design project,** indicates the **qualities of a good designer,** and explains the **need for a systematic approach** that consists of an **eight-step design process.**

The Nature of Engineering Design

In the course of creating new products, engineering design uses available technology to improve performance, lower cost, or reduce risk. For example, if the design of a bridge produces a new structure that is visually stunning with no consideration for its strength, this is design without the engineering. If, on the other hand, the designers of a new-concept car use analysis or experiments to evaluate air drag, structural integrity, and manufacturability in addition to style when coming up with a new exterior design, this is engineering design.

Prior to the 1960s, the preceding description of engineering design might have been adequate, but this is no longer the case. The definition has been broadened to include the systematic thought processes and best practices that define the modern engineering design process. This systematic approach, which has become synonymous with engineering design, forms the heart of these design chapters.

Those of you experienced at tinkering with your own inventions in the basement or garage might say that you got along fine without engineering analysis and without training in the systematic approach. Such a view ignores the realities of engineering design. Usually, designers must search for the best possible design under severe conditions of limited time and limited resources (especially cost). For example, in most design courses, there is barely enough time to produce a single prototype. The idea behind an engineering approach, as embodied in engineering analysis and the systematic approach, is to minimize the number of design iterations required to achieve a successful design. Fewer design iterations means lower cost and shorter development times.

One of the greatest challenges of engineering design is the breadth of knowledge required of the designer. The diversity of topics covered in earlier chapters provides a hint of what a design might entail; some electromechanical designs could conceivably touch upon them all. In addition to those topics, there will be issues related to manufacturing, economics, aesthetics, ethics, teaming, government regulations, and documentation of the design, to name but a few. In the spirit of these design chapters, we propose that the best way to understand the multifaceted nature of engineering design is to experience it for yourself through the following hands-on design exercise.

Hands-on Design Exercise: "The Tower"

The design objective is to build the tallest tower out of the materials supplied. This is both an individual and a team competition.

Setup

- Divide the section into teams of three students per team.
- Provide each team with five sheets of 8.5 × 11 standard copier paper.
- Other materials will be distributed unevenly as follows (assuming eight teams per section):
  - Two teams each receive one roll of Scotch® tape.
  - Two teams each receive a roll of duct tape.
  - Two teams each receive one box of paper clips.
  - Two teams each receive one pair of scissors.

Rules

- Teams will have up to 10 minutes to build a tower.
- Measurement of tower height must be made by the instructor.
- Teams should indicate to the instructor when they are ready for a measurement.
- If the tower is composed of materials other than the supplied paper, Scotch® tape, duct tape, or paper clips, it will be disqualified.
- Tower must be stationary when the measurement is made.
- Tower must be built on a flat surface. Tower cannot lean against or be attached to any other surfaces.

- Any team that intentionally knocks over another team's tower before it has been measured will be disqualified.
- After 3 minutes, one individual on each team will be offered 8 points to jump to the next team.
- There are no other rules.

Scoring

- One point will be awarded for each inch of tower height. (Heights will be rounded to the nearest inch.)
- First team to be measured will receive a bonus of 10 points.
- Winners are (1) the team with the highest point total and (2) the individual with the highest point total.

After the Exercise

Discuss the importance of the following issues to the outcome of the competition:

- Quality of teaming among different teams (were materials shared?).
- Quality of teaming within the team (was everyone within the team given the opportunity to contribute ideas, or did one person dominate the decision making?).
- Ethics (was it ethical to jump to another team, not share materials, or copy the design of another team?).
- Manufacturability (how important was it to have the right materials?).

Benefits of a Hands-on Design Project

A practicing engineer does not have to be an expert in machining or other basic manufacturing operations. Still, a basic understanding of the challenges involved in manufacturing a product is essential for producing a successful design. The best way to appreciate that fact at an early stage in your career is to manufacture a design yourself.

The lessons to be learned are universal. Don't expect your design to work on the first try. Leave a lot of time for testing. Complicated designs take a lot longer to build and have a lower probability of success. If you have a choice of manufacturing a part yourself or buying it, buy it. Many such lessons are foretold by the design principles and design for manufacture guidelines of later chapters. The consequences of violating those principles are best understood by experiencing the results of having done so.

There are other lessons to be learned from a hands-on design experience. For electromechanical systems with moving parts, that experience might be the only way to accurately evaluate the design. Also, students gain a sense of accountability by learning that it is not enough for a design to look good on paper. In order to actually be a good design, it has to lead to an end product that works.

In particular, we recommend that the hands-on design project should be done under a competition format, involving interactions between the machines. It is a natural motivator—the challenge of the task is heightened by having to deal with the unpredictability of your human opponent, and the other designs provide a relative scale against which to assess quality of performance.

Qualities of a Good Designer

These are the qualities of a good designer:

- *Curiosity about how things work.* Seeing other design solutions provides you with a toolbox of ideas that you can draw from when faced with a similar design challenge, so when you come across an unfamiliar device, try to figure out how it works. Take things apart; some companies actually do this and refer to it as "reverse engineering." Visit a toy store; the products there demonstrate creative ideas and new technology.
- *Unselfishness.* A key ingredient to effective teaming is suppressing ego and sacrificing personal comfort to serve the best interests of the team.
- *Fearlessness.* It takes a leap of faith to step into the unknown and create something new.
- *Persistence.* Setbacks are inevitable in the course of a design project. Remain resilient and determined in the face of adversity.
- *Adaptability.* Conditions during the design process are constantly evolving. For example, new facts may surface, or the rules of the design competition may change in some way. Be prepared to take action in response to those new conditions. In other words, if the ship is sinking, don't go down with the ship—redesign it.

The Need for a Systematic Approach

There are two main goals of the systematic approach to engineering design: (1) to eliminate personal bias from the process and (2) to maximize the amount of thinking and information gathering that is done up front, before committing to the final design. The result is fewer costly design changes late in the product development stages.

The engineering design process also provides a blueprint for design of complex systems. For example, you might be able to get along without a formalized design procedure when designing a new paper clip,[1] but when taking on the daunting task of designing a complex system like the space shuttle, brain gridlock can set in. The design process offers a step-by-step procedure for getting started, as well as strategies for breaking down complex problems into smaller manageable parts.

The Steps in the Engineering Design Process

The systematic approach to engineering design may be viewed as consisting of eight steps:

1. Defining the problem
2. Generating alternative concepts
3. Evaluating and selecting concepts
4. Detailing the design
5. Design defense
6. Manufacturing and testing
7. Evaluation of performance
8. Preparing the final design report

The next chapter presents two ground rules for engineering design. Subsequent chapters will treat each of the preceding eight steps in detail. At the end of each chapter there is a suggested milestone for successful completion of the step in the design process described in the chapter. Milestones are crucial for measuring progress toward the eventual goal of a successful design. After the eight steps have been described, this portion of the book concludes with a detailed example of an actual design competition.

[1]But see just how difficult it originally was to perfect the paper clip: *The Evolution of Useful Things: How Everyday Artifacts—from Forks and Pins to Paper Clips and Zippers—Came to Be as They Are*, Henry Petroski (Alfred A. Knopf, Inc., 1992).

Chapter 16

© iStockphoto.com/Aksenov Vasilii

Two Ground Rules for Design

This chapter introduces two ground rules for design: the use of a **design notebook** and **effective teamwork.**

Ground Rule Number 1: Use a Design Notebook

When you are a working on a design project and you want to write something down, the design notebook is the place to do it. There is no need for notepads, reams of paper, or sticky notes. The place to record your thoughts is in a permanently bound volume with numbered pages, a cardboard cover, and a label on the front cover identifying its contents. Every college bookstore has them, though they may be called laboratory notebooks.

As a starting engineer, now is the best time to start a career-long habit. Just how important the design notebook is can be explained in the case of Dr. Gordon Gould.

> On November 9, 1957, a Saturday night just given to Sunday, Gould was unable to sleep. He was 37 years old and a graduate student at Columbia University. For the rest of the ... weekend, without sleep, Gould wrote down descriptions of his idea, sketched its components, projected its future uses.
>
> On Wednesday morning he hustled two blocks to the neighborhood candy store and had the proprietor, a notary, witness and date his notebook.

[1]http://inventors.about.com/gi/dynamic/offsite.htm?site=http://www.inc.com/incmagazine/archives/03891051.html.

> The pages described a way of amplifying light and of using the resulting beam to cut and heat substances and measure distance. . . .
>
> Gould dubbed the process light amplification by stimulated emission of radiation, or laser.

It took the next 30 years to win the patents for his ideas because other scientists had filed for a similar invention, although after Gould. Gould eventually won his patents and received many millions in royalties because he had made a witnessed, clear, and contemporaneous record of his invention.

The lesson is that patents and other matters are frequently settled in court for hundreds of millions of dollars by referring to a notebook that clearly details concepts and results of experiments. You must maintain that notebook in a fashion that will expedite your claim to future inventions and patents.

Another more immediate benefit of using a design notebook is that you will know that everything related to the project is in one place. Finding that key scrap of paper in a pile of books and papers on your desk after working for months on the project can be a rather time-consuming endeavor.

Here are some of the most important guidelines for keeping a design notebook:

- Date and number every page.
- *Never* tear out a page.
- Leave no blank pages between used pages. Draw a slash through any such blank pages.
- Include all your data, descriptions, sketches, calculations, notes, and so forth.
- Put an index on the first page.
- Write everything in real time—that is, do not copy over from scraps of paper in the interests of neatness.
- Write in ink.
- Do not use whiteout; cross out instead.
- Paste in computer output, charts, graphs, and photographs.
- Write as though you know someone else will read it.
- Document team meetings by recording the date, results of discussions, and assigned tasks.

Ground Rule Number 2: Team Effectively

Working in teams on a design project is both a joy and a challenge. While there is a sense of security in knowing that others will be venturing into the unknown alongside you, the unpredictability of human interactions can be as perplexing as the design itself.

To reduce the risk of ineffective teaming, rules of conduct will be presented in this section. These are well-accepted best practices based on observations of effective teams.

There are several advantages to attacking a design project in teams. First, design requires a wide range of skills and areas of knowledge. No one person is experienced enough to pursue every unfamiliar design challenge in isolation. Teaming provides an opportunity to expand the talents and life experiences that will be brought to bear on the design problem. Second, if done right, teaming serves to keep personal biases in check. Third, more people should mean that more will get accomplished in a shorter period of time, although it is puzzling to often see team members standing by politely as one team member does all the work (especially during manufacturing). When best practices are followed, a team will be greater than the sum of its parts.

For design projects done during the freshman and sophomore years, three people are the ideal team size. Teams of two may not experience all of the typical dynamics and so may not learn as much about teaming. With teams of four, it may be too easy for one team member to hide. Design teams at this level are usually not assigned a team leader by the instructor. Leadership typically emerges within the team. If a team leader is assigned, the role is *not* to be the boss, but rather to organize and facilitate participation by all team members.

Here are some teaming best practices:

- **Assign clear roles and work assignments.** A few things are best done as a team, such as brainstorming and evaluation of concepts. Most of the time it will pay off if everyone has his or her own assigned responsibilities and tasks to which they will be held accountable by the team. These tasks should be assigned or updated at the end of each and every team meeting.
- **Foster good communication between team members.** An atmosphere of trust and respect should be maintained in which team members feel free to express their ideas without retribution. That trust extends to allowing for civilized disagreement, delicately done so as not to suppress ideas or discourage participation. Everyone should participate in the discussions. Sometimes this means reaching out with sensitivity to the shy members of the team. If you succeed, you will have a team operating on all cylinders.
- **Share leadership responsibilities.** If there is a designated team leader, that person should empower the other team members with significant leadership responsibilities. This will give those students a strong sense of ownership in the project. At the same time, team members have to be willing to step forward to assume leadership roles.
- **Make team decisions by consensus.** Teams make decisions in one of three ways: (1) the team leader makes the decision, (2) discussions continue until everyone agrees (as in a trial by jury), or

(3) after discussions are exhausted, the team takes a vote. Those who disagree with the outcome of the vote are then asked if they can put their opinions aside and move forward in the best interests of the team. In a college-level design project, the only ways to go are (2) and (3), which are both examples of decision making by consensus.

Chapter 17

© iStockphoto.com/Miroslaw Pieprzyk

Clarification of the Task

The design process begins when somebody, whom we shall refer to as the customer, expresses a need and so enlists the services of an engineer. The customer can be an individual, an organization, or the consuming public. Most customers are not engineers. It is up to the engineer to translate the customer's need into engineering terms. The result is cast in the form of a **problem definition** and a **list of specifications**.

Problem Definition

A problem definition states the design objective in one to three clear, concise sentences. For example, the problem definition addressed by Orville and Wilbur Wright at the turn of the twentieth century was *design a manned machine capable of achieving powered flight.*

This problem definition tells us that they wanted to design a flying machine subject to two constraints. First, it must a carry a person, which rules out model aircraft. Second, an onboard power source must be used to take off, which eliminates the possibilities of leaping off a barn with hand-held wings and lighter-than-air craft such as a hot air balloon.

The problem definition is constructed in response to an expressed need. Failure to identify, understand, and validate the need, prior to designing, is one of the most frequent causes of failure of the entire design process.

The customer's statement of need does not typically take the form of a problem definition. For example, consider the following statement of need from a fictitious client:

Need: People who work at the Empire State Building are complaining about the long waits at the elevator. This situation must be remedied.

An engineer might translate this need into the following problem definition:

> ***Problem Definition:*** Design a new elevator for the Empire State Building.

Is this really a good problem definition? Is the main concern of the management at the Empire State Building to reduce average waiting times or to eliminate the complaints? When turning an expressed need into a problem definition, it is important to eliminate assumptions that unfairly bias the design toward a particular solution. A better, less-biased problem definition might be:

> ***Improved Problem Definition:*** Increase customer satisfaction with the elevators in the Empire State Building.

This would admit such solutions as a mirror on the elevator door or free coffee on the busiest floors.

As another example of an inadequate problem definition, consider the following: *Design a device to eliminate the blind spot in an automobile.* This proposed problem definition also contains an assumption that prematurely limits the designer. The word *device* rules out one solution that achieves the design goal (eliminating the blind spot) by simply repositioning the front and side mirrors.

A third example occurred in a design competition named Blimp Wars (see Figure 1). The goal was to *design a system to retrieve Nerf® balls from an artificial tree and return them to the blimp base.* Inclusion of the word *blimp* in the problem definition biased the students toward blimp designs. The alternative of an extendable arm that would span the distance between blimp base and the target balls was not considered.

List of Specifications

After translating the need into a problem definition, the next step is to prepare a list of specifications. The list of specifications includes both "demanded" design characteristics that must be present for the design to be considered acceptable and "wished for" design characteristics that are desirable but not crucial to the success of the final design. It is the usual practice to classify each specification as either a demand (D) or a wish (W). Don't confuse the two. If you treat a wish as if it was a demand, your design may become more complicated than is necessary.

Whenever possible, use numbers to express specifications. For example, instead of merely requiring that weight must be low, state, "Weight must be less than 10 pounds." Sometimes use of numbers is impossible. A quality such as "aesthetically pleasing" is difficult to quantify. However, use numbers wherever possible, even if at this early stage they seem like guesses. The numbers can be refined later on as the design begins to take shape.

Figure 1

Blimp Returning to Base After Retrieving Ball from Tree on Left

The specifications should be solution independent to avoid bias. For example, if you are designing a small mobile device, requiring that "the wheels must be made of rubber" will bias the design in two respects: in the use of wheels and in the choice of materials. Such decisions are reserved for later in the design process after careful consideration of alternatives.

Specifications come in the following categories:

- Performance
- Geometry
- Materials
- Energy
- Time
- Cost
- Manufacture
- Standards
- Safety
- Transport
- Ergonomics

These categories can also be used as headings by which to organize the list of specifications. Here is an example.

EXAMPLE: The following problem definition was posed to three competing design teams.

Design and build a remote-controlled,[1] portable device that will play nine holes of golf at a local golf course with the fewest possible number of strokes.

The instructor also supplied the following demands. It was left to the students to develop a complete list of specifications.

D or W	SPECIFICATION
D	Must cost less than $600 (not including radio).
D	Must be remotely triggered.
D	Total number of radio-controlled servos[2] is eight.
D	Device cannot be touching the golf ball prior to remote triggering of the shot.
D	Entire device must form a single unit.
D	Must be portable.
D	Design must pass a safety review.
D	Ground supports must fit within a 3-foot circle.

∴Solution

The first step was to organize the demands under each heading. Then, using the headings as a guide, additional specifications were formulated. The results follow:

D or W	SPECIFICATION
PERFORMANCE	
D	Must be remotely triggered.
D	Device cannot be touching the golf ball prior to remote triggering of the shot.
D	Driving distance must be adjustable with a range of between 15 and 250 yards.
D	Putting distance must be adjustable with a range between 0 and 15 yards.
W	Must sink 95% of short putts (<3 feet).
W	Driving accuracy of ± 5 yards.
D	Must operate on inclines of up to 45 degrees.
GEOMETRY	
D	Total number of radio-controlled servos is eight.
D	Entire device must form a single unit.
D	Ground supports must fit in 3-foot circle.
MATERIALS	
W	Materials must not degrade under expected range of weather conditions (including rain, snow, $30^\circ F < T < 90^\circ F$).
TIME	
D	Must be designed and manufactured in less than 14 weeks.

[1]A common abbreviation in design for "remote control" or "remote controlled" is "RC."

[2]A servo is a control system to amplify a small signal into a large response. Typically it is an electric motor controlled by a small voltage.

COST	
D	Must cost less than $600 (not including radio).
MANUFACTURE	
D	Must be manufactured using tools available in the machine shop.
D	Must be manufactured using machining skills available within the team.
W	Off-the-shelf parts and materials should be readily available.
STANDARDS	
D	Radio must adhere to FAA regulations.
SAFETY	
D	Design must pass a safety review.
TRANSPORT	
D	Must be portable.
W	Must fit in a car or small truck (for easy transport to golf course).

Design Milestone: Clarification of the Task

There are two versions of this milestone, depending on the format of the design project. If there is a design competition involved, main responsibility for producing the list of specifications shifts from the students to the instructor, as there is a need for everyone to operate under the same set of constraints. In either case, it is assumed that the instructor provides the problem definition.

For a General Design Project (Version A)

Assignment:

1. Interview the customer. (In the case of a consumer product, conduct a product survey.)
2. Prepare a typed list of specifications.

For Design Competitions (Version B)

Assignment:

1. Review the rules of the competition and ask the instructor for rule clarifications.
2. Prepare a typed list of design requirements to supplement those already appearing in the official rules of the competition. For example, set performance goals for your machine. You do not have to list requirements already appearing in the rules.

Design Competition Tips

- Probe the rules for holes that will allow for concepts not anticipated by the creators of the competition.
- Avoid any temptation to bias the requirements toward a particular solution or strategy.
- Expect the list of supplemental requirements to be very short if the rules are well defined.

Chapter 18

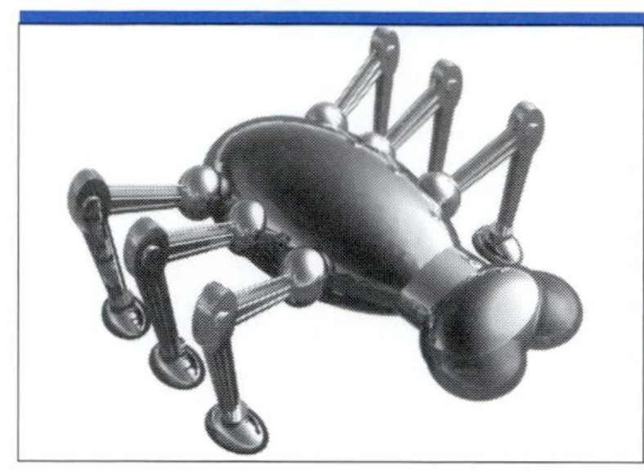

© iStockphoto.com/Michael Knight

Generation of Alternative Concepts

Once the problem statement is in place, and the specifications have been listed, it is time to generate alternative concepts. By a *concept,* we mean an idea as opposed to a detailed design. The representation of the concept, usually in the form of a sketch, contains enough information to understand how the concept works but not enough information to build it. By *alternative,* we are requiring that the various proposed ideas must be fundamentally different in some way. The differences must go beyond appearance or dimensions. The usual rule of thumb in design courses is to generate at least three fundamentally different concepts.

In this chapter, four aspects of concept generation will be discussed: **brainstorming, concept sketching, research-based strategies,** and **functional decomposition.**

Brainstorming

The most common approach for generating ideas is by brainstorming. As the term implies, you rely on your own creativity and memory of past experiences to produce ideas. Usually, team members will generate ideas on their own before meeting with the team for a brainstorming session.

Brainstorming is based on one crucial rule: *criticism of ideas is not allowed.* This enables each team member to put forth ideas without fear of immediate rejection. For example, a professor once recorded the brainstorming session of a small team of students. At one point, a student offered an idea, and another student referred to it as "stupid." The voice of the first student was never heard from again during the session. Instead of a team of four, it had become a team of three.

It is important to devote some of the brainstorming time searching for bold, unconventional ideas. In the case of a design competition, this could mean searching for holes in the rules that could lead to ideas that the creators of the competition had not anticipated.

Only when brainstorming is complete should the team eliminate concepts that are not feasible, not legal or not fundamentally different. After this weeding-out process, at least three concepts should remain. If not, more brainstorming is in order. The following example illustrates this step:

EXAMPLE 1

Assuming the alternative concepts in Figure 1 were generated as part of an effort to design a new bat for Major League baseball, which concepts should be eliminated because they are not feasible, not legal, or not fundamentally different?

Solution

Not feasible: E because it stands no chance of being competitive.
I because it is too difficult to find in nature.
Not legal: F, G, H, and J.
Not fundamentally different from each other: C and D because basic shape is the same; only dimensions differ.

Therefore, the condensed list of viable alternatives consists of concepts A, B, and C.

Concept Sketching

For an idea to be considered a feasible alternative concept, it must be represented in the form of a conceptual sketch. The goal in producing a concept drawing is to convey what the design is and how it works in the clearest possible terms. Any lack of clarity, such as failure to represent one of the subfunctions, will translate into doubts about the feasibility of the concept when it comes time to evaluate it.

At the same time, however, this is not a detailed design drawing. Dimensions and other details not relevant to understanding the basic nature of how the concept will work are left out.

It is best to proceed through two phases when generating a concept drawing. First, in the creative phase, hand-sketching is done freestyle and quickly, without regard for neatness or visual clarity. A few simple lines, incomprehensible to others, might be enough to remind you of your idea. Sketching is a means for both storing ideas and brainstorming others. The final outcome is a rough sketch of the concept. Second, in the documentation phase, the concept is neatly redrawn and labeled to facilitate communication with team members and project sponsors.

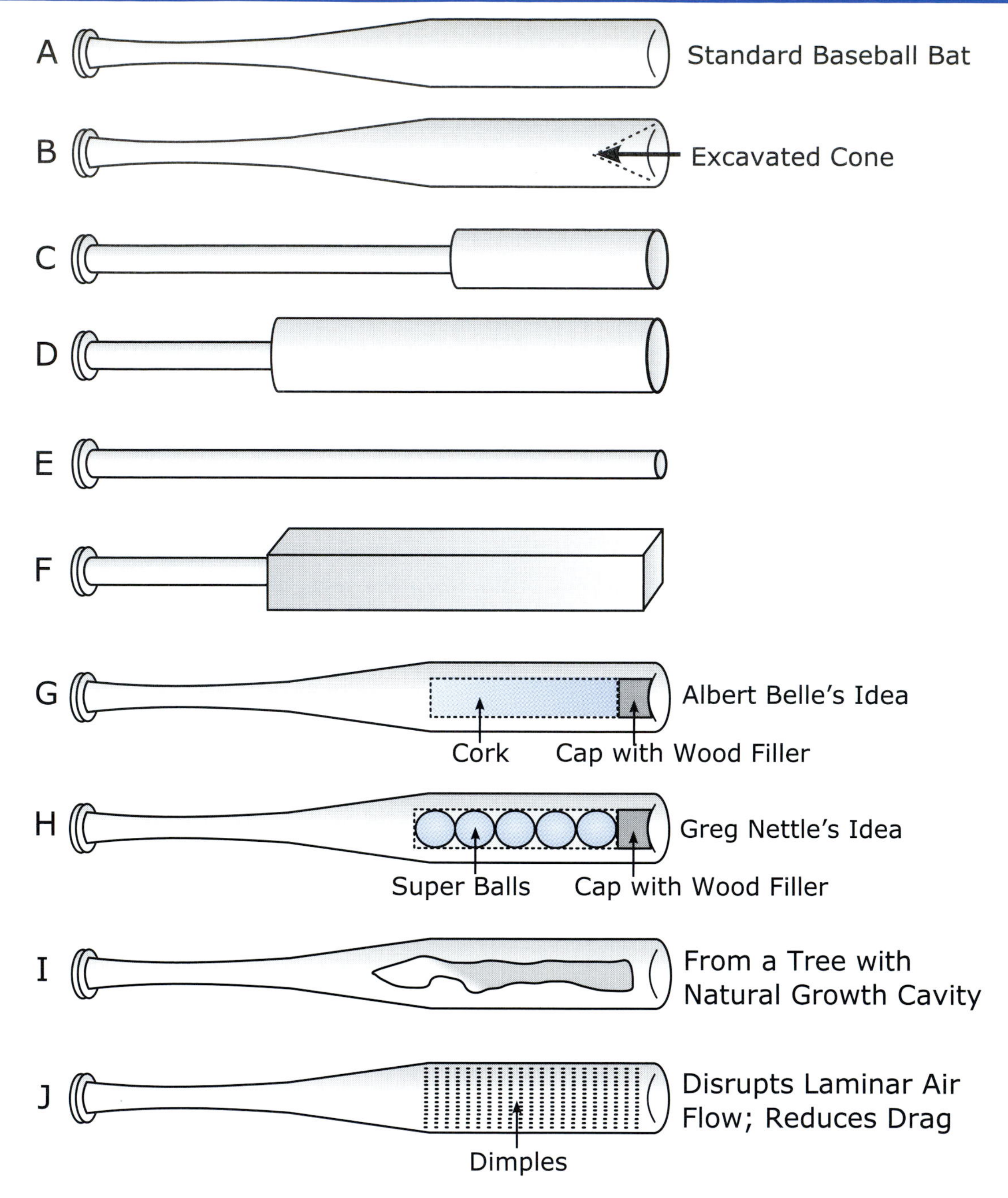

Figure 1 Alternative Concepts for a Major League Baseball Bat

The final outcome is one or more sketches prepared with the following guidelines in mind:

- *Can be hand-sketched or computer generated.*
- *No dimensions.* Remember, this is not a detailed drawing.
- *Label parts and main features.* If the drawing is hand-sketched, handwritten labeling is acceptable.
- *Provide multiple views and/or closeup views if needed to describe how the design works.*

The choice of views is up to you. Isometric views like those shown in Figures 2 and 3 convey a lot of information in a single picture. Most mechanisms can be described effectively using one or more two-dimensional views as in Figure 4. Despite their apparent informality, the quality of these drawings is crucial to fairly representing the designs during the evaluation process. In some cases they are the only source of evidence for judging if a design is likely to work.

Hands-on Design Exercise: "The Tube"

The design objective is to extract a golf ball from the bottom of a free standing, open-ended mailing tube in the shortest possible time.

Setup

- Place a mailing tube vertically on the floor and drop a golf ball in the tube.
- Have a supply of the following materials: string, duct tape, Scotch® tape, 8.5 × 11 standard copier paper, and scissors.

Rules

- Limited to using the supplied materials.
- Scissors can be used for manufacturing.
- Everyone in group can help in manufacturing, but only one person can extract the ball.
- Students are not allowed to handle the materials until it is time to test.
- Must manufacture the design shown on the concept drawing handed to the instructor.
- Time limit of 3 minutes to manufacture concept and extract ball.
- Cannot tip over the tube.
- Cannot touch the outside of the tube with anything.
- No forces can be applied to the inside of the tube in an effort to hold it vertical; accidental contact with inside of tube is okay as long as the tube does not tip over.
- Violation of any of the preceding rules will lead to immediate disqualification.

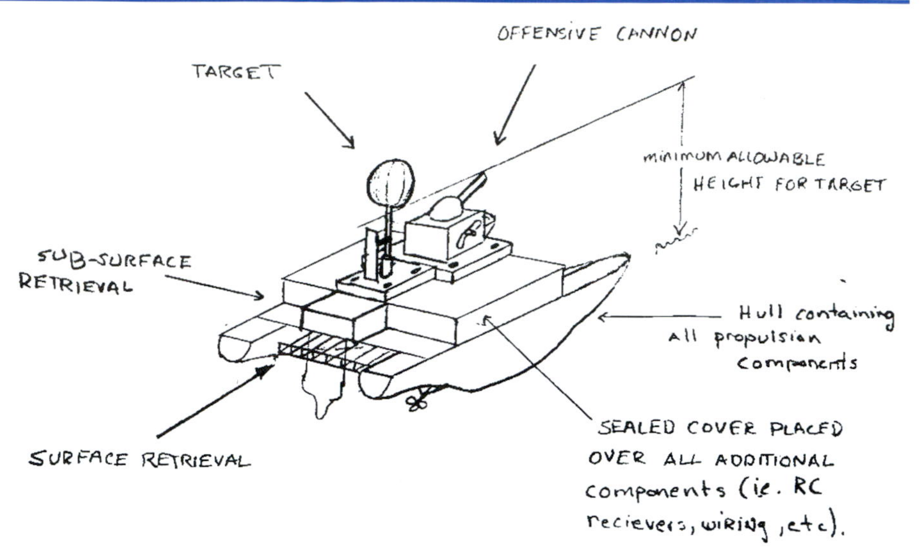

Figure 2

Concept Drawing of an RC Boat for a Design Competition (hand-drawn isometric)

Procedure

1. First allow the students 3 minutes to individually brainstorm (encourage them to draw quick sketches of each of their concepts).
2. Then divide section into teams of four students per team.
3. Allow teams 10 minutes to collect ideas, brainstorm as a team, select their best concept, and give a sketch of their best concept to the instructor.
4. Instructor should walk around during brainstorming to remind teams to (a) generate multiple solutions before selecting one and (b) try to involve everyone in the process.
5. Allow teams 2 minutes to assign responsibilities for manufacture and test.
6. One at a time, give each team 3 minutes to manufacture their concept and attempt to extract the golf ball.
7. Team with the shortest retrieval time wins.

Research-Based Strategies for Promoting Creativity

Some ideas are truly original, but most are drawn from past experience. The following strategies help you to look at old designs in order to generate new ones.

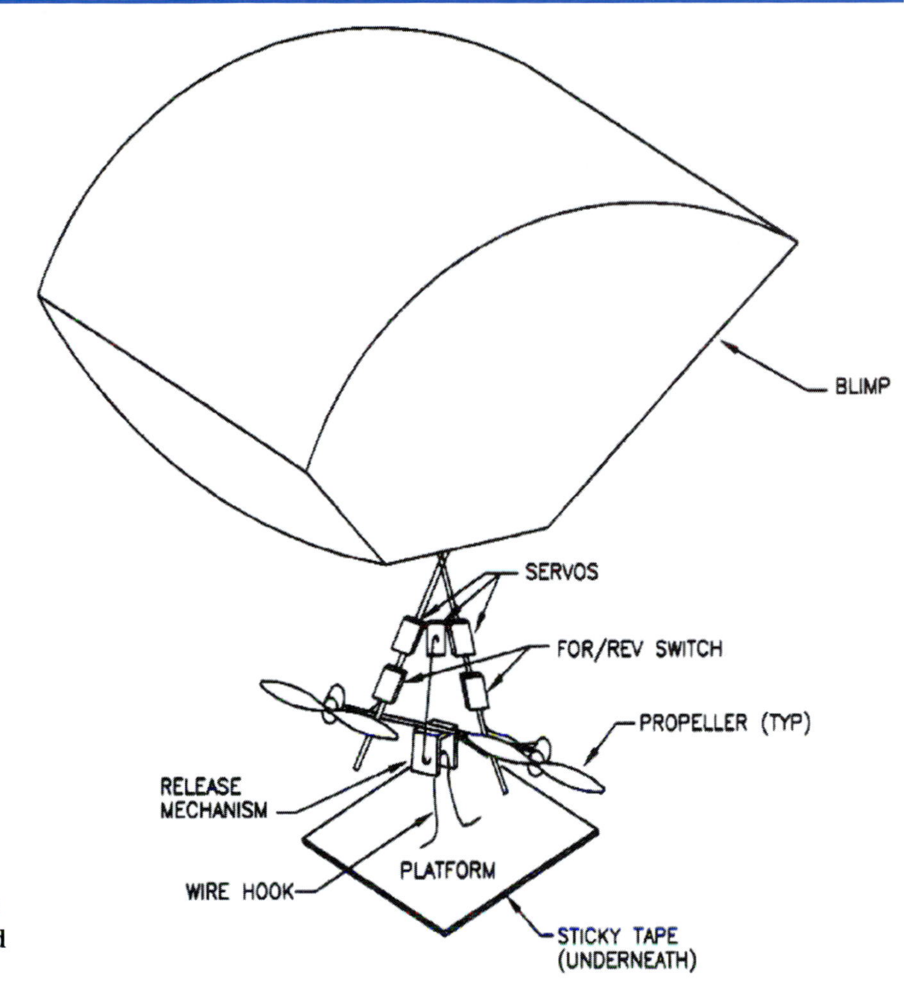

Figure 3

Concept Drawing of an RC Blimp (computer-generated isometric)

Analogies

One often used strategy is to look for analogous design situations in other unrelated fields. To do this, first you have to translate the design objective into an overall function that is general enough to widely apply. For example, you may want to design a system to "climb a vertical wall" or "walk on two legs" or "move efficiently through the water." Nature is filled with solutions to these problems (but because of their complexity, biological solutions usually have to be simplified and adapted before they can be of practical use). If you are designing a system to "throw an object," a survey of ancient artillery could spark ideas.

Reverse Engineering

The basic strategy here is to acquire an existing product that is similar to a design you have in mind, take it apart, figure out how it works, and then

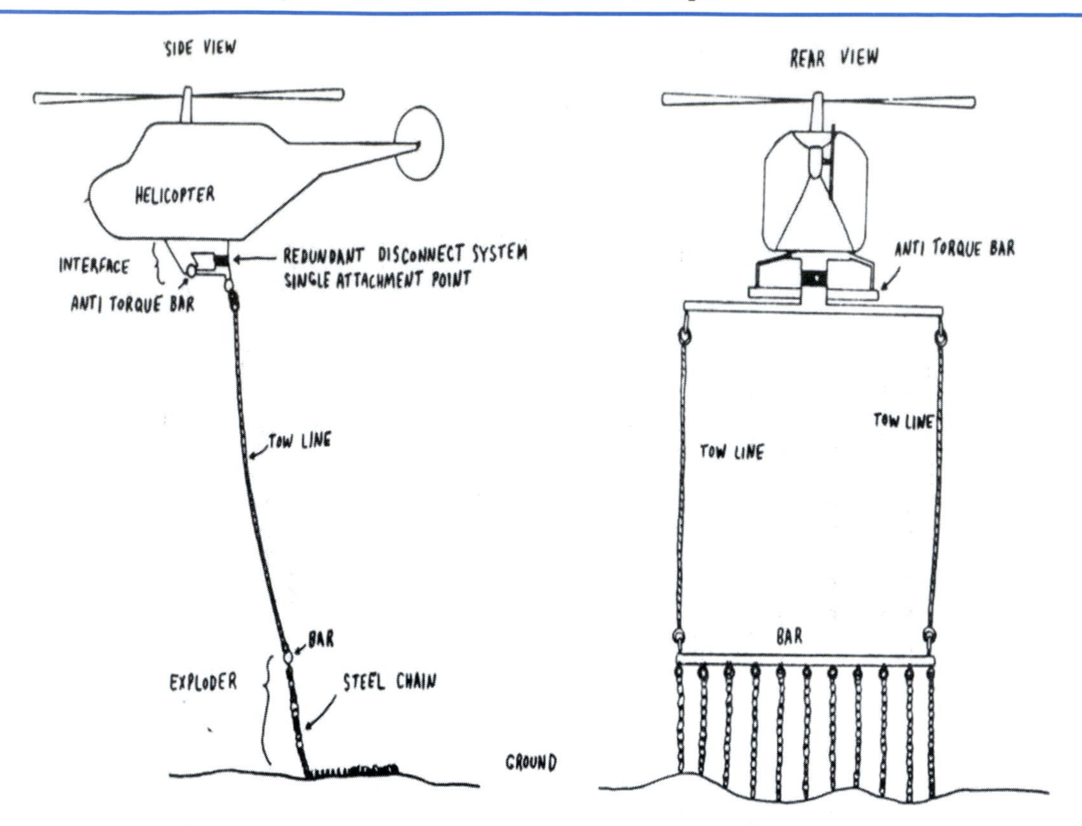

Figure 4 Concept Drawing of an Anti-Personnel Mine Clearing System (two hand-drawn views)

either try to improve on it or adapt some of the ideas to your own design. Toy stores are a great place to search for small electromechanical devices that can be reverse engineered.

Literature Search

Web-based search engines are very effective at finding existing design solutions. For high-tech applications, you should also search books and the electronic databases for technical journals (e.g., the Science Citation Index).

Functional Decomposition for Complex Systems

When confronted with a complex problem, it is frequently advantageous to break it down into smaller, simpler, more manageable parts. In the case of design, those smaller parts usually correspond to the individual functions (or tasks) that must be performed in order to achieve the overall design objective. This approach, known as **functional decomposition,** is the basis of the procedure described below for generating concept alternatives.

Step 1. Decompose the design objective into a series of functions.

Start out by decomposing the overall function into four or five subfunctions. Usually, verbs such as *move*, *lift*, and *control* are used in naming the functions. Figure 5 shows the functional decomposition of the remote-controlled golf machine of earlier examples. It is given in the form of a tree diagram, which is probably the most common form of representation. If more detail was needed, each of the subfunctions could be further broken down into its respective subfunctions. When it is not readily apparent what the subfunctions are, it may help to think in terms of the sequence of tasks that must be performed by the design. The "sequential" functional decomposition for a design to assist disabled people into and out of a bathtub is shown in Figure 6.

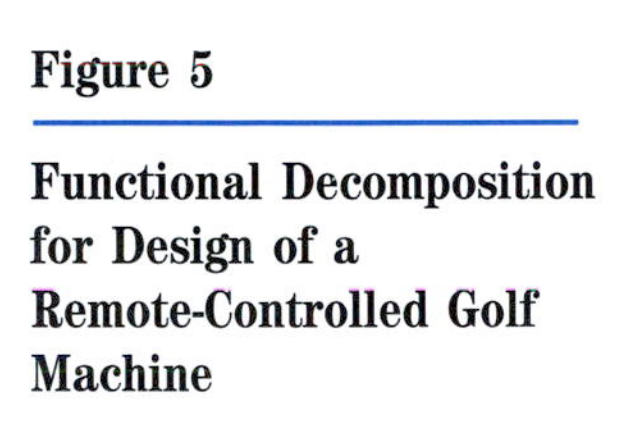

Figure 5

Functional Decomposition for Design of a Remote-Controlled Golf Machine

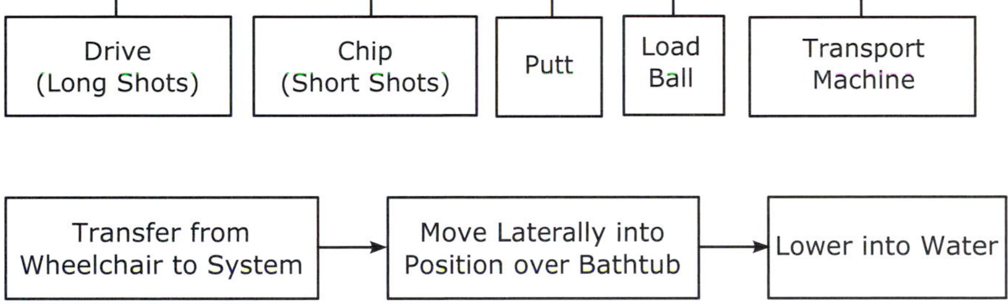

Figure 6

"Sequential" Functional Decomposition for Design of a System to Aid the Disabled into and out of a Standard Bathtub

It is very important that the functional decomposition be general enough to avoid biasing the design solution. For example, the separate drive and chip functions in Figure 5 may cause the design team to overlook the possibility of using the same device to fulfill both functions. If solution bias is unavoidable, introduce multiple functional decompositions.

Step 2. Brainstorm on alternative concepts for each function and assemble the results in a classification scheme.

The classification scheme[1] is a two-dimensional matrix organized as shown in Table 1. The first column lists the functions resulting from the

[1]Pahl, G. and W. Beitz. (1988). *Engineering Design—A Systematic Approach.* New York: Springer-Verlag.

Table 1

Organization of the Classification Scheme

Functions \ Concepts	Concept 1	Concept 2	Concept 3	Concept 4
Function A	A1	A2	A3	A4
Function B	B1	B2	B3	B4
Function C	C1	C2	C3	C4
Function D	D1	D2	D3	D4

functional decomposition. The row of boxes next to each function name contains the corresponding design solutions that have been brainstormed. The design solutions are expressed using a combination of words and pictures, so be careful to draw the boxes large enough to accommodate small illustrations.

Step 3. Combine function concepts to form alternative design concepts.

Table 2 demonstrates how one subfunction concept from each row of the classification scheme is selected to form a total concept. The same sub-function concept can be used with more than one total concept, though keep in mind that the idea is to generate fundamentally different design concepts. The only other rule when deciding upon the best combinations is to be sure that the subfunction concepts being combined are compatible.

Step 4. Sketch each of the most promising combinations.

This is done in accordance with the rules previously presented for concept drawings. Remember that you must end up with drawings for at least three fundamentally different design concepts.

EXAMPLE 2

Use functional decomposition to generate alternative concepts for a proposed remote-controlled blimp, capable of retrieving Nerf® balls from an artificial tree and returning them to blimp base (see Figure 1 of Chapter 17).

Table 2

Combining of Compatible Sub-Function Concepts

Functions \ Concepts	Concept 1	Concept 2	Concept 3	Concept 4
Function A	A1	A2	A3	A4
Function B	B1	B2	B3	B4
Function C	C1	C2	C3	C4
Function D	D1	D2	D3	D4

Total Concept I = A1 + B2 + C2 + D1
Total Concept II = A4 + B2 + C4 + D2

Solution

The first step was to produce the functional decomposition of Figure 7. Then concepts were brainstormed for each of the subfunctions, and the results were assembled in the classification scheme of Figure 8.
Total concepts were formed by combining compatible subfunction concepts.

Three promising total concepts are:

Total Concept I = helium + 2 props + pivot props + sticky tape
Total Concept II = helium + rotating turret + vertical prop + rake
Total Concept III = helium + prop with rudder + string + claw

The final step is to represent each alternative design in the form of a concept drawing. The concept drawing for Total Concept I is shown in Figure 3.

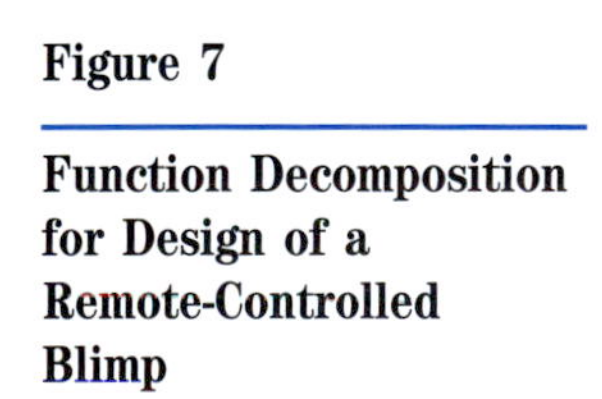

Figure 7

Function Decomposition for Design of a Remote-Controlled Blimp

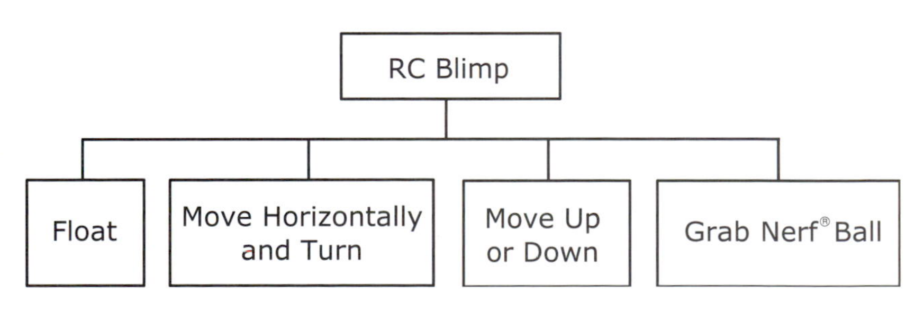

Functions \ Concepts	Concept 1	Concept 2	Concept 3	Concept 4
Float	Hot Air	Helium	Hydrogen	
Move Horizontally and Turn	Prop + Rudder	Two Props	Rotating Turret	2 Props at 90°
Move Up or Down	Pivot Props	Vertical Props	Retract String	Heat or Cool Gas
Grab Nerf Ball	Sticky Tape	Rake	Claw	Pin Cushion

Figure 8

Classification Scheme for a Remote-Controlled Blimp

Design Milestone: Generation of Alternatives

This milestone assumes the system to be designed is sufficiently complex (i.e., at least two subfunctions) to warrant the use of functional decomposition.

Assignment:

1. For the functional decomposition given in class (or a modification of it that you are at liberty to propose), brainstorm to determine at least five feasible alternatives for each subfunction and assemble the results in a classification scheme.
2. Form three promising design concepts by combining compatible subfunction alternatives from your classification scheme.
3. Firm up your three design concepts by sketching them up in the form of concept drawings. Functionality (i.e., how it works) should be clearly indicated in the drawings through the use of labeling and text.

Grading Criteria:

Technical Communication

- Ideas are clearly presented
- Final concept drawings are neatly rendered

Technical Content

- All concepts are feasible, legal, and fundamentally different

- Concepts are presented in sufficient detail
- Requested number of concepts was generated

Design Competition Tips

- The goal is to generate three “strong” concepts.
- Search the boundaries of the rules for unusual ideas that could potentially dominate the competition. If you don’t, someone else will.
- Include strategy as one of the items to be brainstormed in the classification scheme.
- Redraw your concept sketches to enhance clarity and neatness. The quality of the concept drawing, or lack of it, can do much to sway opinions when it comes time to judge the concepts. They should be inked, prior to scanning them in for the oral design defense.

Chapter 19

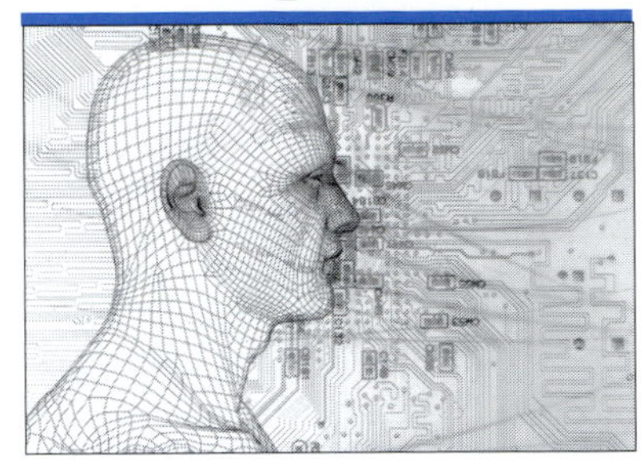

© iStockphoto.com/Emrah Türüdü

Evaluation of Alternatives and Selection of a Concept

Suppose you now have generated three concepts that will meet the problem definition and fulfill the specifications. Which one should you choose as the basis for your final design? There is no magic formula. However, Professor Nam P. Suh of MIT has provided two very helpful design principles for evaluating and improving concepts: **minimize information content** and **maintain the independence of functional requirements.**[1] This chapter adds three additional considerations for evaluating alternatives: **ease of manufacture, robustness,** and **design for adjustability.** It then concludes with a method of pulling together all these ideas: the **decision matrix.**

Minimize the Information Content of the Design

When choosing among promising alternatives, the best design is often the one that can be uniquely specified using the least amount of information or, alternatively, can be manufactured with the shortest list of directions. This idea is sometimes stated as the *KISS* principle: Keep It Simple, Stupid.

[1]Suh, N. P. (1990). *The Principles of Design*. New York: Oxford University Press.

There are a number of design guidelines that naturally follow. A few of the most notable ones are:

- **Minimize the number of parts.**
- **Minimize the number of different kinds of parts.**
- **Buying parts is preferable to manufacturing them yourself.**

Maintain the Independence of Functional Requirements

The functions considered in a functional decomposition provide the basis for Suh's second principle. This principle asserts that these functions should be independent of each other in a good design.

A successful application of this principle is illustrated by the decoupled design in Figure 1. Independence of the functions "lift" and "move" was maintained by designing physically separate mechanisms for each action (scissors jack for lift, wheeled vehicle for move) and by performing the actions in sequence, rather that at the same time. First the scissors jack would lift the vehicle, the vehicle would then slide horizontally onto the next step, and finally the scissors jack would close upward and be pulled back underneath the vehicle. The coupled design employed four articulated arms, tanklike tracks on each arm, and a complicated motion to both lift and move at the same time. Though both machines performed admirably, the decoupled design had a much higher potential payload and was easier to build, since it required half as many motors.

The previous example suggests the following design guideline:

- **Seek a modular design.**

A modular design is one in which the design solutions for each function have been physically isolated. The main advantage of a modular design is that the

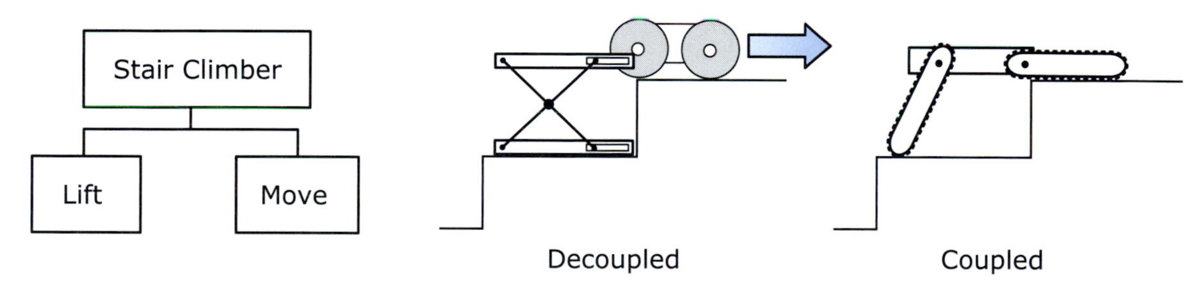

Figure 1 Two Concepts for a Stair-Climbing Machine (The first concept (center) maintains the independence of the functions "lift" and "move" the second concept (right) does not.)

individual modules can be designed, manufactured, and tested in parallel, leading to much shorter product development times.

In looking for opportunities to improve a given design, the situation may arise in which Suh's two principles appear to be in conflict. For example, a design change aimed at increasing the independence of the functional requirements could result in greater complexity. Suh contends that any design change that either increases information or sacrifices the independence of the functional requirements should not be accepted. There always exists a less-coupled design with lower information content.

We now illustrate the application of these two design principles with an example.

EXAMPLE 1

A head-to-head student design competition named Davy Jones's Treasure Trove was based on the following problem definition:

> *Design a system to retrieve surface (Ping-Pong balls) and subsurface (1 lb mass) objects from a swimming pool.*

This was subject to the following major design requirements:

- Must fit in a 2 ft × 2 ft × 3 ft space at start
- Must carry a target that disables boat if struck by opponent

Concept drawings of two of the student designs are shown in Figures 2 and 3. By reference to Figures 2 and 3, evaluate the two student designs by identifying applications and violations of Suh's design principles. (*Hint:* Only relevant features have been labeled in the figures.)

Solution

WATER CANNON DESIGN

- Water cannon might be effective because aiming and steering are independent.
- Catapults will not be as effective because aiming is dependent on steering.
- Boat should be very maneuverable because the two props serve to decouple the move and turn functions; that is, it should be able to turn on a dime.
- Manufacture of the prop systems could be needlessly time-consuming.

TWIN BOAT DESIGN

- Use of two boats will be very effective because it decouples the two retrieval functions. One boat can collect the Ping-Pong balls while the other collects the 1 lb masses.

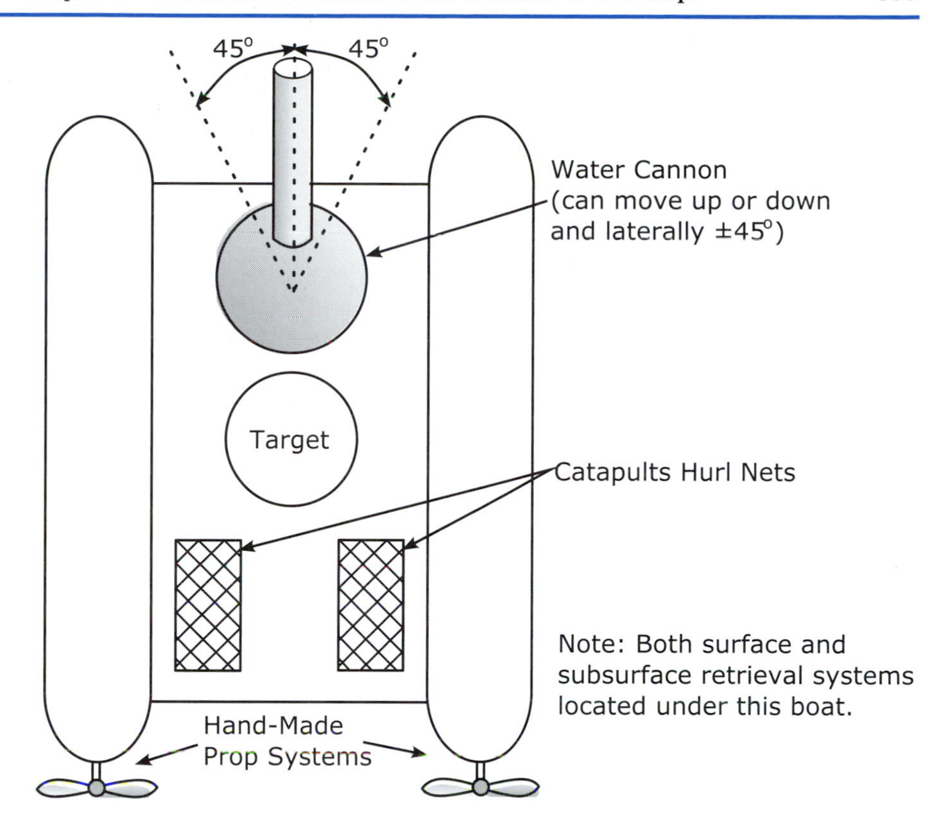

Figure 2

Water Cannon Design (top view)

- Use of store-bought propellers will save time.
- Use of two nearly identical hull designs will simplify both design and manufacturing, thus saving more time.
- The boats will not be as maneuverable as the Water Cannon Design because the move and turn functions are not independent; that is, the single-prop design needs to be moving forward in order to turn.
- Gatling gun will not be as effective as the water cannon because aiming is dependent on steering.

FINAL NOTE

The preceding characteristics provide accurate insight into how the boats actually performed. The Twin Boat Design won the competition largely on the strength of its dual retrieval system. Although the water cannon was far more effective than the Gatling gun, the Water Cannon Design was ultimately at the mercy of its handmade props, which took a lot of time to manufacture, left little time for testing, and proved to be unreliable.

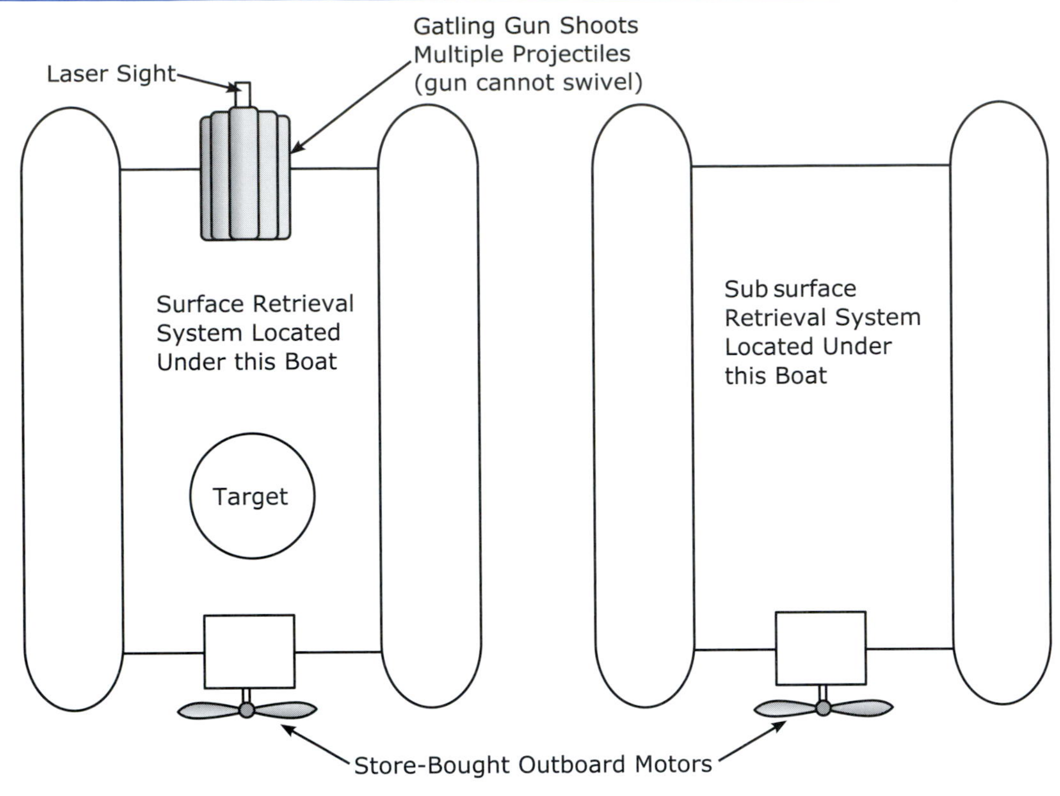

Figure 3 Twin Boat Design (top view)

Design for Ease of Manufacture

There are clear advantages to going with a design that is easy to manufacture. If among competing design teams you are the first to complete manufacture of your design, the extra time can be used to test, debug, and optimize performance. For a commercial enterprise, first-to-market can mean a short-term monopoly in a fiercely competitive marketplace. Often, ease of manufacture goes hand-in-hand with lower costs. Thus, given the choice of two concepts, both of which satisfy the design requirements to the same degree, and where one is more difficult than the other to manufacture, it makes sense to choose the concept that is easier to manufacture.

At this stage in the design process, evaluation of ease of manufacture should be done at a level of abstraction consistent with the concept drawings. The counting of machining operations and assembly steps is reserved for a later time when the requisite level of detail in the design has been attained. Here the emphasis is on developing an impression of ease of manufacture as revealed through Suh's design principles.

A student or team with the formidable task of having to build a complex design themselves should be asking the following questions as each concept is evaluated:

Are there a large number of parts?
If there are a lot of parts that need to be made and assembled, it will take a long time to build.

Are there a large number of different kinds of parts?
For parts of comparable complexity, it takes less time to make two of the same part than it does two different parts because of reduced setup times.

Are there parts with complicated geometry?
These parts will take longer to make.

Can some parts be purchased?
This is not always an option in design competitions, but if it is, the time saved and the proven reliability of the prefabricated part usually justifies the purchase.

Do you have the skills to make all of the parts?
Safest thing to do is to choose a concept that you know you can build.

Is it a modular design?
We noted earlier that modular components can be manufactured in parallel by subgroups within the design team, thus saving time.

Are there opportunities to simplify manufacture of the design by:

- reducing the number of parts?
- reducing the number of different kinds of parts?
- simplifying the shape of some parts?
- purchasing some parts?
- redesigning the parts that are difficult to make?
- modularizing the design?

Design for Robustness

Manufacturing errors, environmental changes, and internal wear can cause unexpected variations in performance. When the designed product is insensitive to these three sources of variability, the design is said to be *robust.* Engineers seek a robust design because performance of such a design can be predicted with greater certainty.

The designer must learn to expect the unexpected. All too often students conceive of a design while assuming ideal operating conditions. Yet, deviations from those ideal conditions can lead to less than ideal performance, as illustrated by the following example.

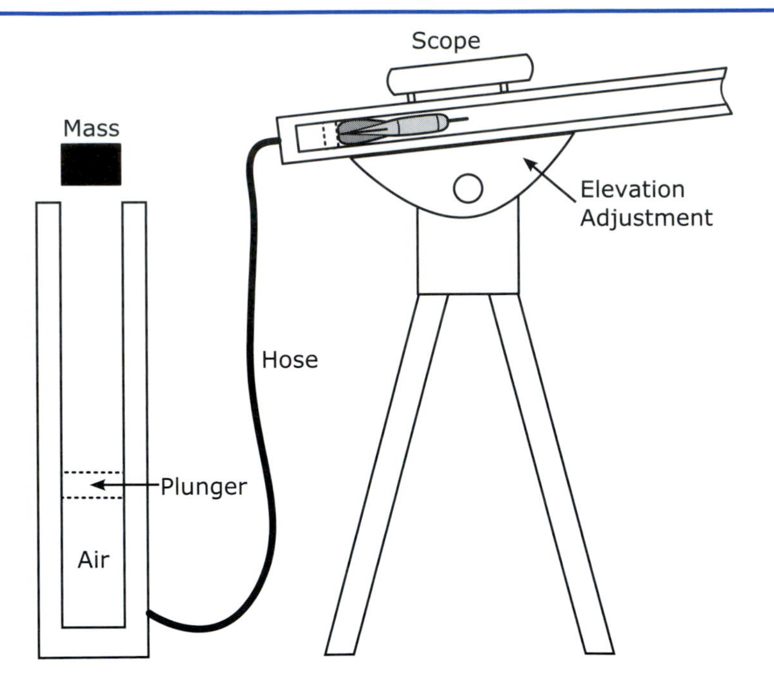

Figure 4

Concept Drawing of a Dart-Throwing Machine

EXAMPLE 2

For a design competition, students had to design machines that could accurately throw darts at a dartboard. The machines were powered by large falling masses. Yet, none of the machines was perfectly repeatable. For example, from 8 feet away, the best that the machine in Figure 4 could do was to keep the darts within a 1-inch circle.
What factors contributed to this loss of dart throwing accuracy?

Solution Relevant manufacturing errors, environmental changes, and internal wear sites were brainstormed. The following list resulted:

MANUFACTURING ERRORS

- Small dimensional differences between darts in the set of three

ENVIRONMENTAL CHANGES

- Small air currents
- Inexact repositioning of the plunger
- Inexact repositioning of the dart within the blow gun
- Inconsistent releases of the falling mass

INTERNAL WEAR

- Damage to the dart fins
- Blunting of the dart tip
- Damage to the dartboard ■

When evaluating concepts with respect to robustness, you should be asking yourself the following questions:

Will small manufacturing errors dramatically impair performance?
If parts have to be manufactured perfectly in order for the design to function properly, you should expect to run into problems. You want a design that will work even when part dimensions are a little off. This is one reason why gear sets are such a popular design choice. Small errors in center distance between mating gears do not change the gear ratio.

Will the design function properly over the full range of environmental conditions?
Environmental conditions subject to variation include applied forces, atmospheric conditions, and roughness of surface terrain. The expected range of relevant environmental conditions should be clearly defined in the list of specifications. If they are not there, now is a good time to include them.

In a head-to-head design competition, have the actions of the opposing teams been anticipated?
Those actions can contribute significantly to the variability of the environmental conditions. For example, an opposing machine can apply forces to your machine or alter the roughness of surface terrain by laying obstacles. Thus, you want to select a strategy/design combination that will perform well irrespective of what the opposing teams may do.

Design for Adjustability

In engineering courses there is usually only enough time and resources to manufacture one design, and that design almost never performs as planned on the first try. Optimizing performance by building several designs is not an option. The only remaining course of action is to design adjustability into the initial implementation.

There are a number of ways to design for adjustability. One way is to design the system with modularity. This can serve to isolate required design changes to a single subsystem. In a mechanical system, dimensional adjustability can be attained by using nonpermanent fastening methods such as screw joints instead of a permanent method like epoxy.

Design for adjustability can be incorporated into the evaluation process by asking the following questions as each concept is reviewed:

What are the main performance variables?
Usually, only one or two of the most important variables need be considered. Typical performance variables are speed, force, and turning radius.

Can those performance variables be easily adjusted?
Common methods were described previously. Others methods are found by brainstorming and by examination of the governing equations.

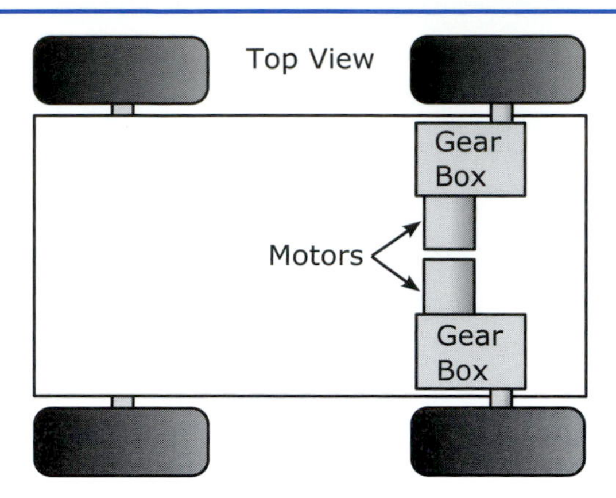

Figure 5

Moving Platform

EXAMPLE 3

A motor-driven moving platform is a common feature of many small-scale vehicle designs. A top view of one such moving platform design is shown in Figure 5.

Once manufacture of this design is complete, what adjustments can be made to:

(a) Increase the speed of the moving platform?
(b) Increase its pushing force?

Solution

(a) Alternative methods for increasing the speed are:

i. Decrease the gear ratio
ii. Increase the radius of the tires
iii. Increase the voltage from the power supply
iv. Switch in motors with higher RPM

(b) Alternative methods for increasing the pushing force are:

v. Increase the gear ratio
vi. Decrease the radius of the tires
vii. Increase the voltage from the power supply
viii. Switch in motors with a higher peak torque ■

Hands-on Design Exercise: "Waste Ball"

Scenario

A company that uses radioactive substances for research sometimes has "spills" of spherical radioactive objects. When this happens, the radioactive objects must be transported to a waste container by the emergency team.

Design Objective

Design a method to transfer the radioactive substance (plastic ball) from the site of the spill to a waste container (small refrigerator) at another location.

Setup

- Divide the team, which consists of the entire class, into subfunctional groups of two or three students each. Each subfunctional group will be responsible for one leg of the transfer.
- Prior to the class, the instructor has to lay out the course. There must be as many different legs as there are subfunctional groups. The room in which the ball is initially placed and another room that contains the small refrigerator account for two of the legs. Other legs can consist of a corridor, a stairwell, an elevator, or an outdoor excursion. Try to make each challenge a little different to promote development of specialized designs by the subfunctional groups.
- Distribute the following materials to each subfunctional group:

    1 daily newspaper (or equivalent)
    1 roll of duct tape
    1 foam plate
    1 plastic cup
    1 pair of scissors (for construction only)

The team will also receive:
  3 balls of string
which must be shared among the groups.

Rules

1. Since the ball is radioactive, no one can be within 8 feet of the ball.
2. You must use only the materials provided.
3. For safety purposes, running is not allowed.
4. If necessary, doors must be safely held open by the teams and then closed immediately after waste passes through.
5. If during transport, the ball accidentally touches something besides the transport container (e.g., floor) a 30 seconds penalty will be imposed and the group carrying the ball must restart at the location where it received the handoff.
6. The team has 3 minutes per group to complete the design for the transport.
7. The class with the minimum transit time wins.

After the Exercise

- Assess team performance by comparing times to other sections.
- Discuss the quality of communication between subfunctional groups.
- What were the lessons learned?

The Decision Matrix

The decision matrix promotes a systematic and exhaustive examination of concept strengths and weaknesses. The entire procedure, from selection of evaluation criteria to filling out the matrix, is designed to remove personal bias from the decision-making process. The results give a numerical measure for ranking alternatives and ultimately selecting the best concept.

Evaluation Criteria

The criteria by which the concepts should be judged are all contained in the list of specifications. To even qualify as a feasible concept, the expectation must be that all of the design requirements designated as "demands" will be satisfied. Therefore, the ranking of the feasible concepts ultimately depends on the degree to which they fulfill the design requirements designated as "wishes." However, at the conceptual level, qualities associated with both demands and wishes are included among the evaluation criteria owing to the uncertainty still associated with estimating their degree of fulfillment.

The design requirements selected to serve as evaluation criteria are usually reworded to indicate the desired quality. For example, instead of weight, cost and manufacture, the corresponding evaluation criteria become *low weight*, *low cost*, and *easy to manufacture*.

Evaluation criteria should be independent of each other to ensure a fair weighting of requirements in the decision matrix discussed below. For example, low cost and ease of manufacture will be redundant and thus double counted if cost of labor is a significant fraction of total cost.

The number of evaluation criteria can vary depending on the situation. We suggest a level of detail consistent with the amount of detailed information available about the concept. For most hands-on student projects, five to seven of the most important evaluation criteria should suffice. *Easy to manufacture* and *low cost* are almost always included in this list.

Procedure for Filling Out a Decision Matrix

Step 1. Identify the evaluation criteria.

This step is described in the previous section.

Step 2. Weight the evaluation criteria.

Weight values are assigned to each evaluation criterion in proportion to its relative importance to the overall success of the design; the larger the weight, the more important the evaluation criterion. Though not a mathematical necessity, it is usually a good idea to define the weights such that their sum is equal to 1—that is;

$$\sum_{n=1}^{N} W_n = 1$$

Table 1

Organization of the Decision Matrix

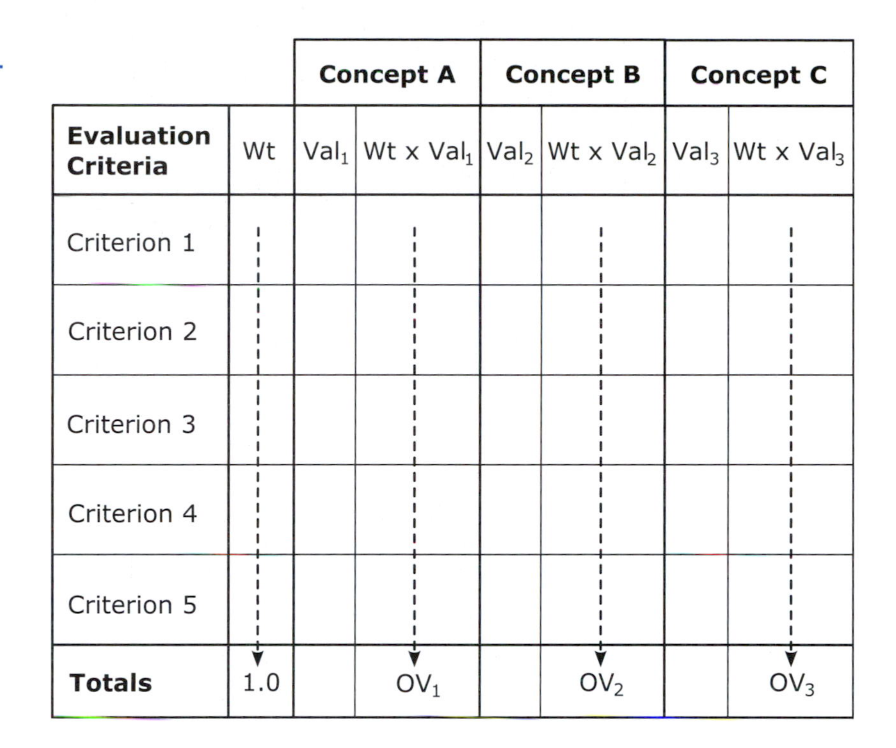

		Concept A		**Concept B**		**Concept C**	
Evaluation Criteria	Wt	Val_1	Wt x Val_1	Val_2	Wt x Val_2	Val_3	Wt x Val_3
Criterion 1							
Criterion 2							
Criterion 3							
Criterion 4							
Criterion 5							
Totals	1.0		OV_1		OV_2		OV_3

in which N is the number of evaluation criteria. This constraint instills the view that weights are being distributed among the criteria and in so doing helps to avoid redundant criteria.

Step 3. Set up the decision matrix.

The organization of the decision matrix is illustrated in Table 1. The names of the concepts being evaluated are filled in at the top of each column. Likewise, the evaluation criteria and their assigned weights are written in the leftmost columns of the matrix. Scoring and intermediate calculations will be recorded within the subcolumns under each concept and then totaled at the bottom of the matrix.

Step 4. Assign values to each concept.

Starting in the first row, each concept is assigned a value between 0 and 10 according to how well it satisfies the evaluation criterion under consideration. The values are assumed to have the following interpretation:

0 = *Totally useless* concept in regard to this criterion
5 = *Average* concept in regard to this criterion
10 = *Perfect* concept in regard to this criterion

and are recorded under the first subcolumn of each concept. This process is repeated for each criterion, going row by row to avoid bias. Usually assignment of values is based on a qualitative assessment, but if quantitative information is available, they can be assigned in proportion to known parameters.

Step 5. Calculate overall value for each concept.

For each concept-criterion combination, the product of the weight and the value is calculated and then recorded in the second subcolumn. After these calculations are completed, the overall value (OV) is computed for each concept using the following expression:

$$OV = \sum_{n=1}^{N} (W_n V_n)$$

which is equivalent to summing the second subcolumn under each concept heading. The OVs are recorded at the bottom of the matrix.

Step 6. Interpret the results.

The highest overall value provides an indication of which design is best. Overall values that are very close in magnitude should be regarded as indicating parity given the uncertainty that went into assignment of weights and values. The final result is nonbinding. Thus, there is no need to bias the ratings so as to obtain the hoped for final result. Rather, the chart should be regarded as a tool aimed at fostering an exhaustive discussion of strengths and weaknesses.

Additional Tips on Using Decision Matrices

1. Every member of the design team should individually fill out a decision matrix prior to engaging in team discussions. This will give everyone a chance to think about strengths and weaknesses ahead of time and thus make it more likely that they will be active participants at the team meeting.
2. Use the matrix to identify and correct weaknesses in a promising design. Give priority to the weaknesses that are most heavily weighted.
3. Feel free to create new alternatives by combining strengths from competing concepts.

EXAMPLE 4

Recalling the golfing machines of earlier examples, the three concepts appearing in Figures 6–8 have been proposed as best fulfilling the design requirements. Evaluate these three concepts by using the previously described procedures for filling out a decision matrix.

Solution The following design requirements were selected to serve as evaluation criteria:

1. Drives well
2. Putts well

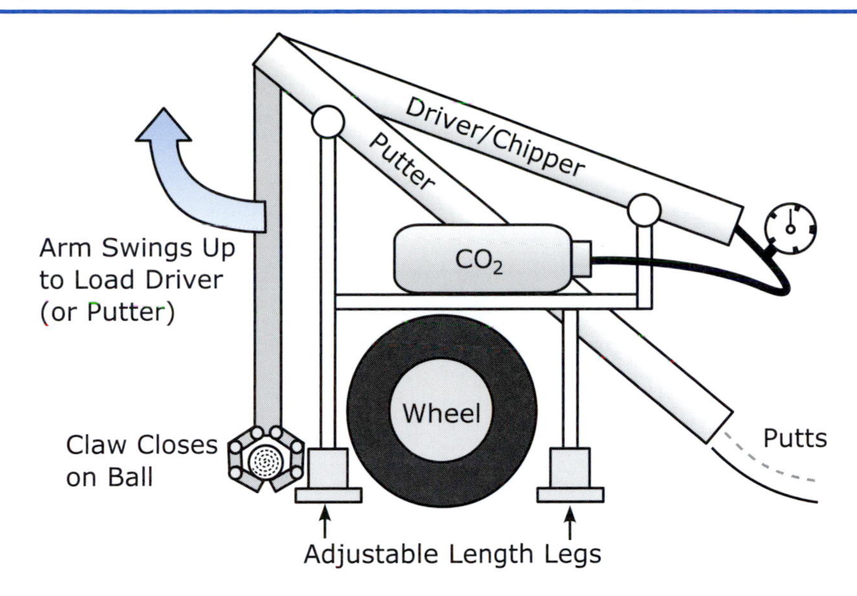

Figure 6

Concept Drawing of the "Cannon"

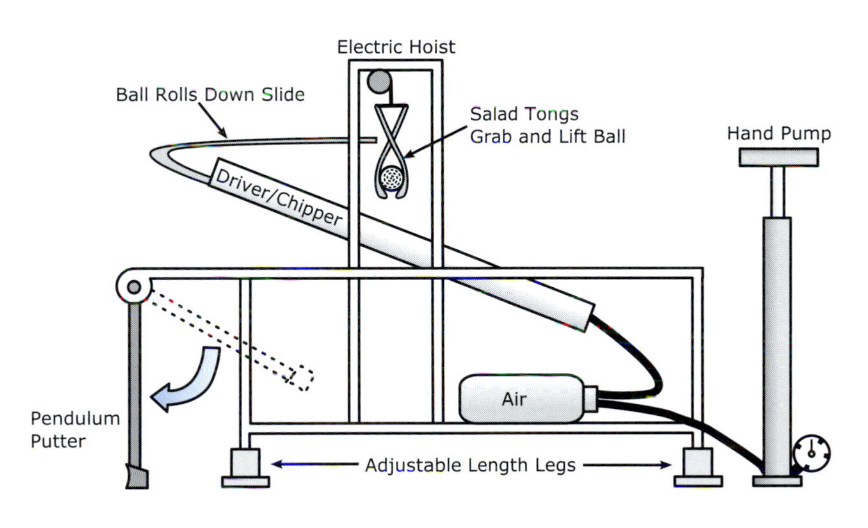

Figure 7

Concept Drawing of the "Original"

3. Ball loader is robust (i.e., picks up ball off of all types of terrain)
4. Easy to transport
5. Easy to manufacture

Low cost, which usually appears, was not selected because all three concepts met the cost requirement and cost was not involved in the design competition scoring.

With these in hand, the decision matrix can be drawn up and weights assigned to each criterion. The drives and ball loader were considered equally important because one cannot work without the other. The drives/chips were weighted slightly higher than the putts because only 43% of all

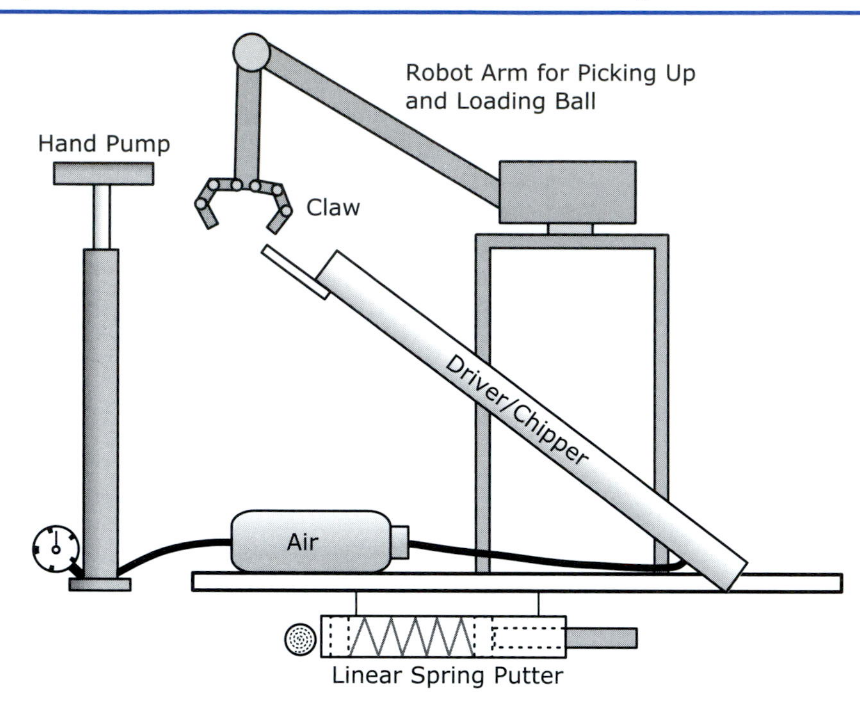

Figure 8

Concept Drawing of the "Robogolfer"

shots taken by golf professionals are putts. Transport is weighted low because it does not factor into scoring. Ease of manufacture is always important because of its impact on development times. The resulting weights are listed in the decision matrix of Table 2.

Then, proceeding one evaluation criterion at a time, the team analyzes the strengths and weaknesses of each concept in the context of the given criterion and assigns corresponding values to each concept in the decision matrix. The results of the analysis are presented following, and the values are recorded in Table 2.

DRIVES WELL

- All three drivers appear to be promising given the effectiveness of the notorious potato gun. However, since the CO_2 tank comes prepressurized and the hand-pumping is subject to a 60 s time limit on preshot preparation, the "Cannon" is likely to be firing the ball at higher pressures and thus should have a greater range.

PUTTS WELL

- The greens at the site of the competition will be severely sloped and slow. Therefore, the machines must be capable of executing long putts. Of the three machines, the Robogolfer is the most adjustable, since the springs can be easily replaced if the range proves inadequate. On the other hand, the potential energy of gravity powers the other two

Table 2

Decision Matrix for the Golf Machine Concepts

		Cannon		Original		Robogolfer	
Evaluation Criteria	Wt	Val_1	Wt x Val_1	Val_2	Wt x Val_2	Val_3	Wt x Val_3
Drives Well	.25	9	2.25	8	2.00	8	2.00
Putts Well	.20	4	0.80	4	0.80	8	1.60
Loader is Robust	.25	6	1.50	6	1.50	9	2.25
Easy to Transport	.05	9	0.45	5	0.25	5	0.25
Easy to Manufacture	.25	5	1.25	7	1.75	3	0.75
Totals	1.0		6.25		6.30		6.85

putters, and it will be difficult to increase starting heights once these machines are built. Therefore, there is greater risk associated with these putters.

LOADER IS ROBUST

- A wide range of lies is possible, from severe slopes to sand and divots. The Cannon and the Original address this issue by using legs that are adjustable in length. The robot arm of the Robogolfer is clearly the most flexible design and requires no setup time.

EASY TO TRANSPORT

- The rules require that only one student from the team may be used to transport the machine to the next shot location. The Cannon is the easiest to transport because only it has wheels.

EASY TO MANUFACTURE

- The robot arm of the Robogolfer stands out as easily the most complicated system on any machine. As a three-degree-of-freedom mechanism, it requires three independently controlled motors.
- The Original's loader should be straightforward to manufacture. The salad tongs and the parts for the electric hoist can be easily purchased.

DISCUSSION OF RESULTS

The Robogolfer is the clear winner on points. But the challenges involved in designing and manufacturing that robot arm should make you nervous (unless you have a robotics expert on your team). The decision matrix also revealed that the putters for the Cannon and the Original are weak concepts. If they are replaced by the Robogolfer's linear spring putter, the Original ends up with the most points.

The three concepts in Figures 6–8 correspond to actual student designs that were designed, manufactured, and tested. The Original team (so named because they were the first team to develop an air cannon) won the design competition. They compensated for their weak putter by chipping the long putts and adding a ramp to make the short putts. The Robogolfer team (who had a robotics expert) took longer than the Original to complete design and manufacture and thus had less time to test. The Cannon had the longest drives and the shortest putts. ■

Design Milestone: Evaluation of Alternatives

Successful completion of this milestone requires three strong design concepts, an open mind, and a lot of careful thought.

Assignment:

1. Decide on five to seven evaluation criteria that will be used with a decision matrix to evaluate the three concepts from the previous milestone.
2. Assign weights to the evaluation criteria.
3. Fill out a decision matrix. One row at a time, discuss the strengths and weaknesses of all of the concepts in the context of the given criterion, and then assign values by consensus before moving on to the next criterion.
4. Analyze the results of the decision matrix. Use the matrix to look for weaknesses and attempt to correct them by combining ideas from different concepts.
5. Select the best concept.
6. Document your evaluation process as per Example 4.

Grading Criteria:

- Are weights and values accurate and fully justified?
- Were the results of the decision matrix interpreted thoughtfully when searching for and selecting the best concept?
- Were all three concepts strong designs?
- Is the documentation typed and clearly written?

Design Competition Tips

- There is no need to rig the results of the decision matrix to come out to the concept you want, as the results are nonbinding.
- Do not blindly obey the results of your decision matrix; the selection of evaluation criteria may have been flawed to begin with.
- Engage everyone in the decision-making process.
- Do not shy away from bold designs just because they are different from everyone else's. Those differences could lead to victory at the final competition.

Chapter 20

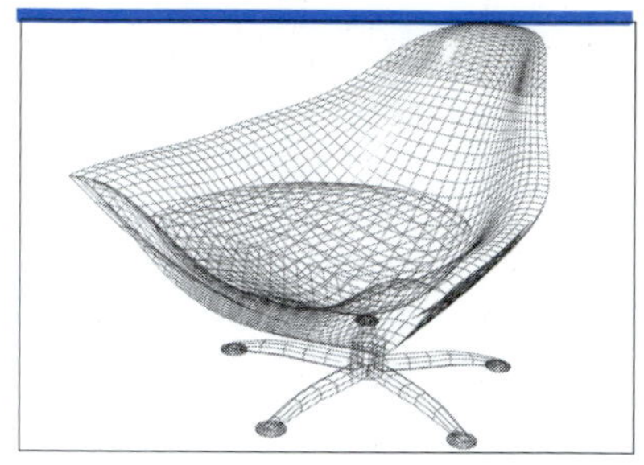

© iStockphoto.com/Rui Jordao

Detailed Design

The goal of this step in the design process is to specify the details of the design so that it can be manufactured. Those details are typically the dimensions and material composition of parts, as well as the methods used to join them. The decisions made during detailed design are guided by **analysis, experiments**, and **models** to reduce the "risk" that additional design changes will be needed later on. The final results are documented in the form of **detailed drawings.**

Analysis

"Analysis" refers to the application of mathematical models to predict performance. The role of analysis in freshman design projects is limited because the analytical capabilities of an engineering student are just starting to develop.

One calculation that may prove to be useful for small electro-mechanical devices is the determination of the optimal gear ratio. In developing this mathematical model we will draw from the equations on gearing and gear ratios in Chapter 6.

Assume that we want to determine the best overall gear ratio for the drive train of Figure 1. In Figure 11 of Chapter 6, this overall gear ratio corresponds to the combined gear ratio of the gear box and the external pair of gears driving the axle. The overall gear ratio (GR) is equal to the reciprocal of the overall velocity ratio as expressed by the following equation:

$$GR = \frac{N_{motor}}{N_{axle}} \tag{1}$$

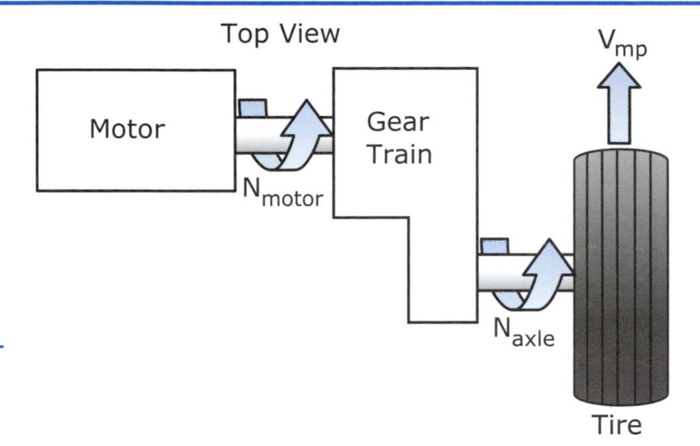

Figure 1

Schematic of the Drive Train for a Moving Platform

in which N_{motor} is the angular speed of the motor shaft (in RPM) and N_{axle} is the angular speed of the axle (in RPM). The linear speed of the moving platform (V_{mp}) is in turn related to the angular speed of the axle through the following relationship:

$$V_{mp} = \omega_{axle} R_{tire} \tag{2}$$

in which ω_{axle} is expressed here in terms of radians/s and R_{tire} is the radius of the driven tire. Changing units on angular speed to RPM in equation 2, we obtain:

$$V_{mp} = \frac{2\pi N_{axle} R_{tire}}{60} \tag{3}$$

Rearranging equation 3 to obtain an expression for N_{axle} and substituting the result into equation 1 leads to:

$$GR = \frac{2\pi N_{motor} R_{tire}}{60\, V_{mp}} \tag{4}$$

where length units on R_{tire} and V_{mp} must be the same, and the time units of V_{mp} are seconds. Equation 4 can be used to calculate the overall gear ratio required to achieve a desired speed V_{mp}, but only if we know the value of N_{motor}.

With small DC motors, the determination of N_{motor} is not always a straightforward matter. To begin with, N_{motor} is linearly dependent on the rotational resistance, or torque, acting on the motor shaft as illustrated in Figure 2. Each point on the motor curve represents a different equilibrium state of the motor. For example, if the motor shaft is allowed to freely spin, it will have an angular speed equal to the "no load angular speed," denoted by N_{noload}. Conversely, if you grab the spinning motor shaft between your fingers and gradually increase the pressure, the shaft will stop spinning when the torque you are applying reaches a value equal to the "stall torque," T_{stall}. Thus, to determine N_{motor}, we need to know N_{noload} and T_{stall} of the given

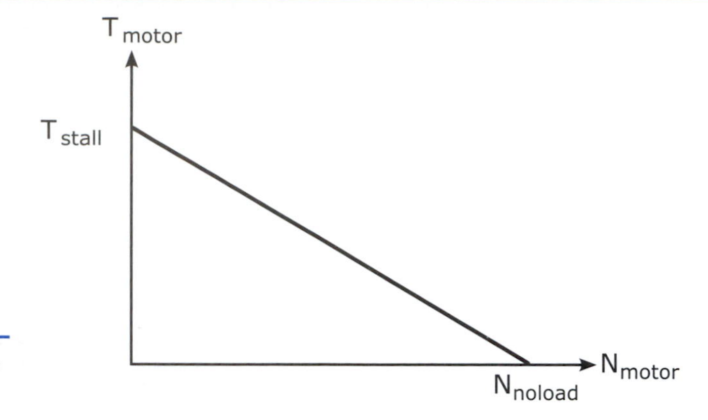

Figure 2

Typical Motor Curve for a Small DC Motor

motor, as well as the torque (T_{motor}) acting on the motor shaft. Herein lies the challenge, for the determination of T_{motor} is considered beyond the scope of this text and N_{noload} and T_{stall} are not always provided with the motor specifications.

Given these constraints, we recommend that you **only calculate gear ratio if T_{motor} is close to zero for the application.** This requires that the vehicle be very small and light (so that frictional losses are negligible) and that it only move on a level plane without pushing against anything. Then, $N_{motor} = N_{noload}$, and equation 4 specializes to:

$$GR = \frac{2\,\pi\,N_{noload}\,R_{tire}}{60\,V_{mp}} \tag{5}$$

Since frictional losses are hard to avoid, you can expect the moving platform to run at a speed that is smaller than the target value. If there are times during operation when T_{motor} is not negligible, such as when climbing a hill or pushing against an opponent, equation 5 is no longer valid. All we can tell you is that the gear ratio will have to be larger than the equation 5 prediction. How much larger will depend on the magnitude of the traction force on the driven tire.

When the value of N_{noload} is not provided, you may (at your own risk) assume an average value of 9000 RPM, given that N_{noload} for most small DC motors is in the range of 6,000 to 12,000 RPM. This will not work if your DC motor is a gearhead motor, which already has a built-in gear box. These will spin at much lower rates, and so your only recourse, if you want to use equation 5, is to try to measure N_{noload}.

EXAMPLE 1

We want to design a moving platform with a top speed of 0.500 ft/s on the flat. The motor, with specifications of $T_{stall} = .210$ oz-in and $N_{noload} = 11{,}600$ RPM, is already in hand, as are the 2.00 inch diameter tires.
Determine the overall gear ratio required to achieve the desired speed.

Solution Substituting into equation 5:

$$GR = \frac{2\pi N_{noload} R_{tire}}{60\, V_{mp}} = \frac{2\pi(11{,}600)(1)}{60\,(6)} = 202$$

where V_{mp} and R_{tire} were expressed using the same length units (inches). ■

Experiments

Physical experiments are a particularly effective way to reduce risk when working with small electromechanical systems. Because of the small scale, materials needed for the experiments can probably be scavenged, or at least obtained at low cost, and realistic forces can easily be applied. Also, physical experiments are often more accurate than idealized mathematical models at this scale.

Since the actual design has not been built yet, the subfunction being investigated may have to be idealized for the purposes of the experiment. For example, you might use cheaper materials or use your hands to create the motion. The errors introduced by these approximations will be tolerable if they are much smaller than the changes in performance being observed.

Knowing when to use experiments requires a keen awareness of the sources of risk in a design. This is no time for overconfidence; you can safely assume that if something can go wrong, it will. Thus, it is vital that you be able to distinguish between the aspects of the design that you are sure about and those aspects that you are not so sure about. The latter are candidates for physical experiments.

The steps for formulating an experimental plan are as follows:

1. Identify aspects of the design and its performance that you are uncertain about.
2. Associate the aspects in step 1 with one or more physical variables that can be varied by means of simple experiments.
3. Carry out the experiments that will do the most to reduce risk within the available time frame.
4. If possible, document the results in the form of graphs or tables.

EXAMPLE 2 A concept for a design competition named Dueling Duffers has been proposed and is shown in Figure 3. The object of this head-to-head competition is to be the first to deposit up to 10 golf balls into a hole in the center of the tabletop playing field of Figure 4.

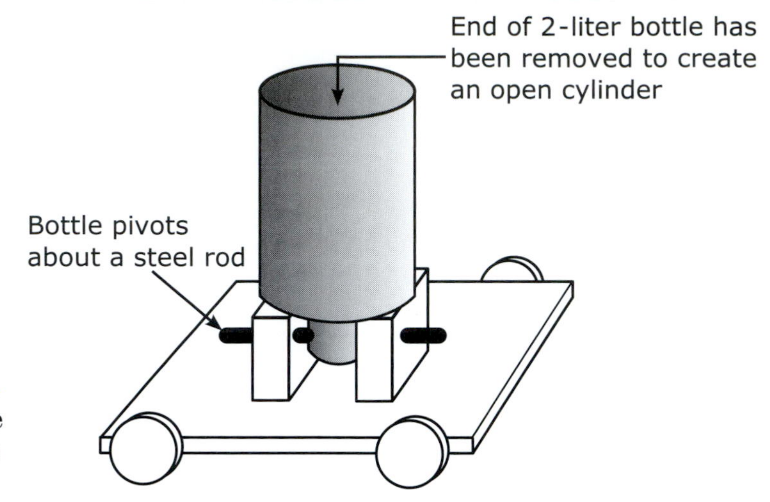

Figure 3

Proposed Concept for the "Dueling Duffers" Design Competition

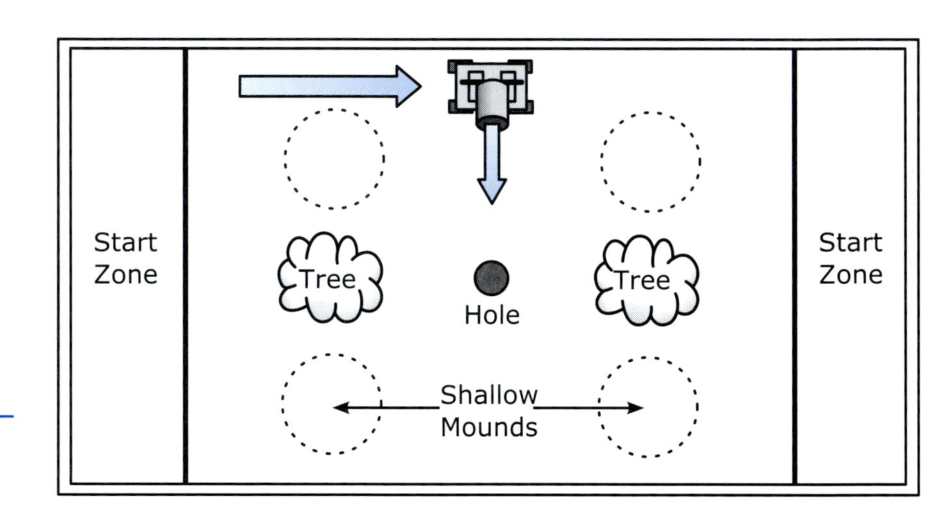

Figure 4

Playing Field for the "Dueling Duffers" Design Competition

These are the main features of the proposed design:

- It holds 10 golf balls in the top half of a 2-liter bottle.
- It uses the side rail to steer.
- When it is even with the hole, it dumps the golf balls in the general direction of the hole.

The path of the vehicle and the direction in which the balls are dumped are indicated in Figure 4.
For this example, you are asked to (1) identify the main sources of risk and (2) propose experiments to address those sources of risk.

Solution

SOURCES OF RISK/UNCERTAINTY

Sources of risk are presented here in the form of questions, the answers to which are currently unknown. These are the same type of questions that will be asked by the jury at the oral design defense.

1. Will all 10 balls fit in the top half of the 2-liter bottle?
2. What is the optimal height from which to dump the balls?
3. Will all 10 balls drop into the hole when dumped in this manner?
4. Is it important to have the vehicle perfectly positioned before dumping the balls?
5. Will the machine travel slowly enough that its position can be easily controlled?

PROPOSED EXPERIMENTS

The experimental setup in Figure 5 can be used to address each of the first four sources of risk just listed. The last item in the list is best handled by computing the overall gear ratio using equation 5.

The bottle, the steel rod and the two blocks of wood are all easily obtained. Assume the playing field is available for testing. The bottom of the plastic bottle will need to be cut off, and the steel rod will need to be inserted through the top of the bottle. There are no other parts that need to be joined. You can use your hands to keep the rod in place above the blocks, while taking care to let the bottle and balls fall under their own weight.

The numbers of the experiments described below correspond to the preceding numbered sources of risk.

1. The control variable is H (refer to Figure 5). Put 10 golf balls into the bottle and measure the minimum H required to hold 10 balls.

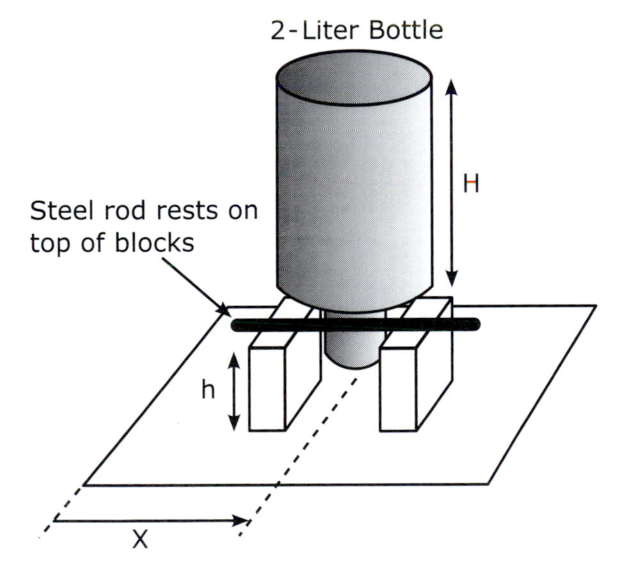

Figure 5

Experimental Setup for Establishing Key Dimensions and Reducing Risk

2. The control variable h determines how fast the balls will be rolling when they pass the hole. Increase h by inserting books under the blocks; decrease h by sawing the ends of the blocks. For each value of h, dump the balls three times and record the number of balls that drop into the hole. Document results by plotting h versus the average number of balls that dropped.
3. Here the concern is less with the speed and more with the distribution of the balls as they pass the hole. Control variables might be H or the manner in which the balls are packed within the bottle (e.g., a vertical divider could be used to keep the balls on the side of the bottle facing the hole). Again, use three trials for each value of the control variable and plot the control variable versus average number of balls that dropped.
4. The control variable is X. Perform three trials for each value of X, and plot X versus the average number of balls that dropped. The shape of this graph will provide the answer to question 4.
5. Calculate the gear ratio using equation 5. ■

Models

Models are scaled replicas constructed out of inexpensive, readily available materials. In the case of small electromechanical devices, they are often constructed out of cardboard or foam board. Models are used to check geometric compatibility, establish key dimensions of moving parts, and to visualize the overall motion.

Typical examples are shown in Figures 6 and 7. The foam board model of the stair climbing device in Figure 6 was used to prove the feasibility of the design. Several other stair climbing concepts, mainly wheel and track designs, were regarded as feasible until models proved otherwise. In Figure 7, a dart-throwing mechanical linkage was modeled in 3D using cardboard. All of the links move, including the grip mechanism that releases the dart when the arm strikes a stopper.

Detailed Drawings

By definition, a detailed drawing will contain all of the information required to manufacture the design. The drawings should be so complete that if you handed them off to someone unfamiliar with the design, that person would be able to build it.

The usual practice is to specify dimensions on multiple orthogonal views of the design. An isometric view is also sometimes provided to assist with

Figure 6

Model of a Stair-Climbing Device

Figure 7

Model of a Dart-Throwing Mechanism

visualization. In all, six orthogonal views are possible: front, back, left, right, top, and bottom. Three views, however, are most common. Figure 8 shows a detailed drawing with five orthogonal views.

Additional information such as material specification, part type, and assembly directions are conveyed through written notes on the drawings. Closeup views can be employed to clarify small features.

While practicing engineers will generate drawings like Figure 8 using computer-aided design (CAD) software, first-year engineering students probably have not taken a CAD course yet. Therefore, we recommend that the usual standards for preparation of detailed drawings should be relaxed somewhat and replaced by the following set of guidelines:

- Drawings can be neatly hand drawn using ruler and compass.
- Drawings must be drawn to scale, though not necessarily full-scale.
- Drawings of at least two orthogonal views of the design should be prepared. An isometric view is not required, but closeup views should be used to clarify small features.

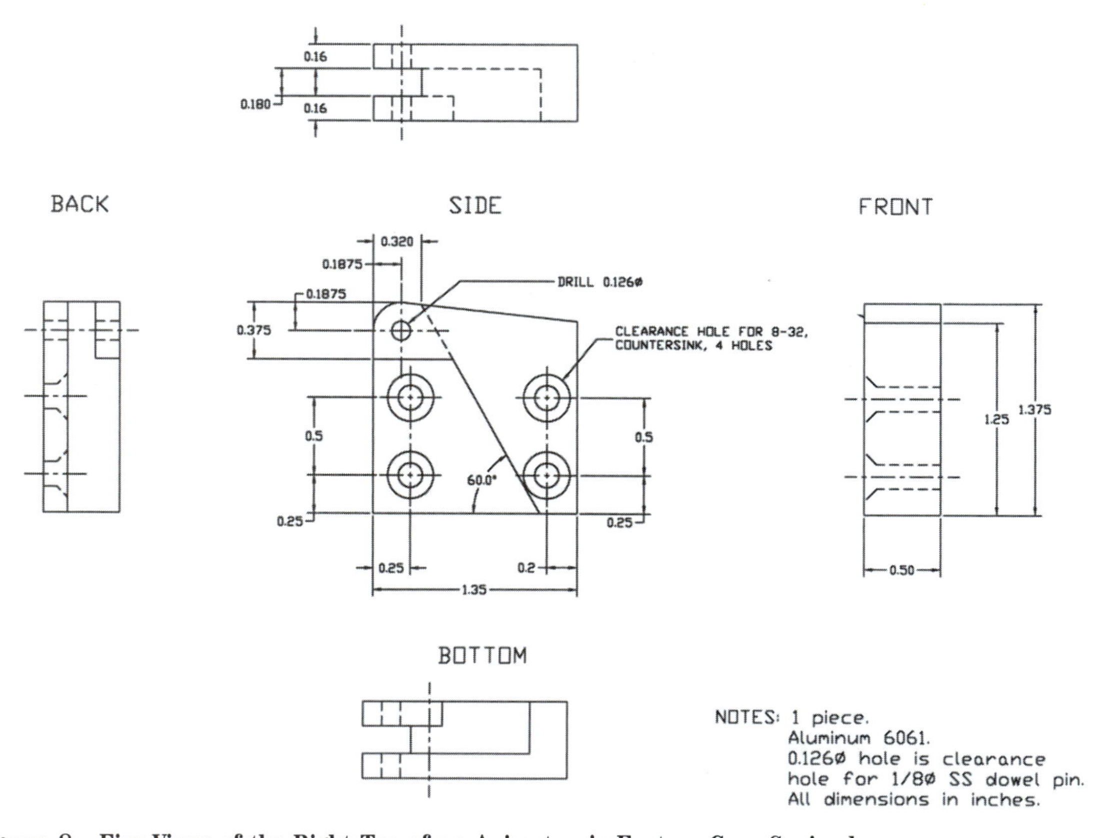

Figure 8 Five Views of the Right Toe of an Animatronic Eastern Gray Squirrel

- Show hidden lines only when they will enhance clarity. These are dashed lines that are used to show edges that are not visible from the viewer's perspective.
- It is acceptable to show only essential dimensions—that is, key dimensions that either have a direct impact on performance or are needed to demonstrate that geometry constraints are satisfied.
- Use notes or labels to indicate material specification, part type, and assembly directions.
- It is acceptable if the manufacturing details are incomplete. Students who lack the experience to fully specify them will have to discover those details by trial and error during building.
- If an electric circuit was designed, show it in the form a neatly hand-drawn but fully specified circuit diagram.

An example of a detailed drawing prepared in accordance with these guidelines is shown in Figure 9.

Although the freedom to leave out some dimensions and manufacturing details has been allowed keep in mind that missing details amount to higher risk in the minds of those being asked to provide resources to the project. Thus, a design with fewer missing details may be viewed by the instructor as having lower risk.

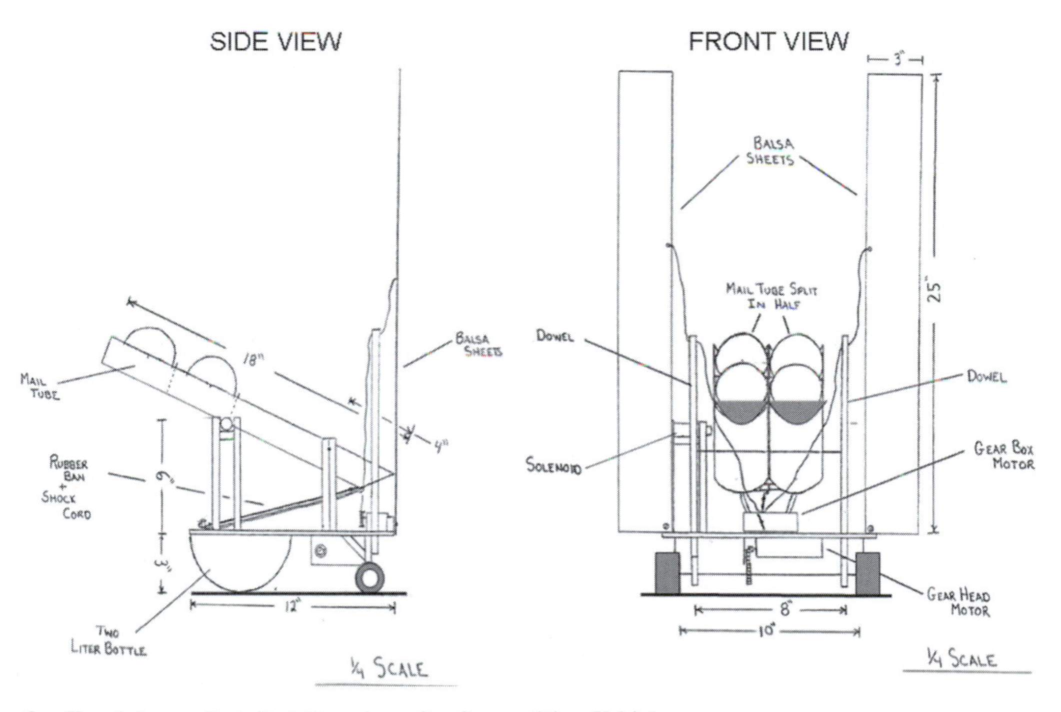

Figure 9 Hand-drawn Detailed Drawing of a Competition Vehicle

Design Milestone: Detailed Design

This milestone is all about reducing risk, not only in your own mind but in the minds of the jury at the upcoming oral design defense.

Assignment:

1. Use analysis, experiments, and models to help establish dimensions and proof of concept.
2. Prepare detailed drawings of the design concept you selected.

Documentation:

- Write up analysis details in the usual format, stating all assumptions.
- For each experiment: (a) state the purpose of the experiment, (b) describe the experimental procedure, (c) present results, and (d) state conclusions.
- Summarize useful information yielded by models, and turn in models.
- Attach hand-drawn detailed drawings.

Grading Criteria:

- From examination of the detailed drawings, does the design have a chance of working?
- Have opportunities to reduce the level of risk (through analysis, experiments, and models) been fully exploited?
- Is there enough information in the detailed drawings to manufacture the design?
- What is the overall quality of the detailed drawings?

Design Competition Tips

- Time spent now on analysis, experiments, and models pays off later in fewer design iterations during manufacturing and testing.
- Of the four methods for reducing risk, good detailed drawings will reap the most rewards in a freshman design project.

Chapter 21

© iStockphoto.com/Jennifer Trenchard

Oral Design Defense

Engineers must convince customers that a design is worth expenditures of money and the time of skilled people. In a student design project, this process of convincing customers is simulated by an **oral design defense.** The goal of this oral presentation is to win the confidence of the project sponsors, henceforth referred to as the *jury*.

In assessing a team's chances for future success, the jury will be searching for answers to the following questions:

1. Did the team adhere to the systematic approach?
2. How does the final concept work?
3. What is the level of risk associated with this design?
4. Do the students appear to be teaming effectively?

The jury's concerns suggest some strategies that should be effective. First, the organization of the presentation should parallel the steps in the design process, as shown in Table 1. This is your way of saying that you followed a systematic approach. Second, you should try to get the jury to understand how your final concept works as quickly as possible. This frees up more time during questions for alleviating concerns about the design. Third, you should anticipate that the jury will ask questions about potential sources of risk, and prepare evidence in advance that will quell those concerns. This evidence should be in the form of quality detailed drawings, results of calculations and experiments, and models. If you built models, bring them; if you conducted experiments, try to bring some evidence that indeed you did them. The strategies that serve to reduce risk, and their counterparts that don't, are summarized in Table 2.

Meanwhile the jury will also be evaluating your teaming. They will base their impressions on the quality of your design and oral presentation (see Table 3 for some tips on delivery and visual aids). There are other

Table 1

Suggested Organization of the Oral Design Defense

ORGANIZATION	Slides
Title	1
Outline of Presentation	1
Problem Definition	1
Important Design Requirements	1
Alternative Concepts Not Selected	2
Final Concept	1–2
• Describe main features	
• Explain why you selected it	
Detailed Design	
• Show main drawings	2
— Explain how it works	
— Explain how you will construct it	
• Zero in on special features with closeup views	1–2
• Present results of analyses, experiments, and models	1–2
Summary	1
• Summarize strengths of the design	
• Quantify performance expectations (e.g., top speed)	
• Describe your strategy at the final competition	
Total =	12–15

Table 2

Strategies That Either Reduce or Amplify Risk

RISK REDUCERS

1. High-quality concept drawings and detailed drawings.
2. Calculations, experiments, and models that help establish proof of concept.
3. Manufacturing details have been explained.
4. Quality visual aids.

RISK AMPLIFIERS THAT COULD DELAY MANUFACTURE

1. Poorly detailed drawings.
2. One or more subfunctions obviously will not work.
3. No thought given to manufacturing.

telltale signs. For example, did everyone contribute equally to the presentation? Was everyone involved in answering questions? Did team members refer to themselves as "we" or "I" when citing accomplishments?

When answering questions, be forthright and honest. Failure to do so will lead to an unending chain of questions. If your response is an opinion

Table 3

Oral Presentation Tips

TIPS ON DELIVERY

1. Do not read sentences directly off the slides.
2. Try to look at the audience.
3. Try to stand next to the screen when speaking.
4. Practice the presentation.
5. Be positive, dynamic.

TIPS ON VISUAL AIDS

1. Avoid using too many words
2. Use a font size that can be seen easily.
3. Ink concept drawings and detailed drawings before scanning.
4. Put a heading on each slide.
5. Show results (not details) of calculations or experiments.
6. View the projected images in advance to make sure all words and pictures will be visible to the audience.

and not a fact, state so, because one erroneous answer can damage your credibility and thus elevate the risk associated with your design.

Design Milestone: Oral Design Defense

To qualify to receive parts and materials for the manufacturing phase of a hands-on design project, a majority of the jurors must be convinced that your design will work. In the event such a consensus is not achieved, teams will be asked to revise their designs and resubmit at a later date.

Assignment:

1. Prepare the visual aids for the oral design defense. You must use Powerpoint or an equivalent software package. Relevant drawings should be inked before scanning them in.
2. Practice the presentation.
3. Deliver the oral presentation to a jury of evaluators.

Typical Format:

- Eight minutes for the oral presentation; 4 minutes for questions.
- All team members participate in the presentation and in responding to questions.

Grading Criteria:

- What is the level of risk associated with the design?
- What was the quality of the drawings and the other visual aids?

Design Competition Tips

- Grading tends to be proportional to the quality of the drawings.
- Evidence of the use of calculations, experiments and models will do much to reduce the level of risk in the minds of the jurors. Visual representations of the results are more effective than just saying you did them.

Chapter 22

© Courtesy of Daimler-Chrysler

Manufacturing and Testing

Manufacturing begins once the detailed design has been approved and ends when the machine is placed in the starting zone of the final competition. In between, the machine will undergo numerous modifications. Few new designs work on the first try. Manufacturing and testing will tend to take much longer than expected—probably from three to five times as long. This chapter begins with a summary of good **manufacturing and testing strategies,** and then moves on to describe **materials, joining methods,** and **hand tools.**

Manufacturing and Testing Strategies

There are strategies that you can employ during manufacturing to minimize the time it takes to get the first prototype ready for initial tests. The extra time freed up for testing and design refinements can prove decisive in a design competition.

The following time-saving manufacturing strategies have been observed in successful teams:

- ***Talk to a machinist.*** Professional machinists are considered partners in the design process. What they lack in knowledge of the engineering design process and analysis, they make up for in manufacturing and practical experience. As a practicing engineer, you will be required to consult with a machinist before finalizing your detailed design.

- ***Don't delay in getting started.*** Take the leap of faith and begin manufacturing as soon as possible. Only then will the team gain a realistic sense of the manufacturing timeline.
- ***Divide up responsibilities so that team members can work in parallel on different subassemblies.*** Otherwise, you may find the entire team standing around waiting for the same glue joint to dry.
- ***Keep detailed drawings up-to-date.*** If they are not up-to-date, only one person will know what the actual design looks like. That one person will end up doing most of the manufacturing while the other team members watch. With accurate drawings, team members can work in parallel.
- ***Set and enforce intermediate deadlines.*** Manufacturing can span several weeks. The instructor will set the big deadlines through the milestones; the teams should set the little ones in between.

Testing will be as important as manufacturing in preparing for a design competition. For example, three design teams were assigned the task of designing machines capable of playing 18 holes of miniature golf at a local course. The first team was stocked with experienced machinists, and so they built their machine out of thick steel parts. The second team had an expert welder, and so they welded together their machine out of steel beams and plates. The third team chose to make their machine out of wood. Three very different manufacturing skill sets, yet all three machines failed at the final competition for the same reason — not enough testing. The first team did not test their machine on synthetic grass before the competition. Their steel machine was heavy, which created large friction forces between its tank-line treads and the synthetic grass. When it attempted to turn, the treads broke, immobilizing their machine. The second team did not have time to test their steering mechanism due to last-minute modifications. As a result, they could not consistently maneuver into position for their putts within the time constraint. The third team completed manufacture and testing of their moving platform two weeks before the final competition. Over the next two weeks, while their putting mechanism was being made, they did not test the moving platform again until about 30 minutes before the start of the competition. Their machine never moved.

The lessons learned by these three teams apply to all design competitions. They are summarized in the following testing tips:

- Always leave a lot of time for testing.
- When conducting tests, try to simulate as closely as possible the conditions at the final competition. If these conditions are not known, test under a variety of conditions to insure robustness.

Materials

When possible, the designs should be made of wood to facilitate manufacture and keep costs down. Manufacturing can be simplified still further by constraining the designs to be small (less than 1 ft^3) and lightly loaded. Under these conditions, balsa can be used as the main structural material.

Recommended materials for a small, lightly loaded electromechanical device are listed in Table 1 along with their relative attributes. Balsa is listed as easy to use with hand tools because it can be cut and shaped easily with a sharp knife. On the other hand, plastic tends to deform rather than shear cleanly under the action of cutting tools, and so it is listed as hard to work with.

When selecting a material from Table 1, the strength and stiffness requirements of the given part also need to be considered. For example, if there are concerns about a part breaking under load, strength considerations will override ease of manufacture, leading one to use plywood instead of balsa. If a small-diameter axle requires high stiffness so that gears can remain engaged, then steel rod may be the best choice.

Table 1

List of Recommended Materials for a Small Electromechanical Design Project. Strength (resistance to breaking) and Stiffness (resistance to deformation) are normalized with respect to values for balsa

Material	Relative Strength	Relative Stiffness	Ease of Manufacture	Useful Forms
Balsa	1	1	Very easy	Sheets, beams, blocks
Woods	5	4	Easy	Plywood sheet, wood dowels
Plastic	5	1	Hard	Prefabricated gears
Steel	50	80	Medium	Thin rods
Rubber	0.5	Very small	Very easy	Long strands

Joining Methods

For small-scale balsa and wood structures, the preferred methods for joining parts are adhesives, wood screws, and machine screws with nuts. Typical joint configurations employing these methods are shown in Figure 1. Use of tapes, especially duct tape, is frowned upon for their nonpermanence and poor aesthetics. Nails are not particularly compatible with balsa because of the large impact forces involved and the possibility of wood splitting.

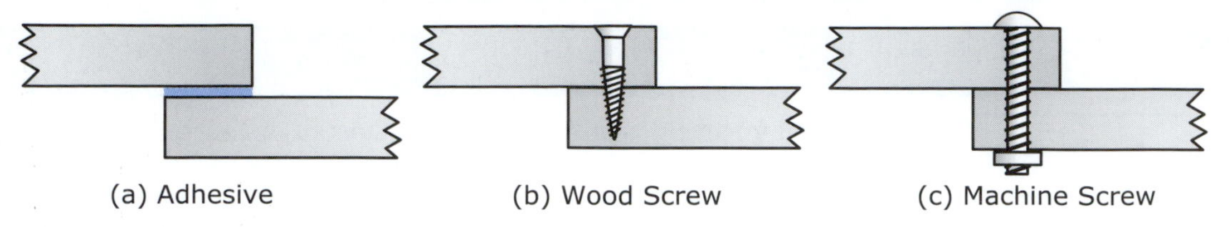

Figure 1 Different Joining Methods

Adhesives, in particular hot glue, are the method of choice when balsa is the predominant structural material. Hot glue cures quickly, and, though essentially permanent for wood-to-wood bonds, metal-to-wood bonds can be adjusted and/or broken by heating. It is so general purpose that it can be used to mount a motor to a plywood base. The main drawback of hot glue is its low strength, but this is usually not an issue for lightly loaded balsa structures. When it does become an issue—for example, when the surface area available for the glue joint is very small—one of the higher strength adhesives listed in Table 2 may be substituted.

Table 2

Common Adhesives

Adhesive	Typical Uses	Setting Time*	Curing Time†	How to Apply	Relative Strength
Wood glue	Wood, paper	8 hr	24 hr	Apply in liquid form direct from bottle.	13
Epoxy	Wood, metal	5 min to 12 hr	3 hr to 3 days	Comes in 2 tubes; mix equal amounts and apply with stick.	14
Hot glue	Almost anything	1 min	2 min	Place glue sticks in heated gun and apply with gun.	1

** Setting time = time to harden*
†Curing time = time to reach maximum strength

On balance, wood screws are a less popular alternative for balsa-to-balsa joints. To reduce the chances of wood splitting, a pilot hole equal in diameter to the screw without the threads should be drilled prior to inserting the screw. This material removal plus the persistent possibility of wood splitting reduces the effective strength of the balsa members. In situations where adjustability is needed, the benefits of easy screw removal may override these costs.

Wood-to-wood joints are a different matter. Wood is much stronger than balsa, and so a higher strength joint is justified. In such cases, use of screws is often preferable to the long curing times of the higher-strength adhesives.

Useful Hand Tools

Small-scale electromechanical devices can be crafted using common hand tools, provided balsa and/or wood are the primary structural materials. In this section, we offer a short compendium of those tools for your easy reference. Our goal in presenting them is to make you aware of the alternative manufacturing operations at your disposal. We will not go into much detail on how to use them. Instead, we advise that you observe your classmates, talk to a machinist, or just give it a try.

For each tool, we will (1) show a picture of it (so that you can find it), (2) name it (so that you can ask for it), (3) describe its use, and (4) provide additional comments on usage.

Tools for Measuring

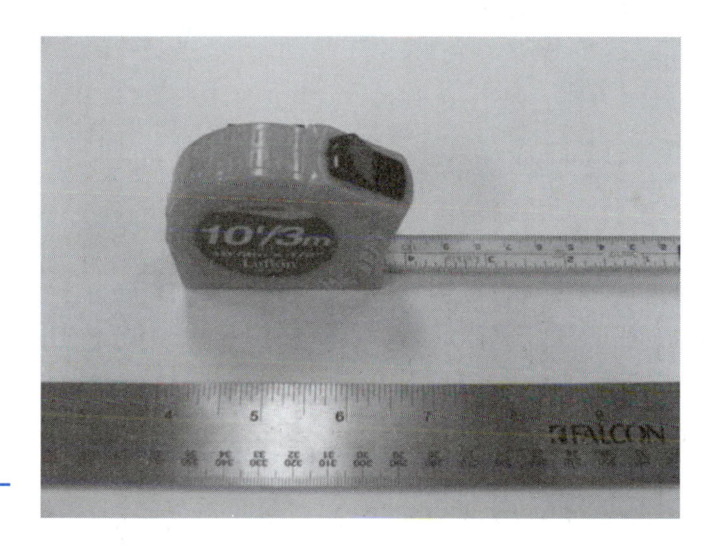

Figure 2

Tape Measure/Ruler

Name: Tape measure; steel ruler.
Use: For measuring length.
Comments: Steel ruler can also be used as a straight edge when making straight cuts in balsa with the X-acto knife.

Tools for Cutting and Shaping

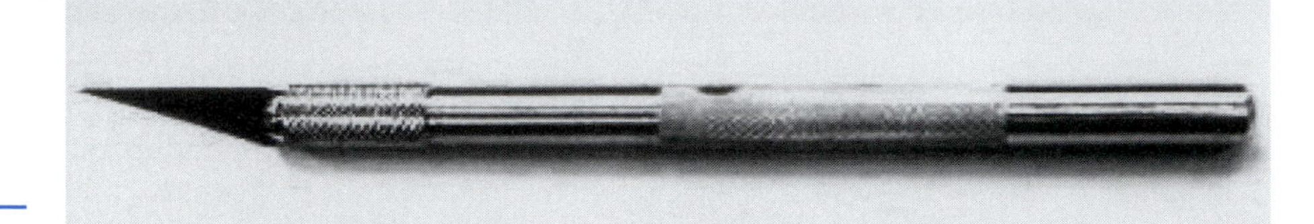

Figure 3

X-acto Knife

Name: X-acto knife (Hunt Manufacturing Co.).
Use: For cutting and shaping balsa.
Comments: Blades come in different shapes and are replaceable.

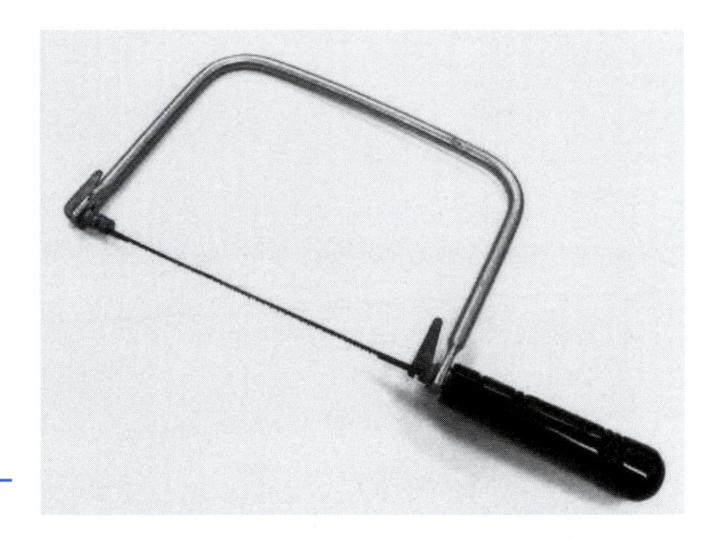

Figure 4

Coping Saw

Name: Coping saw.
Use: For cutting curves in wood.
Comments: The blade can be mounted in the frame to cut on either the push or pull stroke. Unscrewing the handle relieves tension on the blade so that it can be removed, rotated up to 360°, or inserted through a small drilled hole in order to cut out a larger hole.

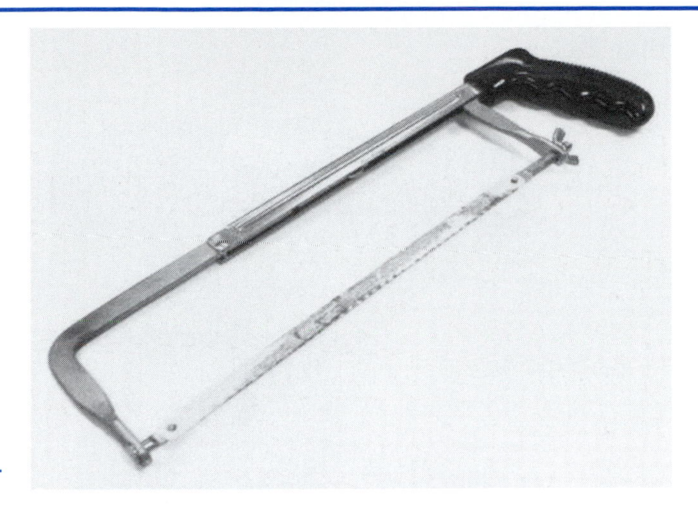

Figure 5

Hacksaw

Name: Hacksaw.
Use: For making straight cuts in metal.
Comments: It cuts on the push stroke, when blade is mounted correctly. Though designed for metal, it can also be used with wood. Blades are detachable and can be rotated to cut in four directions.

Figure 6

Dremel with Bits

Name: Dremel® (Robert Bosch Tool Corporation).
Use: For shaping and drilling holes in wood.
Comments: This is an electric hand drill without the handle. Because of its shape, forces can be applied more easily when grinding/sanding.

Tools for Drilling Holes

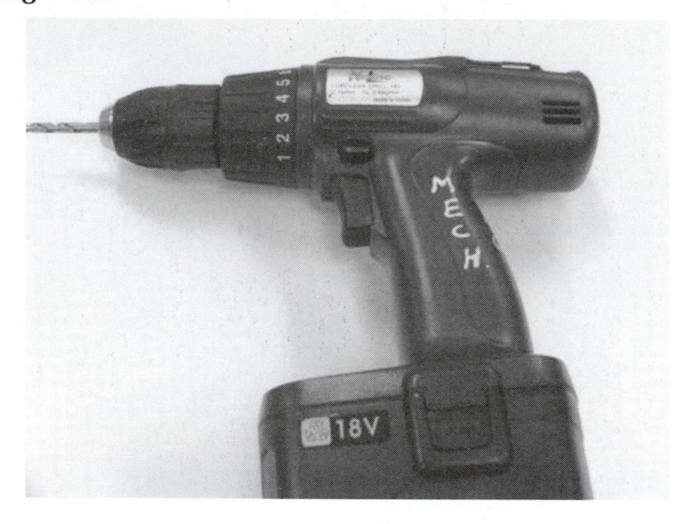

Figure 7

Cordless Hand Drill

Name: Cordless hand drill.
Use: For drilling holes in wood and plastic.
Comments: Release trigger lock and pull trigger to start drilling. Power switch controls the direction of spin.

Tools for Joining Parts

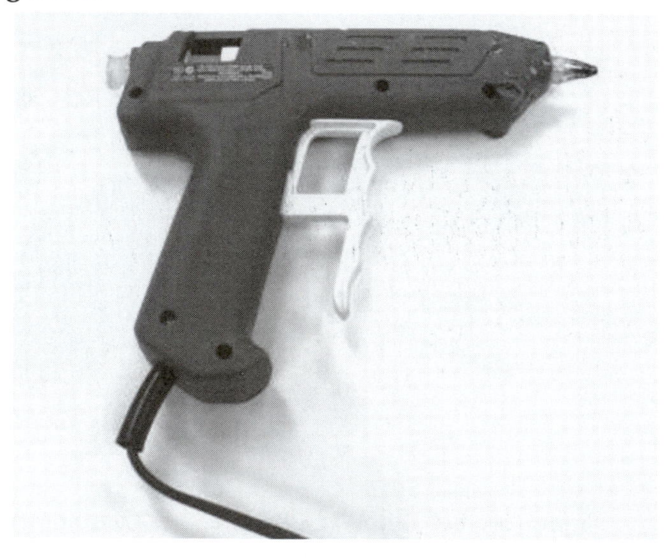

Figure 8

Hot Glue Gun

Name: Hot glue gun.
Use: For applying hot glue.
Comments: Insert a glue stick in the back, wait 3–5 minutes for the gun to heat up, and then apply the glue by pulling the trigger. Handle with care as the glue can reach temperatures of 400°F.

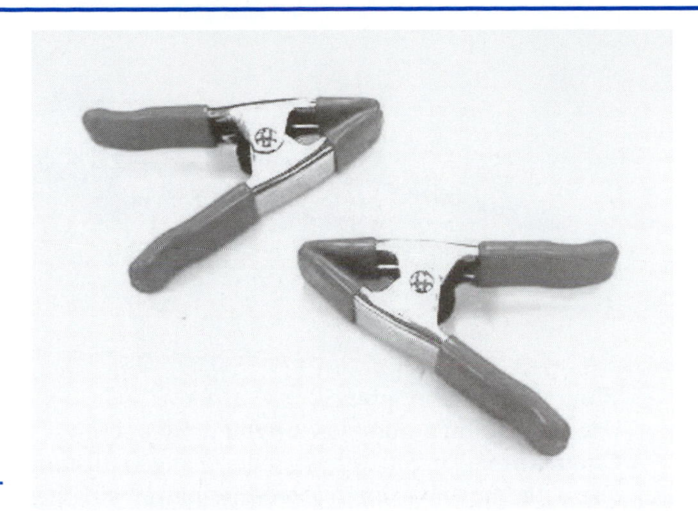

Figure 9

Spring Clamps

Name: Spring clamps.
Use: For holding parts together when waiting for an adhesive to set.
Comments: If the grips do not open wide enough for your application, C-clamps are a good alternative.

Figure 10

Screwdriver

Name: Screwdriver.
Use: For inserting and removing screws.
Comments: The longer and thicker the handle, the easier it is to apply torque. The screwdriver in Figure 10 has a removable tip. You can pull out the tip, rotate it 180°, and reinsert it to switch from a regular screwdriver (⊖) to a Phillips head screwdriver (⊗).

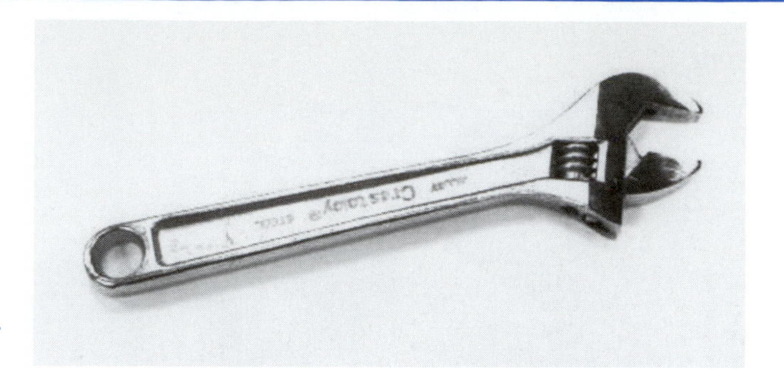

Figure 11

Adjustable Wrench

Name: Adjustable wrench.
Use: For holding nuts when tightening a screw.
Comments: Thumb adjustment changes distance between jaws. Unlike pliers, there is no gripping force, so parts being held must have flat surfaces.

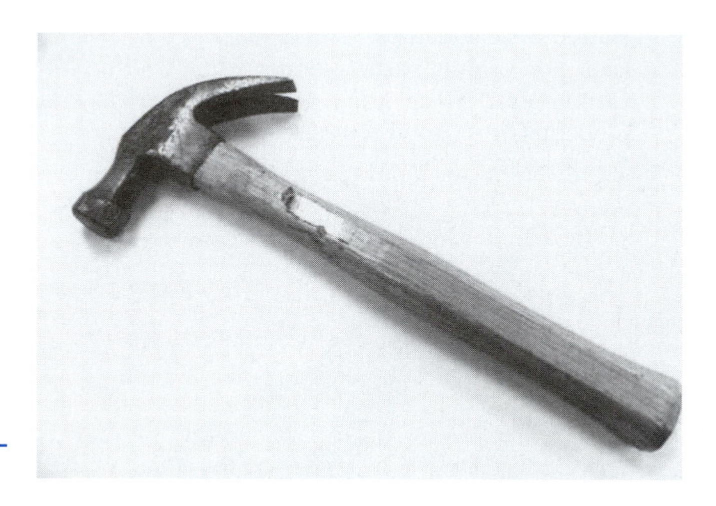

Figure 12

Claw Hammer

Name: Claw hammer.
Use: For driving or removing nails.
Comments: Nails should not be used to join balsa parts.

Tools for Wiring

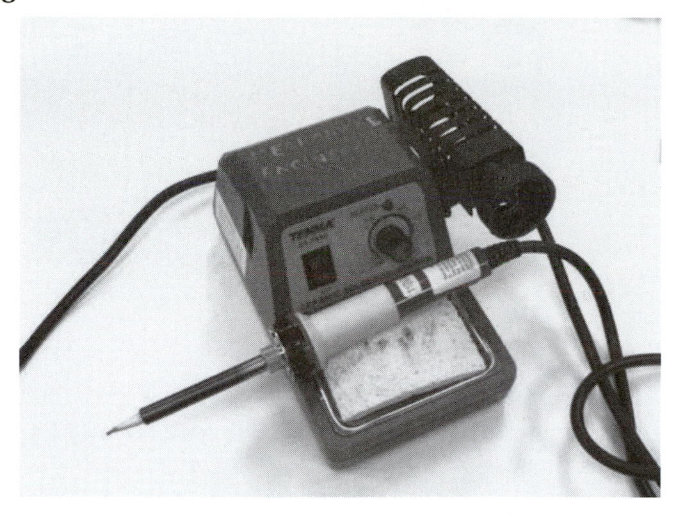

Figure 13

Soldering Iron

Name: Soldering iron.
Use: For attaching or connecting wires with solder.
Procedure: (1) Create mechanical connection (e.g., by twisting wires together); (2) let iron heat up; (3) apply solder to tip of iron (called tinning); (4) heat wires with iron; (5) bring solder in contact with heated wires until solder melts and flows; and (6) let wires cool.

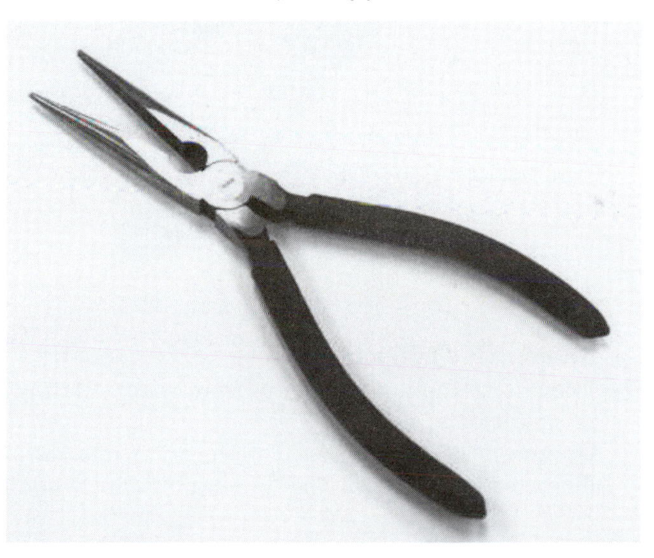

Figure 14

Long-Nosed Pliers

Name: Long-nosed pliers.
Use: For bending wires, grasping parts, cutting wires, and reaching into tight places.
Comments: Clamping forces are highest at the wire cutters and smallest at the tip of the nose.

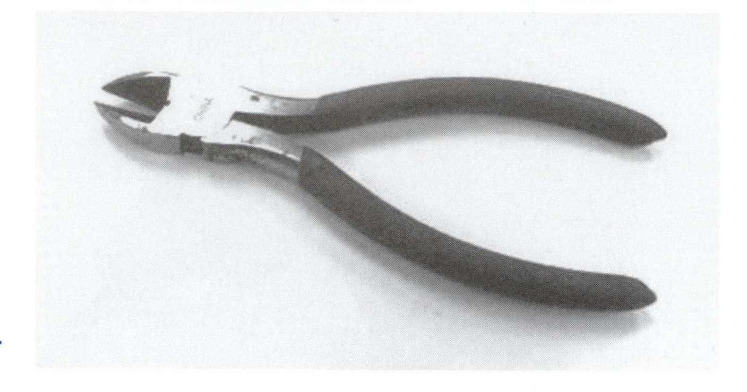

Figure 15

Wire Cutters

Name: Wire cutters.
Use: For cutting or stripping wires.
Comments: You may not need this tool if your long-nosed pliers have a good wire cutter.

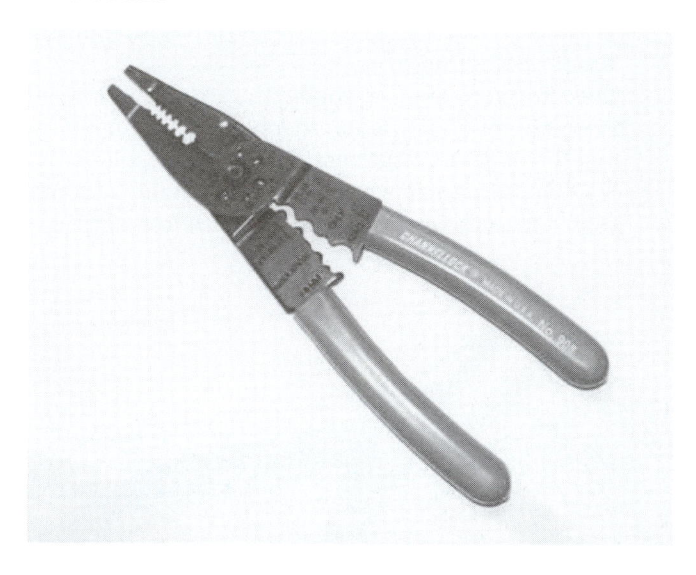

Figure 16

Wire Strippers

Name: Wire strippers.
Use: For stripping insulation from wires, cutting wires, and shearing small diameter bolts.
Comments: To strip insulation, lay wires across jaw in notch of same diameter as metal wire, and then pull wire while holding jaws compressed.

Design Milestone: Design for Manufacture Assessment I

The **Design for Manufacture** (DFM) assessments are so named because the teams that gave serious consideration to design for manufacture principles when first formulating their designs will have the best chance of doing well on these milestones.

This first DFM assessment should occur about halfway through the first manufacturing iteration. For a design project of the scale proposed in these chapters, that's about one week after receiving the box of parts.

Assignment:

Make as much progress as possible toward completion of manufacturing and testing of your design.

Grading:

Teams should bring their materials and a detailed drawing of their machine to the next design studio. Grading will be based on the amount of progress made.

100 = manufacturing is more than 50% completed
80 = manufacturing is between 25% and 50% completed
60 = some progress made (but less than 25%)
0 = no progress

Design Competition Tips

- Team members should work in parallel on different aspects of the design to accelerate progress.
- From here on the design project is a race to begin testing. Teams with the most testing time tend to win competitions.

Design Milestone: Design for Manufacture Assessment II*

By the time this milestone is reached, teams should have completed their first manufacturing iteration and begun testing. For a freshman project, this could be as soon as one week after the previous DFM milestone.

The instructor will evaluate progress by conducting one (ideal) or more well-defined performance tests to determine whether or not the various subfunctions are working. Evaluation of how well they work is reserved for the next milestone. Subfunctions to be evaluated must be carefully selected so as not to be biased toward a particular design or strategy.

Assignment:

Get all subfunctions working by the assigned due date.

*Instructor must specialize this milestone to the design project.

Performance Test:

To be defined by the instructor.

Grading:

Team grades will be based on the level of functionality demonstrated in the best of three trials, and will be computed as follows:

If manufacturing is 95% completed, then:

$$Grade = B + \sum_{i=1}^{I} (W_i s_i)$$

where: $Grade$ = assigned grade on a scale of 100
B = base grade (typically $B = 70$)
I = total number of subfunctions being evaluated
W_i = number of grade points associated with the ith subfunction and defined such that $100 = B + \sum_{i=1}^{I} W_i$
s_i = 1 (if the subfunction worked)
0 (if the subfunction did not work)

If manufacturing is less than 95% completed, then:

$$Grade = f B$$

where: $Grade$ = assigned grade on a scale of 100
B = same base grade as above
f = fraction of manufacturing that was completed

Chapter 23

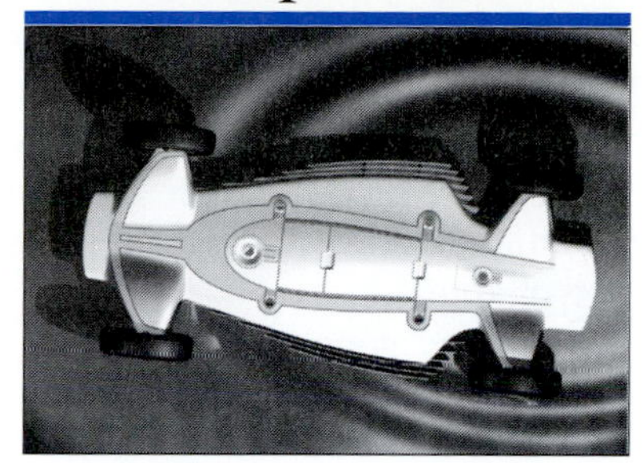

© Courtesy of General Motors Corp.

Performance Evaluation

Once the design has been manufactured, it is time to evaluate performance. If the machines are required to interact with each other, performance should be measured in two stages. First, comes **individual performance testing.** The manufactured device is tested alone, under controlled conditions, to verify that it is capable of doing what the problem definition requires. Then comes the **final competition.** The device is tested against other machines in a series of head-to-head matches to determine the best overall design. The student grade resulting from performance evaluation will typically constitute up to 50% of the project grade.

Individual Performance Testing

Performance of a given machine in head-to-head matches may vary with the opponent. For example, an offensive or defensive strategy that works well against one opponent may not work against another. So the only way to test all of the machines under the same set of conditions is to test them in isolation, without an opponent.

The basic approach is to measure one or more quantities, referred to as "metrics," that are good predictors of success in the head-to-head matches. Typical metrics are time, speed, pushing force, or number of points scored against a stationary obstacle representing the opponent. Some basic rules when selecting metrics are (1) they should be easily measurable, (2) they should be continuously variable to maximize information content, and (3) they should not be biased toward particular design solutions.

The number of different physical tests required to measure all of the metrics will depend on the choice of metrics. Ideally, you want to be able to design a single test that will measure all of the metrics. Sometimes, each metric will require its own test.

The performance grade is the overall measure of performance expressed on a scale of 100. We recommend that it be computed as a weighted sum of the metric values as expressed in the design milestone at the end of this chapter.

The specifics of the performance tests are often not revealed to the students until a week before they take place. This is to prevent students from tailoring their machines to the performance tests instead of to winning the final competition, which is the real design objective.

The Final Competition

The final competition pits the machines against each other, usually in a series of head-to-head matches that may involve direct interactions between the machines. This is the ultimate test of the machines, as it is the only way to accurately evaluate the effectiveness of offensive and defensive strategies, robustness, durability, and the wisdom of past design choices.

However, it is probably best that the results of head-to-head competition not be linked to the performance grade, since each machine will be facing a different set of challenges. For example, one machine may not match up well against a particular opponent, or a prefabricated part may fail unexpectedly due to an accidental collision.

Design Milestone: Individual Performance Testing*

Gradewise, this is the most important milestone. The testing regimen enforced by this milestone and the previous one will prepare the machines for the final competition.

Assignment:

Optimize performance of your machine in preparation for individual performance testing.

Performance Test:

To be defined by the instructor.

---

*Instructor must specialize this milestone to the design project.

Grading:

Team grades will be based on the quality of performance demonstrated in the best of three trials, and will be computed as follows:

$$Grade = B + \sum_{i=1}^{I} W_i \left(\frac{m_i}{M_i}\right)^n - \sum_{j=1}^{J} P_j$$

where $Grade$ = assigned grade on a scale of 100
B = base grade (typically $B = 70$)
I = total number of metrics
J = total number of penalties assessed for rules violations
W_i = number of grade points associated with the ith metric and defined such that $100 = B + \sum_{i=1}^{I} W_i$
m_i = measured value of the ith metric
M_i = best value of m_i recorded by any team in the class
n = 1 (if performance is directly proportional to m_i)
−1 (if performance is inversely proportional to m_i)
P_j = number of grade points associated with the jth penalty

If the *Grade* calculated is less than or equal to *B*, then

$$Grade = f\,B$$

where $Grade$ = assigned grade on a scale of 100
B = same base grade as above
f = fraction of manufacturing that was completed

Chapter 24

© iStockphoto.com/Edyta Pawlowska

Design Report

The design report documents the final design. It enables someone unfamiliar with a design to figure out how it works, evaluate it, and reproduce it. It is the final step in the design process. This chapter summarizes **organization of the report** and provides some **writing guidelines.**

Organization of the Report

Like the oral design defense, the organization of a design report follows the steps of the design process (see Table 1). The report begins with a concise statement of the "Problem Definition" in the student's own words.

The "Design Requirements" section does not need to repeat the competition rules. Instead, it should begin with a short paragraph describing competition strategy. Then list all performance requirements—for example, "Must have a top speed of 1 ft/s"; "Must deposit at least 6 Ping-Pong balls"; "Must be able to steer."

Almost all of the content of the "Conceptual Design" section should be available from previous milestones. In Section 3.1, present the sketches of your alternative design concepts and briefly describe how each one works. In Section 3.2, present your decision matrix and use the matrix as a vehicle to discuss the strengths and weaknesses of each concept. In Section 3.3, indicate the concept you selected and give your rationale.

The "Detailed Design" section describes the design that appeared at the final performance evaluation. New detailed drawings will have to be prepared. Place these new drawings in Section 4.1 along with text describing the operation and main features of the final design. Describe the overall design first and then zero in on the details of special features. If possible, include digital photographs of your machine. To create Section 4.2, retrieve

Table 1

Suggested Organization of the Design Report

ORGANIZATION	Pages
Title and Authors	1
Table of Contents (with page numbers)	1
List of Individual Contributions to the Report	1
1. Problem Definition	0.5
2. Design Requirements	1
3. Conceptual Design	
3.1 Alternative Concepts	2
3.2 Evaluation of Alternatives	1
3.3 Selection of a Concept	0.5
4. Detailed Design	
4.1 Main Features and How It Works	3
4.2 Results of Analysis, Experiments, and Models	1
4.3 Manufacturing Details	1
5. Performance Evaluation	1
6. Lessons Learned	1–2
	Total = 15–16

the results presented at the oral design defense and add text. Section 4.3 is primarily a summary of the joining methods used.

The results of the performance tests are summarized in the "Performance Evaluation" section. Describe how your machine fared both during individual performance testing and at the final competition, being as quantitative about it as you can. Also compare performance predictions to actual results.

The "Lessons Learned" section is an opportunity to reflect back on the design experience. Write it in the form of three paragraphs, with each paragraph dedicated to answering one of the following questions: (1) How would you redesign your machine to improve performance? (2) What general lessons did you learn about the design process? (3) What general lessons did you learn about teaming?

Writing Guidelines

Use double spacing to leave room for instructor comments. Use the section and subsection headings of Table 1 and boldface them so that they stand out. Finally, figures should be embedded within the text (rather than placed at the end of the report) for ease of reference. In addition:

1. **Be concise.**
2. **Begin each paragraph with a topic sentence** that expresses the theme or conclusion of the entire paragraph. The reader should

be able to overview the entire report by reading just the topic sentences.

3. **Generate high-quality concept drawings and detailed drawings** to pass the "flip test." The first thing the instructor will do before reading the report will be to flip through the pages to examine the figures. The figures will form the instructor's first impression of the report.
4. **Give each figure** (i.e., drawing, graph, etc.) **a figure number and a self-explanatory figure caption.** For example: **Fig. 3: Decision matrix.** Figures should be numbered consecutively and should be referenced from the text using the figure numbers. For example: "The results of the comparison are summarized in the decision matrix of Figure 3."
5. **Use a spelling checker.** Spelling mistakes will cause the reader to question technical correctness.
6. **Employ page numbers** both in the text and in the table of contents.

Design Milestone: Design Report

This is a time both for documentation and for reflection.

Assignment:

Prepare a design report in accordance with the guidelines of this chapter.

Grading Criteria:

- Is the report complete?
- Does it pass the flip test?
- Solid writing style?
- Is it clear how the design works?
- Could someone unfamiliar with your machine manufacture it from the drawings and information in the report?

Chapter 25

A Bridge too Far

An Example of a Design Competition: "A Bridge Too Far"

In this chapter, we present a typical design competition along with one team's solutions to the first four milestones. Design competitions like the one described in this chapter have proven to be very successful with first- and second-year engineering students. The rules, tabletop playing field, and list of parts provided may be used as a template in defining similar competitions.

Design Competition Example: "A Bridge Too Far"

The design project begins on the day that the instructor distributes the rules governing the design competition. This package of rules will typically consist of a statement of the design objective, a list of design requirements, a drawing of the playing field, and a list of parts and materials.

In the design competition named "A Bridge Too Far," the objective is *to design and build a device to outscore your opponent in a series of head-to-head matches*. The playing field is shown in Figure 1. A team receives +1 point for every ball resting in its scoring pit at the end of play. In all there are 17 scoring balls. Six of the balls start out in the possession of the 2 teams; the remaining 11 balls have starting positions on the playing field as indicated in Figure 1. Other key rules are:

- The device must fit within a 1-ft by 1-ft by 2-ft high volume.
- Parts and materials are limited to those listed in Table 1.

- Each device can have up to three independent tethered controls.
- There is a one-minute setup time.
- One game lasts 30 seconds.
- If there is a tie, both teams lose.

The parts and materials of Table 1 will not be supplied to the teams until they successfully defend their designs at the oral design defense. The complete set of rules is given in the Appendix at the end this chapter.

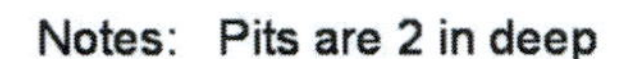

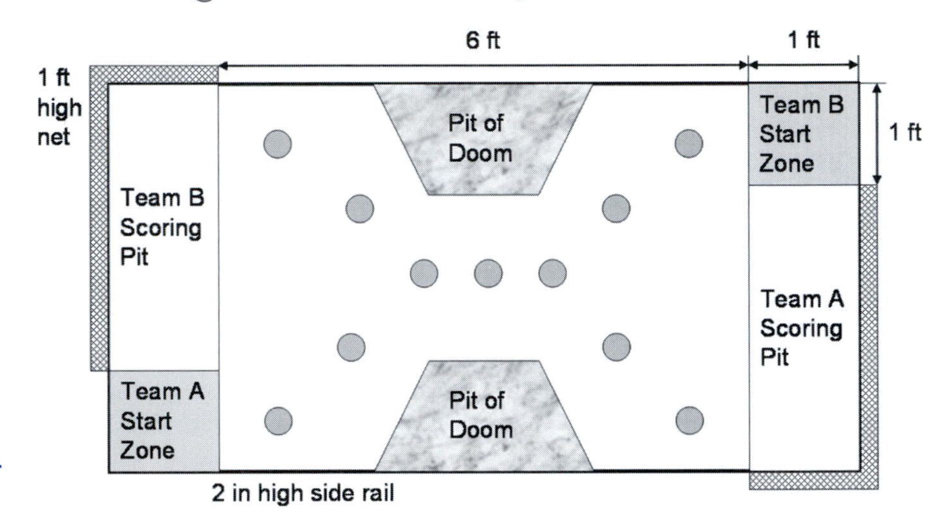

Figure 1

"A Bridge Too Far" Tabletop Playing Field

Table 1

List of Parts and Materials for "A Bridge Too Far"

1	motor with adjustable gear box
1	gearhead motor
1	relay (for use with gearhead motor)
1	diode (for use with gearhead motor)
1	pull solenoid
1	set of 6 gears
1	wooden dowel, 3/8" x 36"
2	balsa sheets, 1/16" x 3" x 36"
2	balsa beams, 3/8" x 3/8" x 36"
1	plywood sheet, 1/4" x 12" x 12"
10	super craft sticks
1	steel rod, 1/8"D x 20"
1	piece of string, 10' long
2	metal hinges, 1" long
1	shock cord with hooks, 1/4"D x 10" long
1	rubber band strip, 3' long
4	wheels with hubs, 2"D (for 1/8" shaft diameter)
1	mailing tube
1	2 liter (plastic) soft drink bottle (provided by team)

Design Milestone Solutions for "A Bridge Too Far"

DM 1: Clarification of the Task

In the case of a design competition, clarification of the task is mostly the responsibility of the instructor. There were just two things left for the design team to do to complete this milestone.

First, the team directed the following questions to the rules committee:

1. Can a machine score from the Pit of Doom?
2. Can you drop obstacles to interfere with the other machine?
3. Can you score points by driving a machine loaded with balls into a scoring pit?
4. Can a machine attach itself to the other machine?
5. Can a machine expand beyond the 1-ft by 1-ft by 2-ft high dimensions once the game begins?

The answer to all of these questions is "yes," since there is nothing against the proposed actions in the rules. Motivation behind the questions is to fully understand the design constraints and to probe for omissions that could lead to a design advantage.

Second, the team compiled a list of performance requirements that were specific to their design and their competition strategy.

D Must score at least four points
W Must score the first point within 10 seconds
D Must hinder the opponent's ability to score

The list is short to avoid solution bias. Later, after a design has been selected, other performance requirements can be added.

DM 2: Generation of Alternative Concepts

The functional decomposition that was settled upon is shown in Figure 2. Consideration of offensive strategy is done through the subfunction "deliver balls."

The results of brainstorming each subfunction were collected in the classification scheme of Figure 3.

Compatible subfunction concepts were combined to form the four promising designs shown in the concept drawings of Figures 4 through 7.

DM 3: Evaluation of Alternative Concepts

The four alternative concepts developed in the previous milestone are illustrated in Figure 8 for easy reference.

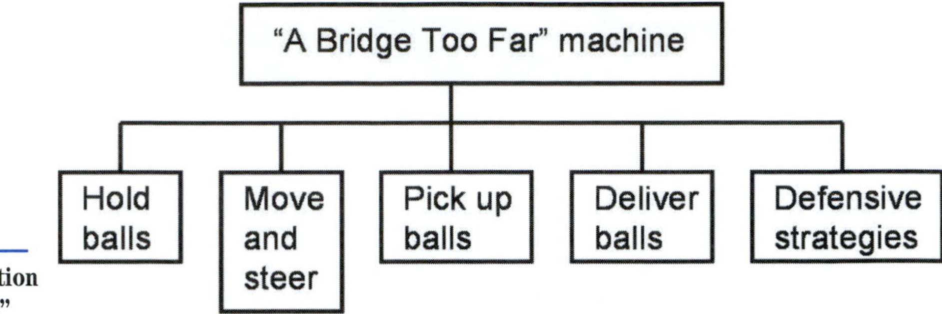

Figure 2

Functional Decomposition for "A Bridge Too Far"

Functions \ Concepts	Concept 1	Concept 2	Concept 3	Concept 4	Concept 5
Hold balls	Place on machine	Keep under machine	Push balls in front	Container to put balls in	Drag behind machine
Move and Steer	2 motors	2 motors	1 motor steers	no steering	Don't move
Pick up balls	Don't pick up balls	wedge	scoop	plow	claw
Deliver balls	Push them into pit	Throw them into pit	Drive into pit with balls	Roll them down a ramp	Push opponent into pit
Defensive strategies	Block bridge	Ignore opponent	Throw extra balls into Pit of Doom	Pick up opponent	Throw obstacles

Figure 3

Classification Scheme for "A Bridge Too Far"

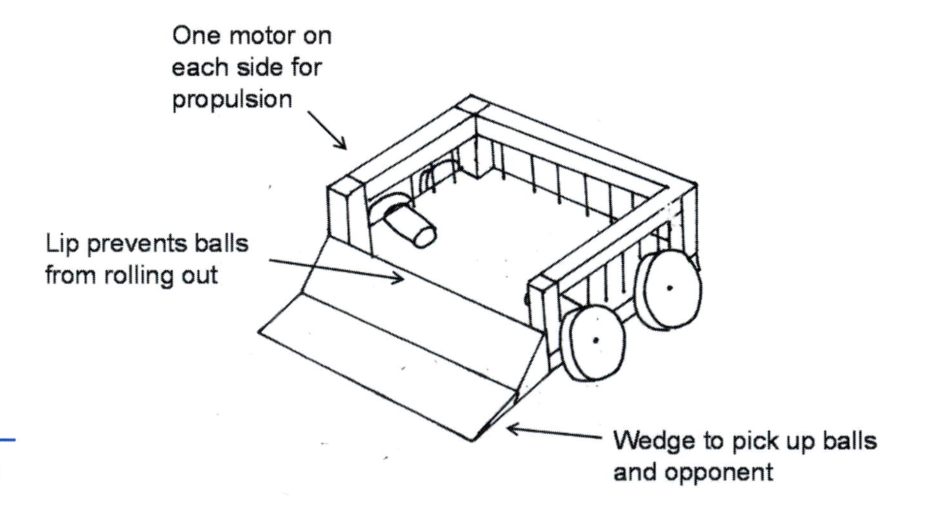

Figure 4

Concept Drawing of the "Wedge"

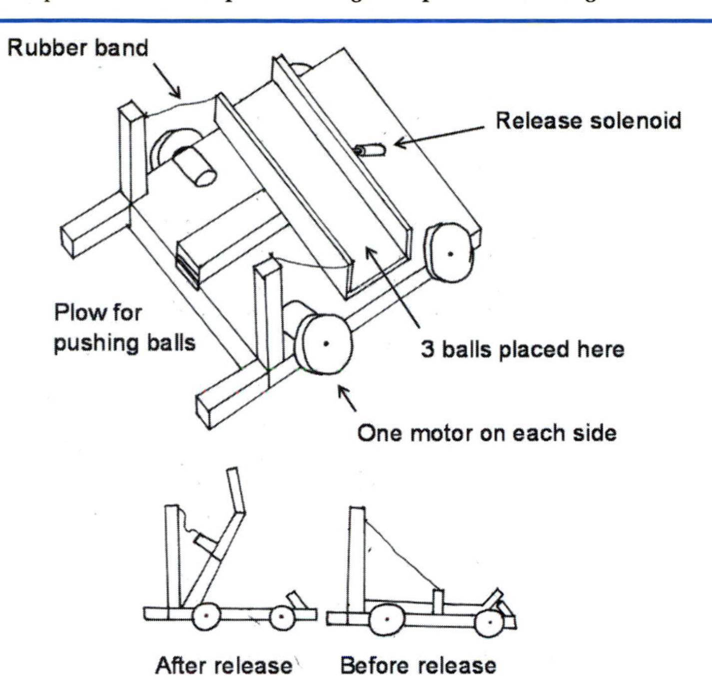

Figure 5

Concept Drawing of the "Catapult"

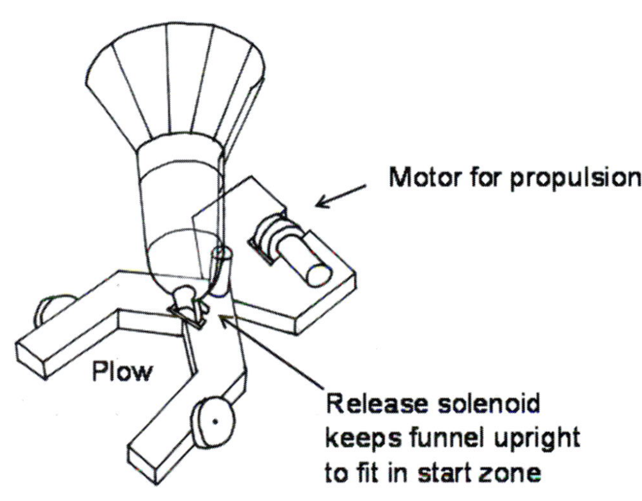
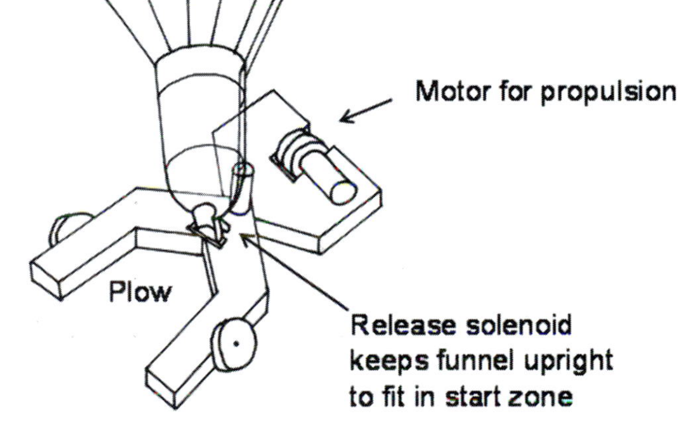

Figure 6

Concept Drawing of the "Snake"

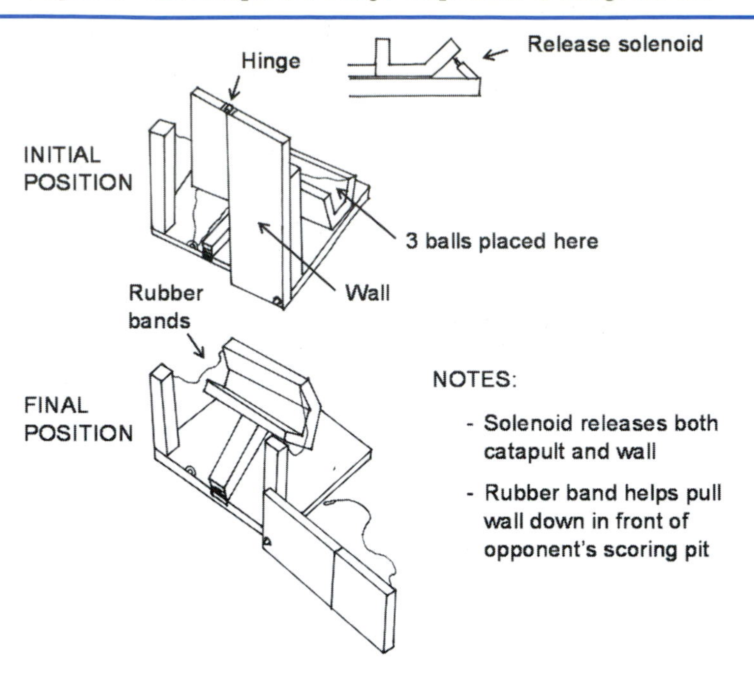

Figure 7

Concept Drawing of the "Wall"

The selection of evaluation criteria was tricky because a high-scoring machine that consistently gathers 9 of the 17 balls should be ranked on the same level as a defensive machine that consistently wins by the score of 1 to 0. The evaluation criteria and their respective weights are shown in Table 2. The first two criteria are both needed to describe scoring potential because it is not enough that you can transport a lot of balls; you must be able to score with them with an opponent in your way. Their combined weighting of 0.4 is slightly higher than the weight of 0.3 assigned to defensive capabilities (third criterion), since you have to score to have a chance of winning. "Easy to manufacture" has its usual importance, but "low cost" was not included because it is not a factor in the competition. "Easy to control" appears because of the potential challenges involved in maneuvering across the bridge. The discussion of concept strengths and weaknesses is organized by evaluation criterion.

HAS A LARGE PAYLOAD

The "Wall" is rated low because it can only score with the three original balls. For the other three concepts, ratings are proportional to carrying capacity. Since the funnel design expands to be much larger than the start zone, it has a much larger carrying capacity than the other two.

ROBUST SCORING CAPABILITY

The two designs that launch the balls score highest here because it is much harder to block a ball that is thrown than one that is pushed. Also, these machines can score without having to cross the bridge.

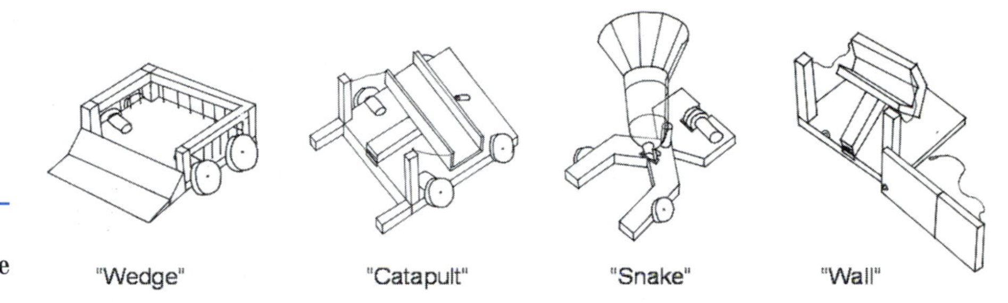

Figure 8

Alternative Design Concepts from "A Bridge Too Far"

Table 2

Decision Matrix for "A Bridge Too Far"

		Wedge		Catapult		Snake		Wall	
Evaluation Criteria	**Wt**	**Val_1**	**$Wt \times Val_1$**	**Val_2**	**$Wt \times Val_2$**	**Val_3**	**$Wt \times Val_3$**	**Val_4**	**$Wt \times Val_4$**
Has a large payload	.2	6	1.2	5	1.0	8	1.6	2	0.4
Robust scoring capability	.2	5	1.0	9	1.8	3	0.6	7	1.4
Can disrupt opponent	.3	7	2.1	5	1.5	4	1.2	8	2.4
Easy to manufacture	.2	7	1.4	5	1.0	5	1.0	7	1.4
Easy to control	.1	7	0.7	6	0.6	3	0.3	9	0.9
Totals	1.0		6.4		5.9		4.7		6.5

CAN DISRUPT OPPONENT

The "Wall" is an obvious favorite for this category, since it is the only one that actively blocks the opponent's goal. The "Wedge" is designed to be able to pick up the opponent and drop it in the appropriate goal to gain more points, so it too can theoretically disrupt the opponent well.

EASY TO MANUFACTURE

The "Wedge" is rated higher than the "Catapult" because it has fewer functions to build. The "Wall" also does well in this category because it does not involve any motors or corresponding drive-trains. In fact, it would be rated even higher were it not for the difficulties anticipated with sequencing the release of the wall and the catapult arm.

EASY TO CONTROL

The "Wall" simply requires flipping one switch, which is almost as easy as doing nothing. The "Snake" is slightly alien to most things people will have controlled and as a result would be hard to drive.

DISCUSSION OF RESULTS

All of the machines demonstrated strength in some areas. Observed weaknesses cannot be corrected without diminishing the attributes that made them strong. For example, the Snake's funnel, which gives it the largest payload, also makes it vulnerable to being pushed around by the opponent.

The "Wall" is the highest-rated design and also the boldest in that it dares to remain stationary. Students tend to shy away from designs like this one because it is different. This design was not selected because it was felt that its defensive capabilities have been overrated. The design may not be able to resist pushing by an opponent at the end of the wall.

The "Wedge" was finally selected as the best design. It is interesting that it is rated high even though it is not a clear winner in any of the categories. It compares very closely with the "Catapult." The "Wedge's" simplicity, lifting potential, and large payload were the decisive factors.

DM 4: Detailed Design

As you may recall, the "Wedge" (shown on the left in Figure 9) was selected as the final concept. The original wedge design was dual purpose: (1) to lift and push the opponent's machine and (2) to allow balls to roll up into the holding bin above the moving platform.

EXPERIMENTS

There were concerns that the forward speed of the vehicle might not be sufficient to allow the balls to roll up high enough into the bin. Experiments (see Figure 10) were conducted to check this out. Results showed that the concerns were justified. Acceptable dimensions for the wedge and holding bin were established based on the results of the experiments. The modified concept is shown on the right in Figure 9.

ANALYSIS

Two identical gearhead motors were available for creating the drive trains for the left and right sides of the vehicle. The no-load angular speed (N_{noload})

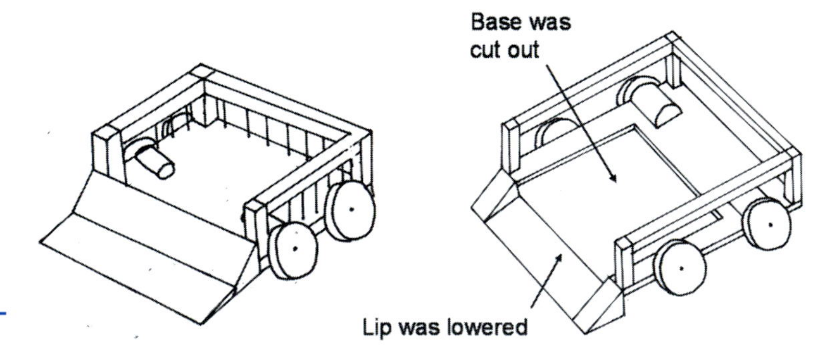

Figure 9

Design Modifications in Response to Experiments

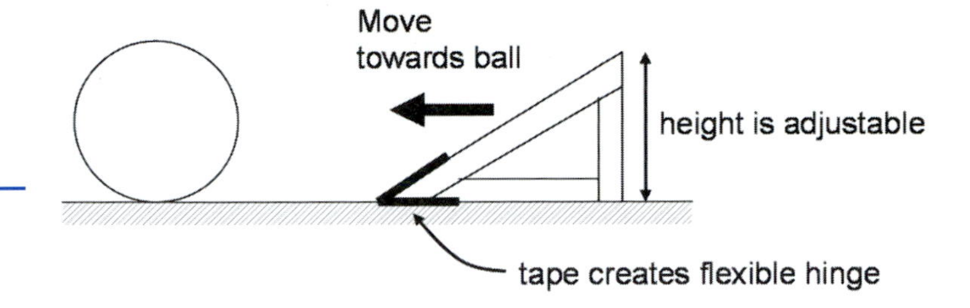

Figure 10

Experiment Used to Determine Optimal Wedge Geometry

at 24V was listed in the catalogue as 145 RPM. To maintain a speed of 1.00 ft/s on the flat with tire diameters of 2.00 inches, the overall gear ratio needs to be (according to equation 5 of Chapter 20):

$$GR = \frac{2\,\pi\,N_{noload}\,R_{tire}}{60\,V_{mp}} = \frac{2\,\pi\,(145)\,(1)}{60\,(12)} = 1.26$$

that we will round up to 2, based on the set of gears that was provided. With the pushing requirement of this design, it might seem that the actual gear ratio needs to be much higher. However, this particular motor is quite powerful; the motor shaft could not be visibly slowed by manually gripping the ends of the shaft. Given the power of the motor and the low weights of the vehicles, this gear ratio was deemed to be reliable even with the pushing requirement. The resulting drive train geometry is shown in Figure 11.

DETAILED DRAWING

Two views of the final design are shown in the hand-drawn (and inked) detailed drawing of Figure 12. Some manufacturing details are provided.

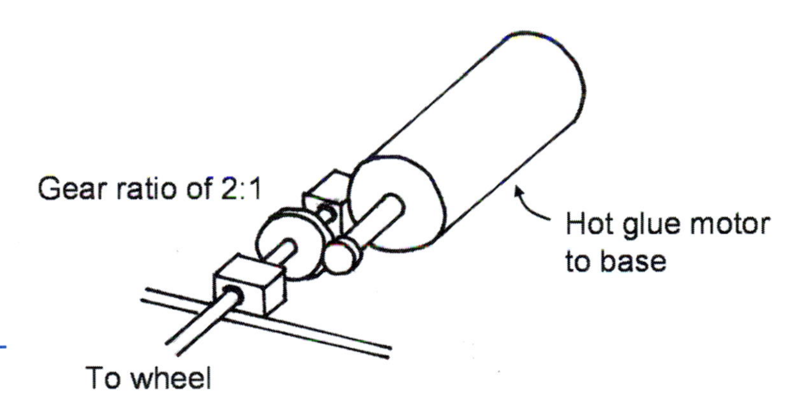

Figure 11

Close-up View of the Drive Train

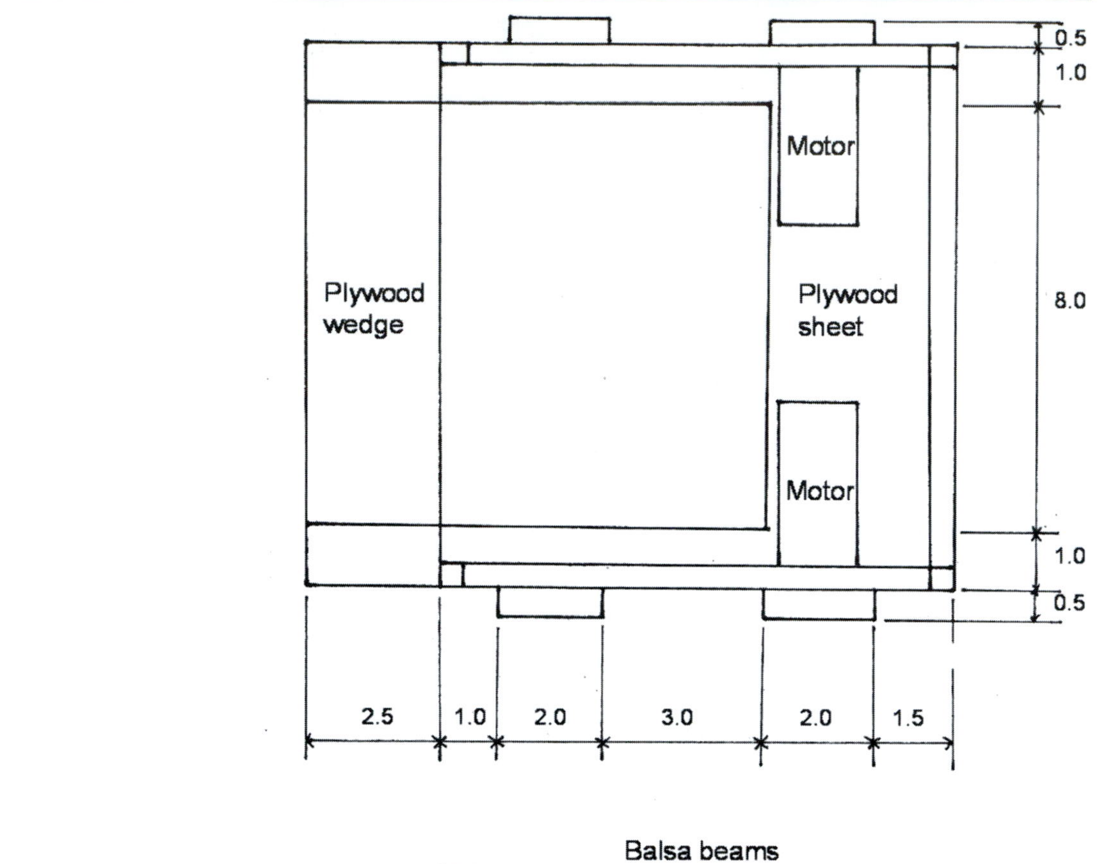

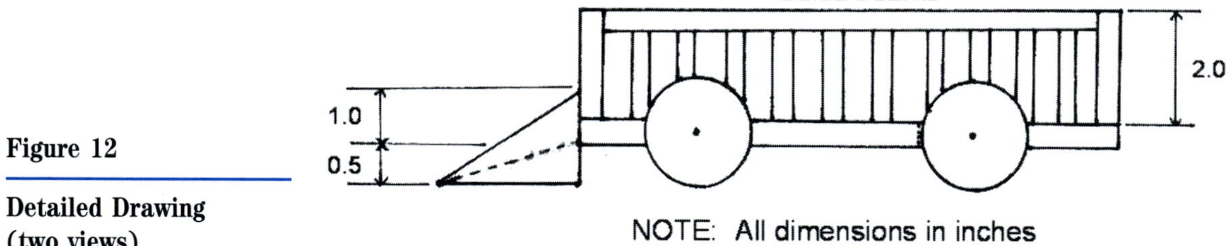

Figure 12

Detailed Drawing (two views)

Appendix

Official Rules for the "A Bridge Too Far" Design Competition

Objective

Design and build a device to out score your opponent in a series of head-to-head matches.

Official Rules

Constraints

1. Parts and materials that may be used in the construction of the device are limited to those defined in Table 1. Each team will be provided with a box containing all legal parts and materials.
2. The devices must be constructed entirely by the members of the team (i.e., if you lack the required expertise or tools to manufacture a certain part of the device, redesign it).
3. Each device can be loaded with up to three scoring balls. If a team chooses not to load all three balls, the discarded balls will be removed from the playing field.
4. When placed on the table at the start of the game, each machine must fit completely within its assigned 1-ft by 1-ft by 2-ft high start zone.

5. An external power source will be made available to each device for use during the matches. It will consist of:

    - overhead wires that connect to the device (*Note:* These wires are considered to be part of the playing field.)
    - a control box with two forward-reverse-off switches and one on-off switch

The Game

1. Just prior to the game, there will be a one-minute setup time during which each team should:

    - place their device in the starting zone
    - attach and check the electrical connections
    - load their device with up to three scoring balls

2. Except for manipulation of the electrical control boxes located at one end of the playing field, no human interaction with the device (or playing field) is allowed once the game begins.
3. The game begins when indicated by the referee and ends 30 seconds later when the power is switched off.
4. The game, and all scoring, ends as soon as one of the following occurs:

    - all movement stops as a result of power being switched off
    - 5 seconds have elapsed since power was switched off

Scoring

1. At the end of the game, each team will receive 1 point for every ball in its scoring pit, irrespective of which team caused the ball to fall into the pit. The team that scores the most points wins.
2. In all there are 17 scoring balls; six of the balls start out in the possession of the two teams, and the remaining 11 balls have starting positions on the playing field as indicated in Figure 1.
3. A ball is counted as lying in a scoring pit if an imaginary vertical line through the center of the ball lies within the boundary of the scoring pit. The referee can ascertain the status of a scoring ball at the end of the game by removing other balls from the scoring pit; if the ball in question then falls into the pit, it will count as lying in the scoring pit.
4. In the event of a tie:

    a. If neither team has scored, both teams lose.
    b. If both teams have scored, the winner will be the machine that has advanced farthest down the field as based on final positions. If this criterion proves indecisive, both teams will advance to the next round.

Other Rules

1. If the tethers (i.e., electrical connections) should entangle as a result of the machines passing each other, time will stop and both machines will be returned to their respective starting zones. Play will then continue at the signal of the referee. This situation can occur because each machine will be tethered to the nearest of two overhead rods running parallel to the length of the table. To avoid entanglement, keep to the right of the opposing machine when passing.
2. Devices that permanently damage the playing field or the balls will be disqualified.
3. Any attempt to intentionally inflict permanent damage upon an opponent's device will result in immediate disqualification. However, devices should be designed to hold up under expected levels of nonaggressive contact. For example, some pushing should be expected, but that pushing should not occur at significant ramming speeds.
4. Any device deemed unsafe will not be allowed to participate in the matches.
5. Implementation of any strategies that are not directly addressed in the rules but that are clearly against the spirit of the rules (e.g., intentionally interfering with the person at the controls) will lead to disqualification.
6. The rules committee has the final word on any interpretation of the rules.

Chapter 26

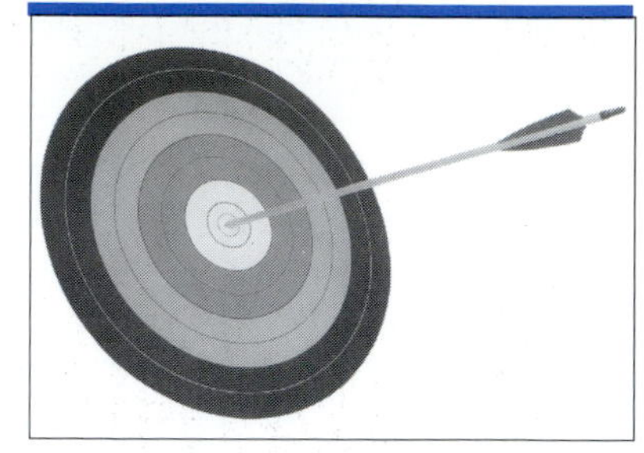

© iStockphoto.com/Malcolm Romain

Closing Remarks on the Important Role of Design Projects

If you ask professors or students why they do design projects in engineering courses, you can expect to hear responses like this:

- "They are a motivational tool."
- "They apply the analytical methods taught in courses."
- "They help to develop written and oral communication skills."
- "They teach teaming."

Indeed, these are all valuable outcomes of a design project, but each one can be achieved by some other means. The answer must lie elsewhere.

Part of the answer is found in the view that every engineering endeavor is ultimately about finding or designing a solution to an expressed need. The analytical methods, the teaming skills, and the rest are tools for achieving that goal—that is, they are the means to the end, not the end itself. Each design project offers a rare opportunity for students who spend most of their time deeply immersed in learning analytical methods to see the big picture.

The rest of the answer has to do with the real purpose behind these design chapters. Engineering design is at its core an unbiased and structured methodology for dissecting and solving complex problems. It is the way engineers should and must think. In contrast to analytical methods that are each limited to their own special class of problems, design methodology has universal applicability—to design, to research, to all fields of study. Design projects are the best way we know to exercise and develop this most fundamental of all engineering methods.

Hands-on design projects come closest to fully realizing these goals. They complete the design process, for as we have seen, it does not end with the detailed design; there will be design modifications to be made during manufacturing, testing, and the final performance evaluation. Students will learn the importance of design for manufacture principles by experiencing the results of having failed to heed them. They will also gain a sense of accountability by learning that it is not enough for a design to look good on paper—it has to *work*.

Index